AIR POWER IN UN OPERATIONS

Military Strategy and Operational Art

Edited by Professor Howard M. Hensel, Air War College, USA

The Ashgate Series on Military Strategy and Operational Art analyzes and assesses the synergistic interrelationship between joint and combined military operations, national military strategy, grand strategy, and national political objectives in peacetime, as well as during periods of armed conflict. In doing so, the series highlights how various patterns of civil–military relations, as well as styles of political and military leadership influence the outcome of armed conflicts. In addition, the series highlights both the advantages and challenges associated with the joint and combined use of military forces involved in humanitarian relief, nation building, and peacekeeping operations, as well as across the spectrum of conflict extending from limited conflicts fought for limited political objectives to total war fought for unlimited objectives. Finally, the series highlights the complexity and challenges associated with insurgency and counter-insurgency operations, as well as conventional operations and operations involving the possible use of weapons of mass destruction.

Also in this series:

Understanding Civil-Military Interaction
Lessons Learned from the Norwegian Model
ISBN 978 1 4094 4966 9

Clausewitz's Timeless Trinity
A Framework For Modern War
ISBN 978 1 4094 4287 5

British Generals in Blair's Wars
Edited by Jonathan Bailey, Richard Iron and Hew Strachan
ISBN 978 1 4094 3735 2

Britain and the War on Terror
Policy, Strategy and Operations
Warren Chin
ISBN 978 0 7546 7780 2

Confrontation, Strategy and War Termination
Britain's Conflict with Indonesia
Christopher Tuck
ISBN 978 1 4094 4630 9

Air Power in UN Operations
Wings for Peace

Edited by

A. WALTER DORN
Royal Military College of Canada

Routledge
Taylor & Francis Group

LONDON AND NEW YORK

First published 2014 by Ashgate Publishing

Published 2016 by Routledge
2 Park Square, Milton Park, Abingdon, Oxon OX14 4RN
711 Third Avenue, New York, NY 10017, USA

Routledge is an imprint of the Taylor & Francis Group, an informa business

British Library Cataloguing in Publication Data
A catalogue record for this book is available from the British Library

The Library of Congress has cataloged the printed edition as follows:
Air power in UN operations : wings for peace / edited by A. Walter Dorn.
    pages cm. -- (Military strategy and operational art)
Includes bibliographical references and index.
ISBN 978-1-4724-3546-0 (hbk) -- ISBN 978-1-4724-3549-1 (pbk) --
1. United Nations--Peacekeeping forces--Case studies. 2. Air power--Case studies.
I. Dorn, A. Walter.
JZ6374.A47 2014
358.4'14--dc23

2014005722

ISBN 9781472435460 (hbk)
ISBN 9781472435491 (pbk)

Contents

PART VI   EVOLVING CAPABILITIES

List of Figures

List of Tables

About the Editor

A. Walter Dorn is Professor of Defence Studies at the Royal Military College of Canada (RMC) and chair of the Department of Security and International Affairs at the Canadian Forces College (CFC). He is also Chair of Canadian Pugwash, an organization of physical, life, and social scientists seeking to reduce the threats to global security. Dr Dorn is a scientist by training whose doctoral research was aimed at chemical sensing for arms control. He now covers both national and human security, especially UN peacekeeping. As an "operational professor" he has visited many UN missions and gained direct experience in field missions. He has served in Ethiopia as a UN Development Programme consultant, at UN headquarters as a training adviser, and as a consultant to the UN Department of Peacekeeping Operations. He has provided guidance to the United Nations on introducing unmanned aerial vehicles to the eastern Congo. He is a member of the UN's High Level Panel of Experts on Technology and Innovation in UN Peacekeeping Operations. He is author of a book on technology and innovation in UN operations.[1]

---

1   Dorn, A.W. *Keeping Watch: Monitoring, Technology, and Innovation in UN Peace Operations* (Tokyo: United Nations University Press, 2011).

About the Contributors

Christian F. Anrig is Deputy Director, Doctrine Research and Education, Swiss Air Force. From 2007 to 2009 he was a Lecturer in Air Power Studies in the Defence Studies Department of King's College London while based at the Royal Air Force (RAF) College. In 2009 he became a member of the RAF Centre for Air Power Studies Academic Advisory Panel. Dr Anrig began his professional career in Defence Studies as a researcher at the Center for Security Studies, Swiss Federal Institute of Technology (ETH Zurich), in 2004. The author of *The Quest for Relevant Air Power*,[1] he has also published various articles and book chapters covering topics from European military transformation to modern air power and its ramifications for small nations. While working in the United Kingdom, he was on the editorial board of the *Royal Air Force Air Power Review*. Dr Anrig spent the first half of his military career in the artillery, Swiss Army. Currently, he is a reserve major assigned to the Air Staff, Swiss Air Force. He is a dual-national Swiss and Liechtensteiner.

William K. Carr is a retired Lieutenant-General of the Royal Canadian Air Force (RCAF). His 39-year RCAF career began in 1939. During World War II he flew Spitfire aircraft in England, Malta, Sicily, and Italy. He later commanded squadrons, while accumulating more than 16,000 flying hours in over 100 different aircraft types. In 1960 he served as Senior Air Advisor to the UN Mission in the Congo. In 1971 he was appointed Deputy Chief of Staff, Operations, with North American Air Defence Command (NORAD). In 1974 he was appointed Deputy Chief of the Defence Staff General. He became the first commander of Air Command in 1975. On retirement from the military in 1978 for the next 14 years he was VP of government and military marketing/sales of the new Challenger aircraft for Canadair and then Bombardier. In 2001, he was inducted into the Canadian Aviation Hall of Fame.

Ryan W. Cross is the lead researcher in the political-psychology and adversarial intention section of the Reactions to Environmental Stress and Trauma Lab in the University of British Columbia's Department of Psychology. Projects include intelligence and conflict early warning using both cognitive and social

---

1   Anrig, C.F. *The Quest for Relevant Air Power: Continental European Responses to the Air Power Challenges of the Post-Cold War Era* (Maxwell Air Force Base, AL: Air University Press, August 2011).

psychological measures. He is also a Research Assistant for Professor Walter Dorn of the Canadian Forces College on peace-support operations research and contributes to ongoing work applying just war theory to contemporary conflicts. Cross has completed graduate studies, lectured or worked at universities in Berlin, Bonn, Zurich, Vancouver, and Lucerne, and is completing Master's degrees in War Studies at the Royal Military College of Canada and in Business Administration at Simon Fraser University. His research has been presented internationally to academic as well defence, security, and intelligence audiences.

Roméo Dallaire is a Canadian Senator and retired Canadian Army Lieutenant-General. He served as Force Commander of the United Nations Assistance Mission in Rwanda (UNAMIR) in 1993–1994. His experiences are described in his book *Shake Hands with the Devil – the Failure of Humanity in Rwanda*,[2] which was also the basis for an Emmy Award-winning documentary and a motion picture of the same name. Lieutenant-General Dallaire has served on the UN Advisory Committee on Genocide Prevention, as Special Advisor to the Minister of Veterans' Affairs Canada, and as Special Advisor to Canadian government on War Affected Children. His most recent book is *They Fight Like Soldiers; They Die Like Children – the Global Quest to Eradicate the Use of Child Soldiers*.[3]

William T. Dean III is an Associate Professor of History at the Air Command and Staff College at Maxwell Air Force Base, Alabama. He received his doctorate degree from the University of Chicago and was a Chateaubriand recipient from the French government. He won the Military Officer of America Association (MOAA) award for civilian educator of the year. He has published on French colonial warfare, intelligence, and air power issues in *Revue historique des armées*, *Penser les ailes françaises*, and *Defense Intelligence Review*, and chapters in various books.

James McKay is a Professor in the Department of Politics and Economics and the current Chair of the undergraduate program in Military and Strategic Studies at the Royal Military College of Canada. He was educated at Bishop's University, the Royal Military College of Canada, and King's College London. He teaches courses in strategic studies, Canadian politics and international relations at the undergraduate level and, on occasion, American military affairs at the graduate level. Being an inherently interdisciplinary sort, he maintains an eclectic range of research interests, from strategic coercion to First Nations land claims in Canada to the Vietnam War.

---

2   Dallaire, R. *Shake Hands with the Devil – the Failure of Humanity in Rwanda* (Toronto: Random House Canada, 2003; London: Arrow, 2012).

3   Dallaire, R. with Humphries, J.L. *They Fight Like Soldiers; They Die Like Children – the Global Quest to Eradicate the Use of Child Soldiers* (Toronto: Random House, 2010).

David Neil is a former maritime helicopter aviator who served 35 years in Canada's Air Force. Following four flying tours, including command of a Maritime Helicopter Squadron, he served as Project Director for both the Maritime Helicopter Replacement Project and the Aurora Long Range Patrol Aircraft Modernization Project. He was subsequently appointed Director of Strategic Planning for the Royal Canadian Air Force during which time he oversaw the development of the Air Force transformation strategy and the stand up of the Aerospace Warfare Centre. After completing his career as Director General Joint Force Development he was responsible for the Canadian Forces' unmanned aerial vehicle (UAV) and Command, Control, Communications, Computers, Intelligence, Surveillance, and Reconnaissance (C4ISR) campaign plans, and Canada's military space portfolio. In 2007, after his retirement from the military, he joined MDA Corporation's UAV team.

Robert C. Owen is a Professor in the Department of Aeronautical Science at Embry–Riddle Aeronautical University, Daytona Beach Campus. He also is an adjunct or contract researcher for several public and private organizations. Professor Owen joined the Embry–Riddle Faculty in 2002, following a 28-year career with the United States Air Force (USAF) and a brief career in private industry. His military career included a mix of operational, staff, and advanced education assignments. He is both a military Command Pilot and a Commercial Pilot with Instrument and Multi-Engine ratings. Professor Owen also served on the HQ Air Force Staff and the HQ Staff of the Air Mobility Command. His academic assignments included tours as an Assistant Professor of History at the U.S. Air Force Academy and as Dean of the USAF's School of Advanced Airpower Studies, the Service's graduate school for strategic planners. His books include the volume on Chronology of the *Gulf War Airpower Survey*, *Deliberate Force: A Case Study in Effective Air Campaigning*, and *Air Mobility: A Brief History of the American Experience*.[4]

Kevin Shelton-Smith is Chief, Aviation Projects and Planning Department of Field Support, UN Headquarters, New York. After joining the Royal Air Force (RAF) as a Halton Apprentice (Airframes and Propulsion) in 1976, he became a technician on the Harrier jump jet at RAF Wittering. He was commissioned as a pilot, flying several aircraft, including the Harrier. He gained an Engineering Honours degree in Aeromechanical Systems from the Cranfield Institute of Technology (now University) at the Royal Military College of Science, Shrivenham. As an Engineering Officer he held several appointments at UK bases and in Support/

---

4   Owen, R.C. *Gulf War Air Power Survey*, Dir. Dr Eliot A. Cohen. Vol. 5, Part II, *Chronology* (Washington, DC: USAF Gulf War Air Power Survey, 1993); Owen, R.C. *Deliberate Force: A Case Study in Effective Air Campaigning* (Air University Press, 1999); Owen, R.C. *Air Mobility: A Brief History of the American Experience* (Dulles, VA: Potomac Books, 2013).

Logistics Command HQ and notably introduced the Boeing E3-D AWACS to service at RAF Waddington. His final role in the RAF was as the Engineering Authority for the Pegasus engine as fitted to his much-loved Harrier II. In 1999, as a Chartered Engineer and Member of the Royal Aeronautical Society, he joined UN headquarters in New York, serving UN peacekeeping aviation in the various roles of Aviation Safety and Standards, aircraft contracting and management for peacekeeping missions in Kosovo, Bosnia, the Democratic Republic of the Congo, Sierra Leone, Georgia, Iraq, Ivory Coast, Mali, and Chad. He also served as Chief Air Operations in the United Nations Mission in Sudan (UNMIS) in 2005–2006. His responsibilities now include overall planning of aviation support and requirements in new peacekeeping missions, training, staffing, and the introduction of new technologies such as unmanned aerial systems and new aircraft types. Since this volume went to press, the United Nations has accepted its first unmanned aerial vehicles, the Selex Falco UAV, into service in the Democratic Republic of the Congo and is planning further military and civil systems in Mali and elsewhere.

Kevin A. Spooner is Associate Professor of North American Studies and History at Wilfrid Laurier University. He is author of *Canada, the Congo Crisis, and UN Peacekeeping, 1960–64*,[5] published by University of British Columbia Press and awarded the 2009 C. P. Stacey Award by the Canadian Committee for the History of the Second World War and the Canadian Commission for Military History. He has addressed the Cold War dimensions of Canadian foreign policy during the Congo Crisis in his paper, "Just West of Neutral: Canadian 'Objectivity' and Peacekeeping during the Congo Crisis, 1960–61" in the *Canadian Journal of African Studies*.[6] Dr Spooner also co-edited *Documents on Canadian External Relations, 1959*, Vol. 26.[7] His current research projects examine Canadian policy towards Africa in the period of decolonization and the development of Canadian foreign policy autonomy, within the British Empire, during the Laurier years.

Robert David Steele is CEO Earth Intelligence Network. He is a retired Marine Corps infantry, intelligence, and administrative officer who also served as the senior civilian responsible for creating the Marine Corps Intelligence Center and as the Study Director for the flagship study, *Planning and Programming Factors for Expeditionary Operations in the Third World* (MCCDC, 1990). A 30-year veteran of defence and national intelligence, since 1993 he has been a key international proponent for Open Source Intelligence (OSINT), Information Peacekeeping, Peacekeeping Intelligence, and – more recently – Multinational, Multiagency,

---

5   Spooner, K.A. *Canada, the Congo Crisis, and UN Peacekeeping, 1960–64* (Vancouver: UBC Press, 2009).

6   Spooner, K.A. "Just West of Neutral: Canadian 'Objectivity' and Peacekeeping during the Congo Crisis, 1960–61", *Canadian Journal of African Studies* 43(2) (2009), 303–36.

7   Spooner, K.A., Cavell, J. and Stevenson, M.D. (eds). *Documents on Canadian External Relations, 1959*, Vol. 26 (Ottawa: Canadian Government Publishing, 2006).

Multidisciplinary, Multidomain Information-Sharing and Sense-Making (M4IS2). As Chief Executive Officer (CEO) (pro bono) for the Earth Intelligence Network since 2006, he has led 23 others in creating a Strategic Analytic Model for M4IS2 and hybrid Stabilization and Reconstruction Operations. He served as the first international security intelligence analyst for the International Commission Against Impunity in Guatemala (CICIG), a hybrid UN organization. Most recently he has published three seminal articles on the future of Human Intelligence (HUMINT), and a book, *Intelligence for Earth: Clarity, Diversity, Integrity, and Sustainability*.[8]

F. Roy Thomas is a retired Armoured officer with United Nations service in Cyprus, the Golan Heights, Jerusalem, Afghanistan, Macedonia, Sarajevo, and Haiti. He was hijacked in South Lebanon and taken hostage in Bosnia. Roy was the Senior United Nations Military Observer, Sector Sarajevo. He is a recipient of the Meritorious Service Cross, a United Nations Protection Force (UNPROFOR) Force Commander's Commendation, a UN Mission in Haiti (UNMIH) Force Commander's Commendation and a US Army Commendation Medal. His last appointment in the Canadian Forces was as the *first* Chief Instructor of the Canadian military's Peace Support Training Centre (PSTC), Kingston.

Matthew Trudgen is a fellow at the John Sloan Dickey Center for International Understanding at Dartmouth College. He received his PhD in history from Queen's University in 2011. His dissertation examined how different conceptions of the Canadian national interest held by different groups within the Canadian political establishment and armed forces influenced the development of the North American air defence system in the 1950s. He was also a research assistant at the Department of Foreign Affairs, Trade and Development Historical Section from December 2013 to March 2014.

---

8   Steele, R.D. *Intelligence for Earth: Clarity, Diversity, Integrity, and Sustainability* (Earth Intelligence Network, 2010).

Foreword

Lieutenant-General The Hon. Roméo A. Dallaire (Retired)

The operations of the United Nations, no matter how flawed, understaffed or ill equipped, are a vital expression of human concern, needing both national and international support. All nations should have a strong interest in using peace operations to prevent conflict, minimize human suffering, preserve human dignity, and reduce radicalization and extremism in their many forms. No nation can be completely unaffected by the war-torn parts of our interconnected world. War zones export violence and terrorism. They also affect trade – licit and illicit. For example, minerals from Africa are used in everyday products like cell phones. Blood diamonds and conflict coltan are but two of the minerals fueling conflicts in Africa, where illegally exploited natural resources are exchanged for weapons. The world must deal with the contraband that fuels wars and the wars that fuel contraband – a vicious cycle that has to be stopped, in part with international forces on the ground. Raging conflicts anywhere impact global immigration, refugee flows, and national diasporas, with relatives stuck in conflict zones. Wars also deprive children of education, and perpetuate poverty and economic dislocation, even as the young are manipulated into fighting as soldiers. Thus, there is a strong *humanitarian imperative* to support UN operations. It is ethically impossible to stand by as people are slaughtered, as war-affected children are murdered or orphaned or forced to murder. The UN's peace operations, with their twenty-first century mandates to protect civilians and support human rights, are an expression of this humanitarian imperative. Yet these peace operations remain poorly equipped to do the job. Among the many needed capabilities, air power is vital.

For peacekeepers in distant war-torn parts of the world, aircraft often serve as the lifeline for survival and sanity. In Rwanda, in 1994, as genocide was perpetrated all around the peacekeeping mission, it is hard to describe the joy and relief we felt from the sound of incoming aircraft landing with essential supplies, new personnel, and packages from home. The aircrew risked their lives to save ours. The Canadian Forces' Hercules, the only aircraft which flew regularly into the mission during the genocide, took fire from the ground and had to land on unsecured, hazardous airfields. But the courage and skill of the aircrews made our work possible so that we, in turn, could save thousands who would otherwise have perished.

Though air power was limited in Rwanda, it can potentially serve many functions. As force commander responsible for the Rwanda mission, I was deeply troubled by the lack of air support for UNAMIR ground patrols, which regularly sustained fire. Ideally, we would have had close air support. More basically, there was no aerial reconnaissance to regularly monitor, from above, the locations we sought to protect. I longed for the capacity to jam from the air or ground the hate broadcasts of *Radio Télévision Libre des Mille Collines*, which was urging Tutsis to kill the *inyenzi* (literally, cockroaches), a hateful term designating Hutus. Air power and so many other essential capabilities were lacking within that mission.

Fortunately, the United Nations has made progress since the mid-1990s, including building an aviation fleet and expanding its aerial capabilities. For instance, the ongoing eastern Congo mission includes attack helicopters that have helped to keep rebel groups at bay. But there is still much to do. In comparison with advanced militaries and the North Atlantic Treaty Organization (NATO) alliance, the United Nations still operates only at a basic level. Whereas NATO has spent decades dealing with ground–air interoperability, common-funded aircraft (like AWACS, the Airborne Warning and Control System), compatible air training and doctrines, and technological integration, the United Nations still has not developed simple air–ground tactics, techniques, and procedures. For instance, missions are plagued with problems of simple communications between aircraft and ground troops, unless the aircraft and troops originate from the same nation.

As NATO's combat mission in Afghanistan comes to a close, there is hope that Western countries may once again re-engage in UN peace operations and help furnish the necessary air power. It would be unfair to continue the West's current abandonment of peacekeeping, leaving the important job solely to the global South, whose ground and air capabilities are limited. The North has contributions to make in all manner of air power assets. Western nations supported the NATO stabilization operations in Bosnia and Kosovo with a multitude of aircraft. There is a similar need in contemporary UN operations. Successful prevention, mitigation and resolution of conflicts all involve air power.

There is a common but mistaken view of peacekeeping as only an army activity; air forces and civilian aircrews also have a key role to play. Modern militaries rightly stress the importance of joint operations, bringing together components from land, air, and naval forces. Achieving "jointness" is also important for the United Nations. Its operations are made up of disparate nations, with soldiers and aircrews who have not trained together or worked together before. Furthermore, air assets must be managed by a mix of UN civilian and military personnel. This means that there is much room for improvement in integrating forces – air, sea, and ground, as well as military and civilian.

One thing is certain. In this era of "protection of civilian" mandates in complex multidimensional missions, where ambiguity and complexity are the norm, air power will continue to be a required part of the solution. Air power is essential to develop an "environment of security". But it remains an under-used and under-studied tool for peace operations.

The United Nations needs a conceptual base to examine joint air–ground operations. It needs to explore new ways to integrate land and air forces. It has to learn to manage the complexities of modern technological operations. The challenge remains how to achieve integration with many nations and other actors in multidimensional, multiagency, multinational environments, covering air and water as well as land.

I welcome this unique volume on air power in UN operations. It provides a close look at the ways peacekeeping and enforcement can be facilitated from the air. It provides an impressive and wide-ranging examination of air power applications from the past and points to how these can be made more effective in the future. As peace operations gradually catch up to other military operations in technological resourcefulness, such studies will play an important role in illuminating the past to brighten the future.

Preface

Most people think of peacekeeping in terms of ground operations performed by soldiers. In fact, peacekeeping has evolved considerably beyond the two dimensions of space to cover the third as well: airspace. The peacekeepers of the air also have a story worth telling. As in conventional warfare, the air campaign is a vital adjunct to the ground campaign; the two are intrinsically bound together. But the air power story in peacekeeping has hardly been told. To students and practitioners of UN operations, it appears as a major gap in the public, professional and academic literature – one that needs to be filled so all can benefit.

This eclectic volume is the first book to treat the UN's aviation experience, doing so both descriptively and critically, covering the organization's needs and means, its challenges and weaknesses. The book examines the air systems employed for UN operations – humanitarian, peacekeeping and enforcement. It illustrates the lessons with poignant historical case studies. In addition to many UN peace operations, it covers actions by UN-authorized enforcers like the North Atlantic Treaty Organization (NATO) in Bosnia and Libya and supporters like the United States in Haiti.

The book's coverage is based on the core capabilities that air power provides. As in military operations generally, these capabilities are: transportation, observation, and firepower.[1] Simply put, aircraft provide means to *carry*, *see*, and *shoot*. Aircraft are also a means to show presence, though the value of the presence lies in the ability to carry, to observe, or to apply force. Another capability, though less important and less used, is to relay communications.[2] Almost all air power

---

1 There are various ways to consider the applications of air power. The three core capabilities given here provide a simple but accurate description of the most basic capabilities, which can then be combined to carry out the vast majority of applications. Other ways of looking at military air power are found in military doctrine. For instance, the US Air Force's Doctrine Document 1 (14 October 2011) specifies 13 "core functions", while the British Air and Space Power Doctrine (AP3000, Fourth Edition, 2013) gives four "fundamental roles" – namely, control of the air and space, air mobility, intelligence and situational awareness, and attack. By contrast Canadian Forces Aerospace Doctrine (B-GA-400-000/FP-000, December 2010) recognizes six "functions" for air forces – namely, command, sense, act, shield, sustain, and generate, functions which are common to the army and navy. The author asserts that almost all of the applications and roles of air power can be constructed from the three core capabilities cited here.

2 In a rough comparison to the human body, the capabilities of transportation, observation, firepower, and communication, are equivalent to the legs, eyes/ears, arms, and mouth, given that the legs are used to carry materials, the eyes and ears to observe

functions derive from the three basic capabilities, which are sometimes combined during a single flight. For instance, an armed helicopter might carry troops to a conflict zone, observe the movements of opponents, and fire missiles against those who attack the UN forces.

Each of these functions is vital, intriguing, and worth studying in detail. The first, transportation, involves more than deploying peacekeepers into the host country and inserting/extracting them into precise conflict zones (maybe called the "battle space" or even the "peace space"). It also means moving vast quantities of equipment and supplies to sustain not only the peacekeepers but also the "peacekept" – the local population and displaced persons whom the United Nations seeks to save and help. In addition, aircraft can transport and drop leaflets to educate and inform the local population and, in emergencies, provide medical evacuation (air "medevac") for fast transport of peacekeepers and local civilians to hospitals.

Aerial observation, the second capability, can be as simple as a pilot viewing the ground while transporting personnel and goods. But to verify complex peace agreements and to prevent the spread of deadly conflict, the United Nations needs dedicated surveillance flights, sometimes observing raging battles from above. Since many of the violations and atrocities in armed conflicts are carried out at night, the United Nations also must overcome the night barrier by using airborne night vision equipment, which few missions have done. Such devices can spot and help stop night attacks and the smuggling of arms, precious minerals, and human beings.

While peacekeeping is meant to de-escalate violence, it is sometimes necessary to *use force to stop force*. When attacked, UN peacekeepers have a right to defend themselves, including the right to call in close air support. Furthermore, in the twenty-first century, UN missions have a responsibility to protect civilians under imminent attack or threat, requiring rapid and forceful responses, sometimes delivered by air. Such a combat capability is sometimes called "kinetic air power" or aerial firepower; this is the third of the core capabilities. The armed helicopter, the Mi-35, has become an iconic and somewhat ironic symbol of robust peace operations. Once an instrument of suppression and dictatorship, the Russian-made helicopter is now used by the United Nations as an instrument to *prevent* aggression and oppression, proving its worth in the Democratic Republic of the Congo (DRC), Liberia and the Côte d'Ivoire. The combat capability of the Mi-35 is usually applied only when a firefight erupts or an attack is under way, but the mere presence or sound of the heavily armed helicopter can serve as a powerful deterrent. That is the power of presence. Parties are less likely to violate peace agreements if they know that violations will be met with UN resistance backed up by robust UN air power.

---

the surroundings, the arms to push or punch (and much more, of course), and the mouth to communicate.

The mission of peacekeepers is, however, very different from that of warfighters. Rather than gain victory on the battlefield, the United Nations seeks a negotiated settlement so that the conflicting parties can live in peace for the long term. In his article "Peacekeeping at the Speed of Sound", John Hillen observes that UN peacekeeping emphasizes "restraint, perseverance and legitimacy as opposed to offense, surprise and mass".[3] Using all the facets of air power can facilitate negotiations and a sustainable peace, just as it can the fighting of a war.

Aircraft are sometimes used for relaying communications, bouncing signals from the ground to locations much further from their origin. Of course, aircraft also need to communicate their own information, including what they observe from the air and a host of flight details. In addition, aircraft can broadcast messages electronically to the wider public through radio, television and the Internet. Alternatively, they can jam unwanted signals, such as hate radio broadcasts that inflame conflict. (This is usually done by saturating the particular radio frequency with white noise.) Sometimes aircraft are used as mobile relay stations to pass communications to other aircraft or ground forces.

From these core capabilities a host of UN air functions are developed. For example, UN commanders sometimes place themselves aboard helicopters to oversee the movement of their troops and to observe any hostile or opposing forces. In another example, airborne search and rescue crews use aerial surveillance to locate lost persons and air transport to bring them quickly to a hospital or back to base. Similarly, the interdiction of illegally trafficked people and contraband involves surveillance (that is, spotting the illegal traffickers or goods) and the transport of troops to bring traffickers in to custody and seize their ill-gotten gains. It can also involve combat, if the traffickers put up a fight.

Admittedly, the operation of aircraft in peacekeeping has some drawbacks and disadvantages. First, they are *expensive* to operate: US$1,000 to $3,000 per flying hour is typical (personnel included). But this relatively high cost must be measured against the time savings from rapid air transport and, in some cases, the impossibility of moving personnel or equipment into remote areas by ground transportation. Second, the use of aircraft can be *dangerous*, as terrible crashes in UN history have illustrated. One of the UN's most prominent Secretaries-General, Dag Hammarskjöld, lost his life in a plane crash in the Congo in 1961 – according to official reports the crash was caused by pilot error. Ground fire can also down aircraft or strike UN personnel aboard aircraft. In Sarajevo in 1992–1995, the UN peacekeepers in C-130 Hercules aircraft were told to sit on their helmets because of the risk of hostile ground fire that could easily pierce the air frame. In Haiti in 2009, the UN's CASA-212 accidentally crashed into a mountainside, killing all 11 peacekeepers aboard. Overall, however, the UN's flying record is

---

3   Hillen, J. "Peacekeeping at the Speed of Sound: The Relevancy of Airpower Doctrine in Operations Other Than War", *Airpower Journal* 12(4) (Winter 1998), 8. Also available at http://www.airpower.au.af.mil/airchronicles/apj/apj98/win98/hillen.pdf [accessed 27 January 2013].

impressive, given that flights are made in some of the most conflict-ridden parts of the world, and that many more fatalities have occurred on the ground than in the air. Impressively, the UN mission in the Democratic Republic of the Congo (DRC) has the largest aircraft fleet in Africa and an enviable air safety record compared to others operating in the dangerous conditions of the continent.

This description of air power capabilities and challenges only scratches the surface. Answering further questions needs a much deeper study and a higher level of expertise. How were the core capabilities used in different UN missions? What has been the UN experience with air power over its history? When were combat aircraft used? With what effect? More generally, how can the UN make the most effective use of the third dimension of space?

Again, the current paucity of literature on peacekeeping does not allow a fulsome answer to these questions. To seek a fuller understanding, a workshop was held at Canadian Forces Base Trenton, Canada's largest military air base, in June 2011. It brought together military officers (mostly but not exclusively from air forces), UN officials, academics, and industry representatives. Their papers were updated after the conference to include cutting-edge developments, such as the UN's contracting of unmanned aerial vehicles (UAVs) for the DRC. These papers form the basis of the current volume. Where some gaps were found in the coverage of issues, the editor brought additional authors on board to make the book wider in coverage and deeper in depth.

The Editor was very fortunate to have Lieutenant-General (ret'd) and now Senator Roméo Dallaire, the former head of the United Nations Assistance Mission for Rwanda, provide the Foreword, in addition to the keynote address at the workshop where the general's direct experience with air power during a horrendous genocide enlightened us: he described how even the sound of incoming UN aircraft provided immense reassurance and motivation, well before its life-saving supplies were provided. Unfortunately, the combat side of air power was not applied to help General Dallaire stop the genocide and possibly save countless lives. There were only a few precedents in UN history where the United Nations used combat air power.

The first part of this book considers an early, important, and fascinating case study involving combat: the leap in air power made by the United Nations in the Congo (1960–1964). The Congo operation proved irresistible as a prime case study for the development of UN air power. In some ways the mission carried out activities unsurpassed by any peacekeeping mission to the present day. For instance, it was the only mission (so far) to use bomber aircraft. In its multidimensional application of air power it was a forerunner of the many peace operations in the post-Cold War world. The mission saw the creation of the UN's first "Air Force", which expanded in number and type of aircraft as the world organization became embroiled in a battle to maintain law and order, and prevent secession in that new-born country. How the world organization "established this air force from nothing" is told in Chapter 1 by a key participant of the operation, Lieutenant-General (ret'd) William K. Carr, who was in charge of organizing the early air mission and who would go on later to become the Commander of Canada's Air Command. Ironically for the United Nations, the Congo mission soon became embroiled in an aerial arms race and in air-to-air combat with the secessionist Katanga province, as described by A. Walter Dorn in Chapter 2. The politics of contributing to this international adventure in the heart of Africa is told in Chapter 3 by a historian–expert on the mission, Professor Kevin A. Spooner, who uses the Canadian experience to show the challenges, politics, and dilemmas facing national contributors to the difficult and controversial mission.

After reviewing this remarkable case, with its abundance of lessons, the book explores other cases according to the three core capabilities of air power: airlift, aerial reconnaissance, and air combat. Airlift has served as the lifeline for UN missions, bringing supplies and new personnel to peacekeepers sometimes caught in battle zones or enduring emergencies. A classic UN operation, still in existence, which requires airlift at high altitude in the Himalayas, is the UN observer mission in Kashmir, which is considered in detail in Chapter 4 by historian Matthew Trudgen. He looks at Canadian decision-making to provide aircraft for transport and observation. A more recent case, with much greater sophistication, occurred after the devastating earthquake in Haiti on 10 January 2010. The UN's close work with the US Air Force is described in detail in Chapter 5 by Colonel (ret'd) Robert C. Owen, now a professor at Embry–Riddle Aeronautical University in Florida. The tremendous work of the UN Humanitarian Air Service (UNHAS) in Haiti and in other hot spots around the world is described in Chapter 6 by A. Walter Dorn and Ryan W. Cross, showing that cooperation across diverse UN agencies is possible, however difficult. For effective peacekeeping, however, coordinating contributions from UNHAS, troop-contributing countries and private contractors remains a challenge.

The second capability provided by air power is surveillance, to keep a watch over conflict-ridden areas. The case for aerial surveillance, complementary to ground observation and action, is made by A. Walter Dorn in Chapter 7. He also offers, in Chapter 8, a short case study of UN aerial observation during the Lebanese civil war of 1958. Looking to the future, the expanding use of UAVs is advocated in Chapter 9 by Colonel (ret'd) David Neil, who serves MacDonald Dettwiler and Associates Ltd., a pioneering aerospace company that has operated UAVs in Afghanistan.

Moving beyond surveillance, the United Nations must often take direct action, sometimes applying sanctions on national government and leaders. No-fly zones (NFZs) are a particular form of UN sanction imposed on recalcitrant nations such as the former Yugoslavia, Iraq, and Libya in order to prevent those nations from using aircraft to suppress or bomb civilians. The enforcement of this special form of restriction necessitates both observation and a combat capability, and is often left to particular UN member states or coalitions. The Southern NFZ, imposed on Iraq after its disastrous 1990 invasion of Kuwait and its 1992 suppression of the Marsh Arabs, is examined in Chapter 10 by James McKay, a former Canadian military officer now teaching at the Royal Military College of Canada. The NFZ in the former Yugoslavia is covered in Chapter 11 by F. Roy Thomas, a career solider, who locks at air power from the ground during his tour as a UN military observer in Sector Sarajevo in 1993–1994.

Combat aircraft can be used in close support of ground troops or make gun, missile, or bombing runs in a standalone fashion. Close air support by the operation in Somalia (1992–1993) is examined in detail in Chapter 12 by William T. Dean III of Air University in Alabama. A thorough scholarly overview of Operation Deliberate Force in Bosnia 1995 is provided in Chapter 13 by Robert C. Owen, who shows how NATO applied *force for peace* to back the contemporaneous and future peace operations run by the United Nations and NATO, respectively. Close air support by UN attack helicopters in the DRC (2003 to the present) is described in Chapter 14 by A. Walter Dorn using UN archival records. The most powerful application of combat capability under UN mandate was carried out by NATO in Libya during Operation Unified Protector, as documented in Chapter 15 by Swiss Air Force scholar Christian F. Anrig.

Having studied all these applications, the challenge is to be forward-looking while drawing on the lessons of the past. This is done in Chapter 16 by Kevin Shelton-Smith, the Chief of Aviation Projects at UN Headquarters and a former pilot with the Royal Air Force (United Kingdom), who has also served in industry and in UN field missions. Shelton-Smith gives practitioner insights into the United Nations of today and the possibilities for tomorrow. A further creative exploration, with bold recommendations, is offered in Chapter 17 by Robert D. Steele, who looks at how UN air power can be an innovative tool for peace. While not exactly "winged angels", the aerial UN peacekeepers are important agents of protection and support. They are an attempt to bring the better angels of human nature to the fore. UN air power is a celestial and material representation of humanity's concern for humanity. This book shows how air power can save lives, alleviate suffering, and build global security. But these aerial applications can be as complicated and as challenging as they are fascinating.

A. Walter Dorn
Toronto, Canada

List of Abbreviations

A/C	Air Commodore
AA	Anti-Aircraft
AD	Air Defence
AF	Air Force
AFB	Air Force Base
AFDD	Air Force Doctrine Document
AFNORTH	[US] Northern Command Air Component
AFSOC	[US] Air Force Special Operations Command
AFSOUTH	[NATO] Allied Forces Southern Europe [Allied Joint Force Command Naples, from March 2004]
	[US] Southern Command Air Component
AI	Aerial Interdiction
AMC	Air Mobility Command
AMR	After Mission Report
ANC	Armée nationale Congolaise / Congolese National Army
AOC	Air Operations Center
AOR	Area of Responsibility
APC	Armoured Personnel Carrier
APOD	Aerial Port of Debarkation
ARL	Airborne Reconnaissance Low
ASOC	Air and Space Operations Center
ASU	[World Food Programme] Aviation Safety Unit
ATO	Air Tasking Order
ATOL	Automatic Takeoff and Landing
ATU	Air Transport Unit
AWACS	Airborne Warning and Control System
BHC	[UN] Bosnia-Herzegovina Command
BiH	Bosnia-Herzegovina
BSA	Bosnian Serb Army
C2	Command and Control
C4ISR	Command, Control, Communications, Computers, Intelligence, Surveillance, and Reconnaissance
CAS	Close Air Support
CCOS	[Canada] Chairman, Chiefs of Staff (position created in 1951, abolished by 1964)
Celtel	Cellphone Tower
CENTCOM	[US] Central Command

CF	Canadian Forces
CIA	[US] Central Intelligence Agency
CINCSOUTH	Commander-in-Chief, Allied Forces Southern Europe
CMOC	Civil–Military Operations Centers
CNDP	Congrès National pour la Défense du People / National Congress for the Defence of the People
COB	Contingency Operating Base
COIN	Counter-Insurgency
CRE	Contingency Response Element
CRG	Contingency Response Group
CSAR	Combat Search and Rescue
DCOS	Deputy Chief Of Staff
DFS	[UN] Department of Field Support
DHH	[Canada] Directorate of History and Heritage
DMZ	Demilitarized Zone
DND	[Canada] Department of National Defence
DOD	[US] Department of Defense
DOS	[US] Department of State
DPKO	[UN] Department of Peacekeeping Operations
DRC	Democratic Republic of the Congo
DZ	Drop Zone
ECR	Electronic Countermeasures / Reconnaissance
EO	Electro-Optic
EUFOR	European Union Force
EW	Electronic Warfare
FAC	Forward Air Controller
FAK	Force Aérienne Katangaise / Katangan Air Force
FARDC	Forces Armées de la République Démocratique du Congo / Armed Forces of the Democratic Republic of the Congo
FLIR	Forward-Looking Infrared
FW	Fixed Wing
GoH	Government of Haiti
GPS	Global Positioning System
HARM™	High-speed Anti-Radiation Missile
HFOCC	[Provisional] Haiti Flight Operations Coordination Center
HIRI	High-Intensity-Restricted-Infrastructure
HOCC	[UN] Humanitarian Operation and Coordination Centre
HR	Humanitarian Relief
HUMINT	Human Intelligence
IADS	Integrated Air Defence System
IAI	Israel Aerospace Industries
ICAO	International Civil Aviation Organization
ICTY	International Criminal Tribunal for the former Yugoslavia
IED	Improvised Explosive Device

IR	Infrared
ISAF	International Security Assistance Force
ISR	Intelligence, Surveillance and Reconnaissance
ITB	Initiation to Bid
JAM	Joint Assessment Mission
JCO	Joint Commission Officer
JFACC	Joint Forces Air Component Commander
JFC	Joint Force Commander
JMAC	Joint Military Analysis Centre
JOC	Joint Operations Centres
JP	[US Joint Chiefs of Staff] Joint Publication
JSTARS	Joint Surveillance Targeting Attach Radar System
JTF-H	[US] Joint Task Force – Haiti
JTF-SWA	Joint Task Force – Southwest Asia
LAC	Library and Archives Canada
LMC	Logistics Management Center
M4IS2	Multinational, Multiagency, Multidisciplinary, Multidomain Information-sharing and Sense-making
MALE (UAV)	Medium Altitude Long Endurance (Unmanned Aerial Vehicle)
MANPAD	MAN-Portable Air Defense missile
MCDA	Military and Civil Defense Assets
MDA	MacDonald, Dettwiler and Associates Ltd
MIB	Military Information Branch
MINURCAT	[UN] Mission in the Central African Republic and Chad
MINUSTAH	[UN] Stabilization Mission in Haiti
MIPONUH	[UN] Civilian Police Mission in Haiti
MONUA	[UN] Observer Mission in Angola
MONUC	Mission de l'Organisation des Nations Unies en République démocratique du Congo / UN Organization Mission in the Democratic Republic of the Congo
MONUSCO	Mission de l'Organisation des Nations Unies pour la Stabilisation en République démocratique du Congo / UN Organization Stabilization Mission in the Democratic Republic of the Congo
MPs	Military Police
MTPP	ICAO airport identifier: Toussaint Louverture International Airport, Haiti
NAEW&CF	NATO Airborne Early Warning and Control Force
NATO	North Atlantic Treaty Organization
NEO	Non-combatant Evacuation Operation
NFZ	No-Fly Zone
NGO	Non-Governmental Organization
NSC	National Security Council
OAS	Offensive Air Support

OCHA	[UN] Office for the Coordination of Humanitarian Affairs
ONUC	Opération des Nations Unies au Congo / The United Nations Operation in the Congo [originally the United Nations Organization in the Congo]
OP	Observation Post
OSINT	Open Source Intelligence
PGM	Precision Guided Munitions
PSYOPS	Psychological Operations
PVO	Private Volunteer Organization
QRF	Quick Reaction Force
RAF	[UK] Royal Air Force
RAMCC	Regional Air Mobility Control Center
RCAF	Royal Canadian Air Force
RFP	Request for Proposals
ROE	Rules of Engagement
ROVER	Remotely Operated Video Enhanced Receiver
RW	Rotary Wing
SA	Small Arms
SA	Specialized Agencies
SAR	Synthetic Aperture Radar
SCR	[UN] Security Council Resolution
SEAD	Suppression of Enemy Air Defences
SIDM	Système intérimaire de drone male
SLO	Senior Liaison Officer
SMO	[UN] Senior Military Observer
SNFZ	Southern No Fly Zone
SOUTHCOM	[US] Southern Command
SSEA	[Canada] Secretary of State for External Affairs
STOL	Short Take Off and Landing
TACC	Tanker Airlift Control Center
TCC	Troop Contributing Countries
TEZ	Total Exclusion Zone
TLAM	Tomahawk Land Attack Missiles
TWCF	[US Department of Defense] Transportation Working Capital Fund
UAR	United Arab Republic
UAS	Unmanned Aerial System
UAV	Unmanned Aerial Vehicle
UN	United Nations
UNAMI	United Nations Assistance Mission for Iraq
UNAMID	African Union / United Nations Hybrid Operation in Darfur
UNAMIR	United Nations Assistance Mission in Rwanda
UNAMSIL	United Nations Mission in Sierra Leone
UNATF	United Nations Air Transport Forces

UNCIP	United Nations Commission for India and Pakistan
UNEF	United Nations Emergency Force
UNHAS	United Nations Humanitarian Air Service
UNHCR	United Nations High Commissioner for Refugees
UNIKBDC	United Nations Iraq–Kuwait Boundary Demarcation Commission
UNIKOM	United Nations Iraq–Kuwait Observer Mission
UNIPOM	United Nations India–Pakistan Observation Mission
UNITAF	[US] Unified Task Force
UNMIL	United Nations Mission in Liberia
UNMIS	United Nations Mission in Sudan
UNMISS	United Nations Mission in the Republic of South Sudan
UNMO	United Nations Military Observer
UNMOGIP	United Nations Military Observer Group India–Pakistan
UNOCI	United Nations Operation in Côte d'Ivoire
UNOGIL	United Nations Observer Group in Lebanon
UNOMIG	United Nations Mission in the Republic of Georgia
UNOSOM	United Nations Operation in Somalia
UNPROFOR	United Nations Protection Force
UNSC	United Nations Security Council
UNSCOM	United Nations Special Commission (for disarmament in Iraq)
UNSCR	United Nations Security Council resolution
UNTAET	United Nations Transitional Administration in East Timor
USAF	United States Air Force
USAID	United States Agency for International Development
USSEA	[Canada] Under-Secretary of State for External Affairs
USSOUTHCOM	United States Southern Command
USTRANSCOM	United States Transportation Command
W/C	Wing Commander
WCP	Weapons Collection Point
WFP	World Food Programme
WMD	Weapons of Mass Destruction

Acknowledgments

The Editor thanks Major Bill March, the Editor of the *Royal Canadian Air Force Journal*, for recognizing the large gap in air power studies on UN operations and his realization that a workshop and an edited volume of the best papers could help fill that gap. The authors of the papers are to be congratulated for the diversity and depth of their studies. Because the volume needed to cover a wide range of UN-mandated operations, the Editor sought a paper on NATO's operation in Libya 2011. Fortunately, Dr Christian F. Anrig had published an excellent preliminary assessment on the Libya operation and was willing to update it for this volume on short notice in a timely manner (as one would expect from someone from Switzerland!). It was a pleasure to work with Kirstin Howgate and the staff at Ashgate Publishing in the United Kingdom. They showed initiative and speed in reviewing, accepting, providing feedback, copy-editing, and improving the text. The collation, editing and updating of the papers would not have been possible without the diligent assistance of Ryan W. Cross and Matt Trudgen, who provided conscientious and dependable help. They also helped draft papers, section summaries and the Afterword. Some of the draft papers were kindly reviewed by Leif Hellström, Josh Libben and Mohammed Masoodi, who provided insightful commentary and corrections. Finally, many UN officials shared experiences and provided information and insights that greatly enriched this volume. To them, the Editor extends both his admiration and his gratitude.

PART I
The UN's First "Air Force"

The peacekeeping operation in the Congo, from 1960 to 1964, was the UN's baptism by fire in nasty internal (intrastate) conflicts. The United Nations had to deal with coups d'état, secessionist provinces, tribal wars, ethnic massacres, and very real threats to its own personnel, including from air attacks. Notably, a lone fighter jet flown by a mercenary pilot against the nascent mission was able to paralyze UN efforts and embarrass the international community. The United Nations was obliged to participate in an aerial arms race with the secessionist Katangan province in order both to protect itself and prevent the breakup of the newborn country. Aerial reconnaissance, provided by Swedish jets, was essential to predict and pre-empt Katangan attacks on UN forces. Bombers provided by India were able to destroy airfields used by the mercenaries. In Operation Grand Slam of December 1962/January 1963, close air support from Swedish jets assisted ground forces to assert the UN's freedom of movement and to capture key airfields and centres in Katanga, finally winning both the war and the peace. But the air effort began much earlier, starting in July 1960 when the United Nations had to bring over 20,000 troops into the vast Congolese territory, requiring a powerful airlift capacity, originally provided by the US Air Force. Soon over a dozen nations contributed. Thus, the mission made use of all three main elements of air power – that is, transport, surveillance, and combat. For this reason, UN personnel rightly boasted that they created the UN's first "air force", despite the use of aircraft for transport and surveillance in previous UN missions.

The mission was in many ways a precursor of the robust multidimensional missions of the twenty-first century. While the UN's experience in the Congo was an overall success, it came at a great cost in human lives and in funds. Over 200 peacekeepers died in the mission; and the financial cost of the mission taxed the resources of the international community, almost driving the United Nations into bankruptcy. For several reasons, it was the first and only UN peacekeeping initiative in Africa until the end of the Cold War in 1988–1989. It continues to provide rich lessons for modern-day peacekeeping as the world deals with many complex conflicts, especially in Africa and in the Congo again.

Fortunately, one of the senior participants in the Congo mission was able to describe his experiences in setting up the UN's first "air force". Then, Group Captain (later Lieutenant-General) William K. Carr from Canada oversaw the aircraft and crew from a host of nations around the world working together to achieve a challenging goal. In Chapter 1, William K. Carr shows how the United

Nations used practical improvisation and creativity born of necessity to keep the force moving and equipped, even before it acquired its combat capability, as it had never before attempted to create and move such a large force. In 1961, after the deaths of Prime Minister Patrice Lumumba and UN Secretary-General Dag Hammarskjöld, the United Nations adopted a much more robust stance. Combat was authorized not only for self-defence but also for the broader defence of the mission, which now included preventing the secession of the mineral-rich Katanga province. Chapter 2, by A. Walter Dorn, describes the challenges of "fighting for peace". Mission leaders took a defensive posture until the opportune moment when they used combined air and ground power to nullify the military arm of the secessionist government. This showed that combat could be successful in bringing about Katangan peace in a unified country. However, the operations raised many dilemmas. The contributing nation's (in)decision to support the air mission is typical of peacekeeping, as showcased in Chapter 3 by Kevin Spooner, an expert on the operation. For example, how did Canada maintain national support for the beleaguered mission, even when tough or impossible UN requests were made? More generally, how were considerations of Cold War politics balanced? When to support the use of force? The chapters in Part I help answer these important questions using the fascinating case of the Congo in the first half of the 1960s.

Chapter 1

Planning, Organizing, and Commanding the Air Operation in the Congo, 1960

William K. Carr[1]

Until Somalia and Bosnia in the 1990s, the United Nations Operation in the Congo (known by its French acronym ONUC: Opération des Nations Unies au Congo) was by far the largest peacekeeping operation ever conducted by the United Nations. The mission was authorized on 14 July 1960 and finally wrapped up officially on 30 June 1964. The weaponry and firepower employed by ONUC's military component included jet fighter aircraft, artillery, armored personnel carriers, and tanks. At its peak, the Force consisted of almost 20,000 troops from 28 countries. Over its lifetime 93,000 troops served in the force; 127 military personnel died in action and 133 were wounded, along with scores of European expatriates and tens of thousands of Congolese.

ONUC began as a conventional peacekeeping mission modeled on the United Nations Emergency Force (UNEF) based in the Sinai. Like UNEF, ONUC was mandated initially only to use force in self-defence. This idea was considerably extended as, for example, the need arose to protect civilians at risk. By robustly asserting its freedom of movement in Katanga ONUC was able to detain and expel foreign mercenaries and prevent civil war. By the time ONUC ceased to operate on 30 June 1964, UN expenditures amounted to over US$400,000,000.[2]

The aim of this chapter is to tell the tale of Royal Canadian Air Force (RCAF) involvement at the beginning of the Congo operation in 1960 and to recall some of the things which stick in my memory over 50 years later.[3]

---

1 An earlier version of this text appeared as Carr, W.K. "The RCAF in the Congo, 1960: Among the Most Challenging Assignments Undertaken by Canadian Forces in the Peace Keeping Role", *Canadian Aviation Historical Society Journal* 43(1) (2005), 4–11, 31. It is republished here with permission from the Canadian Aviation Historical Society, with major updates.

2 This figure represents only what is termed "incremental costs", that is, those costs billed by contributing nations as being direct out-of-pocket expenses to them.

3 Article originally written in 2004; phrasing adjusted to reflect this.

Why Canada Became Involved and How the Operation Grew

The Congo, a country relatively unknown by Canadians until 1960, was granted independence that year, though it was ill prepared to assume the mantle of nationhood. For nearly 100 years it had been the private domain of the King of Belgium and later a totally dependent colony of Belgium. One factor that sped the decision to grant independence in 1960 was the example of no fewer than 17 former African colonies having recently won self-government.

The first government of the Congo was formed on 24 June 1960, with Joseph Kasavubu as Head of State and Patrice Lumumba as Prime Minister. On 29 June, in Leopoldville – modern day Kinshasa – they signed a Treaty of Friendship with Belgium. At the same time the Belgian King Baudouin proclaimed Congolese independence. Almost at once, a breakdown occurred in what had previously been a system of militarily imposed law and order.

The more than 200 tribes, speaking a myriad of languages, had never viewed Belgian colonization as a benefit, or a stabilizing influence on historic enmities. On 5 July, parts of the 25,000 member indigenous army/police "Force Publique" mutinied against their Belgian officers. This led to the widespread unrest. Belgium reacted by sending in troops to provide protection for its more than 100,000 nationals. Belgium was unable to gain legitimacy for this move by failing to convince Lumumba to invoke the Treaty of Friendship and seek help from the now more than 10,000 Belgian soldiers in the country.

During the second week of July more trouble and violence arose as the mutiny spread. After evacuating all Belgian nationals from the area, Belgian soldiers and warships attacked the port city of Matadi with a considerable loss of life among the local population. Hyped-up reports of this action carried on the Congolese army radio network, sparked new rounds of violence even in areas that previously had been quiet. Far from stabilizing the situation, the appearance of Belgian paratroops at widely separated locations resulted in even more unrest. Increasing numbers of attacks on the remaining Europeans followed.

In the midst of all this turmoil, Moïse Tshombé, the governor of mineral-rich Katanga announced the secession of the province. Lumumba flew to the provincial capital, Elisabethville (now Lubumbashi), to seek conciliation, but his aircraft was prevented from landing. The incident led to a further breakdown of relations with the Belgian government, which supported Tshombé for financial reasons from behind the scenes.

Confronted with a situation beyond his control, Lumumba asked the United Nations for help on 12 July 1960. After Dag Hammarskjöld, the Secretary-General, offered a plan, the United Nations Security Council gave unanimous approval for a security force to be sent to the Congo.[4] A Swedish General, Carl von Horn, then Chief of Staff of the United Nations Truce Supervisory Organization in the Middle East, was appointed to command the force and arrived on the scene on 18 July.

---

4   United Nations Security Council, Resolution 143 (1960), 14 July.

Figure 1.1   Emperor Haile Selassie thanks US Air Force C-130
crewmembers before they airlift Ethiopian troops to the Congo
Source: UN Photo 183490, 25 July 1960.

The buildup of troops was rapid and within a month more than 14,000 military personnel were located throughout the country. They had been delivered directly to their final destinations within the Congo, mainly by aircraft of the United States Air Force (USAF) and the RCAF. Figure 1.1 above shows a USAF C-130 aircraft and its crew, who are about to ferry Ethiopian troops to the Congo.

Canada's Key Role

Because of its already well-earned reputation in UN peacekeeping, and having played a key role in every UN peace mission to that date, Canada became involved at the outset in the planning for the Congo operation. Specifically, the Secretary-General asked Canada to take on the job of running all air operations throughout the Congo and, in addition, to provide a long-range radio network for ONUC, which would be located at key centres. Canada agreed.

The Air Officer Commanding RCAF Air Transport Command, Air Commodore Fred Carpenter, accompanied by Wing Commander Jack Maitland, the Commanding

Officer of 426 Transport Squadron, which flew the long-range Canadair North Star planes, were dispatched immediately to survey the needs and make recommendations as to how they could be satisfied.[5] Carpenter's recommendations were approved and, within days, a small air staff to implement the decisions was assembled and sent on its way to Leopoldville. Canada also agreed to establish and operate the UN forces' radio network as requested by the Secretary-General and, coincidently, took on the task of reactivating and operating the civilian systems which had collapsed with the departure of the Belgians.

While this was happening, my family and I were holidaying at a lake west of Ottawa. I was the Wing Commander of the RCAF's 412 VIP Squadron. Early one morning in late July, the manager of a nearby airport drove up in his pickup and told me I was wanted on the phone by "some big-shot" at RCAF Station Trenton (a large Canadian military base)![6] I went to the phone and spoke to my boss; Air Commodore Carpenter. His words were "You're to go to the Congo tomorrow". Naturally, I politely asked why, and for how long:

> You're to set up and run an air transport operation for the UN operations in the Congo. ... You're to jump on a plane and head for New York, where someone from UN Headquarters will meet you and brief you in more detail. From there you will head for Brussels where you'll get a detailed briefing on the situation in the Congo, and then you'll head by Sabena Airlines to Leopoldville. You should be away for a few weeks and, by the way, you're promoted to Group Captain as of today.

I did as I was told and arrived in New York – where no one met me. I phoned UN Headquarters and spoke to an advisor to Secretary-General Dag Hammarskjöld, Brian Urquhart (later Sir Brian), whom I had met before, and went on to Brussels by commercial air. There, the RCAF air attaché met me, gave me a bottle of Scotch and wished me luck, having informed me he had no idea what was going on. The Belgians were too busy to brief me. The next morning I arrived in Leopoldville and was met by Jack Maitland, whom Air Commodore Carpenter had left behind to help out until the small air staff group from RCAF Station Trenton and I arrived on the scene.

The Role of the UN Air Transport Forces in ONUC

The press release from UN Headquarters stated that I was – to use their phrase – "to command all UN air forces in the Congo". Obviously, this was a

---

5 The Canadair North Star was a 1940s Canadian development of the Douglas C-54/DC-4 long-range transport aircraft.

6 RCAF Station Trenton became Canadian Forces Base Trenton in 1968 with the establishment of the Canadian Forces (the merging of the RCAF, the Royal Canadian Navy, and the Canadian Army). It is a large Canadian military base several hours east of Toronto, on Lake Ontario (detail not in original; provided for readers unfamiliar with the Canadian military).

further endorsement of Canada's reputation and had little to do, I suspected, with my particular talents. The role of the United Nations Air Transport Force (UNATF) was to operate and control aircraft, air traffic, and the facilities needed to support the ONUC commander in the effective execution of his mandate. Our arrival within days of the receipt of the request by Ottawa saw our crew of 10 Canadian airmen undertake an operation which had no precedent in UN peacekeeping history.

On arrival, I had met with General Von Horn, ONUC's "Supreme Commander" (as he liked to be called) and came away with a vague understanding of what the mission would need by way of air support. I found Von Horn a warm, smart, and dedicated UN commander put into the most difficult role the UN peacekeepers had seen to date. He fought for his troops and did well for them. The fact that he may have lacked experience that would have better equipped him for the job is a moot point as there simply was no precedent for ONUC.[7] My first job was to write my Terms of Reference (list of duties) and define our role as precisely as could be done. The General immediately approved what I put in front of him.

The air transport job would include the control of External Airlift and the operation and control of Internal Airlift. The External Airlift involved the movement of military units and equipment, and ingoing logistic support from overseas to the Congo. The Internal Air Transport would include the movement of UN military and civilian personnel and materiel throughout the Congo. In addition it was to provide the resources to be able to deploy by air a battalion group of infantry to trouble spots as might be required to help local UN commanders re-establish stability in their particular region.

We soon discovered that this not only involved operating numbers of different kinds of aircraft over a very wide area but also that it would require the operation of an air traffic control system and the airfields which would be used. To cap it all, the air navigation and communications systems, as modern as any in Europe, had been abandoned by the Belgians and no local Congolese had been trained to the level necessary to put them back in operation. In some cases the equipment had been sabotaged while in others it had been vandalized.

With more than 15,000 UN troops already on location at many widely spread points, we obviously could not wait to produce a nice neat plan to put the whole project together. The troops had to be fed and supported. The limited road and rail and very expansive river transportation systems used by the Belgians were

---

7   "I liked Von Horn. I respected him and was loyal to him. I felt sad and resented the fact that his UN bosses, aided by input from a very ambitious Secretary-General Military Advisor who yearned for the CINC appointment, on occasion openly chose to ignore Von Horn's counsel. I was greatly honoured a couple of years later to help host Von Horn during his official visit to Canada and made sure he knew we thought he had done a first-class job". (This footnote and the preceding three sentences in the body of the text appears in the author's review of fellow chapter contributor Kevin Spooner's book on ONUC: Carr, W.K. "'Canada, the Congo Crisis and UN Peacekeeping, 1960–64' by Kevin Spooner". Book Review. *Royal Canadian Air Force Journal* 1(1) (2012), 84–7.)

no longer in operation. Simultaneously, we would have many activities on the go. All of these, hopefully, would lead eventually to the neat (and very expensive) package we could see down the road, but had neither the time nor the information to create in the rush. Inundated with demands on their talents and time never before experienced, our ten intrepid members managed it with aplomb and perhaps many shortcuts. We did have help from a Pakistan Army motor transport company in assembling loads and dispatching aircraft. And we increasingly commandeered people and equipment from the various headquarters and on the road to get the job done.

Evolution of the UN Air Transport Forces

With the Security Council's decision to create the mission, Dag Hammarskjöld's staff had immediately appealed to selected member nations for the resources they believed to be necessary to meet the mandate. With total confidence in their infallibility, and some limited advice from an eclectic array of ex-military UN employees, infantry units and air force personnel and aeroplanes were requested from different sources. In the army case it worked out well, as some expertise was evident in the staff and useful offers were made and accepted. In the air force case, however, no such knowhow seemed to be on hand when the non-specific requests for airmen and aeroplanes went out. And, unbelievably, before the requirement could be defined in detail, numbers of each appeared on the scene.

On our arrival, we discovered 17 C-47s (military DC-3s) had arrived, along with five C-119s. These had apparently been dug out of the North Atlantic Treaty Organization (NATO) war reserve in Europe and, until their delivery, some had not flown for upwards of 15 years. A mixed bag of helicopters and several Beaver and Otter fixed-wing aircraft as well, had been generously shucked off from Middle East UN missions and US Army units in Europe. While this raised our eyebrows, it was nothing compared to the surprise we got when we discovered that we had, or would shortly have on hand, aircrew and ground crew in uncoordinated lots from 11 different nations! The one encouraging offer of assistance was an Indian Air Force C-119 Squadron which would come as a formed unit. It was followed a short while later by an Italian Air Force C-119 Squadron.

In this confused atmosphere, we soon discovered we had pilots who had never flown the types of aircraft we had inherited and mechanics who were not qualified to fix them. To make matters even more difficult, there was the Yugoslav contingent of mechanics, real experts on their Russian version of the DC-3, but who had no facility in either of the languages that our Brazilian and Argentinian pilots spoke. The pilots spoke good English but the Yugoslavs spoke only Serbo-Croatian, with their sergeant able to speak some French. Initially this, too, created a problem, but the expertise of the Yugoslav mechanics soon convinced our South American pilots that these foreigners were as good, or better, than any they had worked with at home.

UNATF, with a fleet of obsolescent aircraft, many aircrew unqualified to fly them and mechanics of questionable skills and knowhow, was not off to a very impressive start – from the outset it would be expected to logistically support a field force of upwards of 20,000 troops widely dispersed over an area of nearly 1,000,000 sq mi. However, with the unqualified dedication and ability of a few key members, we were soon running conversion courses to qualify personnel and were routinely doing pilot check rides on all our pilots, whether they liked it or not. Flight safety, if nothing else, required it. We were responsible for the safe results of our efforts and had to make sure an acceptable standard could be met.

Flight checks were done using RCAF check standards. There was no sitting in the office and having someone else do it. After the initial run-through we were able to recruit others to accept some of the responsibility. Obviously the Indian Air Force and Italian Air Force squadrons maintained their own air force standards, even though the first time I flew with an Italian C-119 crew, I was a bit surprised to see a wicker-covered bottle of Italian vino in an especially fitted holder between the two pilot seats. (The explanation that potable water simply was not available in the Congo, and seldom in Italy, left me a bit uncomfortable, despite its purported logic!)

Progress

By the end of August, the dust had begun to settle. We were running regular flights to the main UN troop locations and had a better sense of safety regarding the situation on the ground at the airfields we were using. Not only did we operate the airlift but we also inherited several main airfields and their facilities. However, we lacked the expertise to fill the necessary air traffic control slots vacated by the departing Belgians, so we brashly contacted the International Civil Aviation Organization (headquartered in Montreal). Surprisingly and fortuitously, this generated a quick supply of the several professional air traffic controllers we urgently needed, to augment the few bilingual air traffic control tradesmen the RCAF had been able to provide.[8] At the outset we had asked much of the aircraft crews in having them operate into insecure and uncontrolled airfields where the local political situation was uncertain.

The Canadian Army signallers at most of these sites were of inestimable help and our crews went out of their way to make sure that the needs of the signal detachments and our tradesmen took priority. Much innovation was involved in acquiring vehicles, accommodation, and such amenities as we could locate. The UN support system was simply not geared for this kind of operation. However, their ignorance was our bliss!

---

8   The preceding three sentences appeared in Carr, W.K. (2012). "'Canada, the Congo Crisis and UN Peacekeeping, 1960–64' by Kevin Spooner" (see Note 7). The original text was modified to accommodate this additional insight.

By mid-September we had aircrew of 11 nationalities flying 78 aircraft of 13 different types. Despite the language barriers, inadequate training and lack of supplies, we were getting results. We still required the backup of Air Congo (a politically less offensive name invented for the Belgian air carrier Sabena) charter C-54s, as well as the maintenance and repair of our aircraft.

Air Marshal Hugh Campbell, the incomparable Canadian Chief of the Air Staff, was well known for keeping his ear close to the ground on all matters affecting his RCAF members and what they were doing. He called one day from Ottawa, via the long-range radio, and asked how things were going and whether there was anything we needed: I briefed him on some of our aircraft serviceability and aircrew proficiency problems and mentioned that we had a very large backlog of vehicles and equipment urgently needed in the field. He asked whether a couple of C-119s on temporary duty (with crews) would help. I, of course, said "yes" and within three days they were in Leopoldville! (I wonder if today's brass could react so quickly and completely.) In the two months we really needed them, these two borrowed aircraft and crews moved 386,000 lb of freight backlog and hundreds of passengers. Also, during the ONUC operation, the RCAF's 426 Squadron North Stars, in 392 flights, had airlifted more than 2,000 t of freight and 11,476 UN passengers into and within the Congo.

To further give body to UNATF, Canada had purchased four de Havilland Caribou aircraft and offered them to the United Nations for internal airlift. The RCAF in Canada was busily training the aircrews and ground crews to ferry them to the Congo and operate them as part of the UNATF when the Secretary-General, bowing to Russian pressure, refused the offer. This was a blow to our hopes. The Russian pressure was reportedly because of Canada's strong position in NATO, as well as its membership in the Commonwealth, which had a history in African affairs not necessarily covered with glory. The Russians also openly supported Lumumba, even after he had been fired by Kasavubu.

The records show that by the middle of October our "mixed bag" UNATF had actually moved more 10,000 t of freight and hundreds of passengers in its military aircraft. We had also met the UN's voracious need for paperwork by having issued Organization Orders, Air Staff Instructions, Supply Demand forms, job descriptions, and other "useful" documents. (They were all actually modified RCAF Forms printed locally with ONUC letterhead.) The world wondered how such a small and busy bunch of airmen could produce this stuff and still run an air operation. I still marvel at it!

An interesting political situation existed during this period when, for a while, no one knew who the government was. Lumumba claimed the job, of prime minister because he had been appointed into it, even though Kasavubu, with outside encouragement, had fired him, and Joseph Mobutu, recently promoted officer in the Belgian-officered Force Publique (renamed on independence Armée Nationale Congolaise or ANC), further promoted himself and led a coup d'état. Dr Bunche, the Secretary-General's Representative on the spot recognized Mobutu as the point man to deal with. The Russians objected strongly, but the Secretary-

General supported this position. Russian support for Lumumba was strong, even before Congolese independence had evolved. They saw him as the means to get re-established in Africa, having lost their footholds in Egypt and Tanzania.

The Russians tried to pressure the United Nations into allowing them to participate in the provision of aid and, despite the denial of overflight clearances by NATO nations as a result of a timely Canadian recommendation, the Russians did try. At one stage a dozen Il-12 aircraft loaded with "equipment" for Lumumba arrived in Stanleyville, via a very circuitous route, intending to proceed further to Leopoldville. The Ethiopian commander at the airport called on the radio and told me the aircraft were loaded with arms and ammunition. General Von Horn agreed that we should try to prevent this from being delivered. Since we airmen controlled the airfields, we ordered the UN detachments at the usable airfields to block their runways by parking vehicles or 45-gal. gas barrels on them as soon as we gave them word that the Russians were airborne. Our UN commanders did as requested, and the Il-12 pilots, on learning there was no Congolese destination open, had no alternative but to return to Khartoum. Not a word appeared in the press nor was heard from the Russians later.

In retrospect, it is amazing how easy it was to get things done when one judiciously avoided being trapped by the UN bureaucratic network. A few times I was chastised for not seeking authority ahead of time. But when the results looked good and all could take a bow, shorting the system was overlooked. The fact was, we were far too busy to waste time on details when the course of action was obvious. Again, their ignorance was our bliss.

Observations

How to Ground a Russian Tu-104

Late one afternoon in October a Russian Tu-104 military transport jet landed in Leopoldville. General Von Horn informed me that its likely purpose was to lift Lumumba out of the country, and this was not what the United Nations wanted to happen. He wondered if we airmen could quietly arrange to delay his departure for a few hours. I met with two of our intrepid, innovating airmen, stated the problem and was reminded that a high performance jet could not taxi or take off with flat tires. Since we controlled the airport and our good buddies the ANC guards were now very friendly with us (because we had arranged that they be paid their overdue wages), in the dark of night the deed was done.

The Russians were most upset when, late the following morning, they explained to us why they needed to borrow our air compressor. They departed Leopoldville that afternoon and Lumumba was not on board. The Secretary-General's Representative would later mention how lucky ONUC had been that the Russian aircraft had had a problem and Lumumba was unable to get away as he had planned. We choose not to enlighten the UN staff on what happened. With hindsight, had we sought CYA authority,[9] it would never have been granted and Lumumba would not have been stymied.

Official Dinner Guests

While Lumumba was in power, he hosted a black tie dinner to which some UN staff members were invited. Lieutenant-Colonel "Johnnie" Berthiaume, a Canadian Army Officer in the Royal 22nd Regiment – the Van Doos – were seated at the head table. Berthiaume also was one of the ablest and best officers I have ever met in the Canadian Forces. He was an incredibly supportive and loyal aide to General Von Horn, who trusted him completely.[10] At the appropriate time our host decided to speak to his guests, including the US representative, and update them. His speech soon developed into an anti-Western harangue in which Canada in particular was vilified. Colonel Berthiaume and I listened for a while and with Lumumba still in full flow decided simultaneously, I think, to depart the gathering in protest. We were featured in the local press the following day. While we felt some political upsets might follow our actions, I personally was more worried that one of Lumumba's AK47-armed and highly visible guards might shoot us in the back as we left.

The UN Supply System

In ONUC, the UN civilian staff handled all logistics and this included accommodation and ground transportation and there were official forms for everything, including for the bits and pieces we needed to repair aircraft. The UN supply system, though, was hopelessly overloaded, out of its depth, and was virtually impotent when it came to aircraft support. To say it was a slow process is being generous. Another case in point concerns the lack of vehicles for getting

---

9   CYA is defined here as "Cover Your Ass". A perfectly good and frequently used expression to suggest there was rational thought in the activity referred to!

10   "Berthiaume was superbly politically sensitive and he could sway even the most ardent UN bureaucrat to act! He and Colonel Joseph-Désiré [later changed to Sese Seko] Mobuto, a central character in the chaotic Congolese political situation, became close friends. The UN brass did not take to Berthiaume because they knew he knew more about the Congolese political situation than they did. When the Secretary-General's UN representative ordered the closure of the airports to forestall some perceived Lumumba – the first legally elected Prime Minister of the Republic of the Congo – exploit, we of course said 'yes' and ignored it. After the fact, Berthiaume told General Von Horn, who laughed loudly and warned us he had not heard what he had just been told! We had to feed the troops and we had to allow the inflow of external airlift by Canada and the United States, not just 'knee-jerk' react to some inane political solution to a perceived problem". (This footnote and the preceding two sentences in the main body of the text appeared in Carr, W.K. (2012). "'Canada, the Congo Crisis and UN Peacekeeping, 1960–64' by Kevin Spooner". The original text included here was minimally edited at this point.).

crews from their accommodation to the airport and for other administrative purposes. No civilian transportation systems were operating either locally on the few highways and rail spurs, or on the river system. Early on, this shortage hamstrung our efforts, so having had no response from the UN system one can-do RCAF Flight Lieutenant conned his UN civilian supply friends into giving him the forms to requisition the vehicles we needed. He did, and four years later I received a query from the United Nations in New York asking the whereabouts of a dozen or so vehicles I had signed for in August 1960.

On the suggestion of an RCAF supply sergeant in our crew, we fashioned with his buddies back in Canada at RCAF Station Trenton an arrangement whereby we could request bits and pieces for all our aircraft directly from the RCAF's Air Materiel Command, even for the Italian and Indian aircraft, and RCAF Supply would meet our demands and then bill the United Nations for repayment. It worked beautifully and amazed many, including the foreign aircrew and the out-of-depth UN logistics staff. The UN civilian staff was more than pleased and soon became very cooperative in things Canadian originated by the RCAF and the Canadian Army signallers.[11]

Other Tales, Other Times

The cultural and political sensitivities of contingents from 28 different national sources created many headaches for staffs. For example, bivouacking flip-flop-shod Guinean troops alongside American infantry-booted Liberian troops caused the Guineans to demand that they be kitted just as well as their neighbours. The United Nations complied and, in passing, had real trouble rounding up boots big enough to fit previously unshod feet. A similar problem arose over UN service allowances. Egyptian soldiers claimed US$6.00 per day, Canadians US$0.30 per day!

An example of politics entering day-to-day affairs was the case of the Israeli-packed and labelled canned-pork products doled out to the Egyptian contingent by the UN quartermaster. A political crisis ensued, with the Egyptians accusing all and sundry of a deliberate attempt to embarrass them. While members of a UN force, they were still enemies at home!

The chasm between officers and other ranks in some contingents were eye-openers also. For example, we had one group whose Wing Commander complained that their officers were expected to ride on the same bus as their mechanics and this was unacceptable. He wanted separate buses. I suggested to him that if his government would indicate its willingness to buy an additional crew bus and supply a driver, we would have no objection. But in the meantime, perhaps they

---

11   Much of the text in this part of the chapter on the UN supply system has been modified and supplemented with material from Carr, W.K. (2012). "'Canada, the Congo Crisis and UN Peacekeeping, 1960–64' by Kevin Spooner".

could arrange to share the bus, with some sitting up front and the others at the back. I heard no more from that source.

African military personnel, especially those from former colonies, seemed prone to respect the authority of us foreigners more than they did that of other Africans. I could cite many examples of tribal attitudes being basic behaviour driver. but one sticks out. The Force Publique/ANC detested the Ghanaian officers, who bossed them in the provision of airport security. When one group of French-speaking Canadian troops arrived and were mistaken for Belgians, it was an on-site RCAF officer who stopped the mayhem, with absolutely no help from the Ghanaian officers, who stood by and watched. These officers claimed the Congolese soldiers paid no attention to the orders given them.

Compared with the Commander-in-Chief's job of keeping his troops from 28 nations fed and happy, our job to help, while critical to his courses of action, had few of the political and sociological factors to distract us. We had untrained personnel, but they were being trained, and our multinational air force was making good progress. Safety of our crews and passengers was paramount and for the most part, luckily, we were successful.

The Royal Canadian Air Force North Star Lifeline

During the deployment phase of the ONUC operation, 426 Squadron's 13 North Stars were flown at the – until then – unequalled rate of 180 hrs per month each, for a total monthly flying rate of 2,340 hrs. The unit played a critical role in terms of support to the whole UN Congo buildup.

Being unpressurized, the aircraft usually operated below 10,000 ft, especially with passengers on board. There was no passenger oxygen installed. Consequently, and unlike today's high-performance passenger jets, the North Star crews spent much of their time flying on instrument flying rules rather than above the weather. It was hard, tiring work and poorly paid, but a challenge these professionals accepted and relished as their duty. A round trip for a North Star crew from RCAF Station Trenton to the Congo was approximately 70 flight hrs. Through the use of en route "slipcrews" the aircraft could be back in Trenton in a little less than four days. The crews, however, limited to twelve-hour duty days, could be on the route for eight days or more.[12]

The initial Canadian deployment route was from Trenton via The Azores to Dakar, thence to Accra and on to Leopoldville. After these deployments were finished a twice-weekly scheduled flight via Pisa, Italy, normally used for the RCAF Middle East shuttles, was instituted at the request of the United Nations. These flights continued until the wind-up of ONUC in 1964 (from late 1960

---

12   "Slipcrews" is a term denoting aircrews established in situ on the way to replace arriving crews who will have run out of "duty time", or would do so were they to proceed further. The arriving crew rests and replaces the next incoming crew, and so on. So crews "slip" back to the next flight.

onwards they were carried out by the Yukon, the new and much more operationally capable North Star replacement).

The UN Air Staff

Our airmen from eleven nations, speaking six languages, were nothing short of amazing in what they were able to do. They needed little supervision or direction and, regardless of nationality, seemed to be blessed with the knowhow and understanding which led to the on-the-spot innovation and action that produced the results needed and normally would not have been seen to be possible, even from personnel of much higher rank and experience. Biased I may be, but in retrospect the glue in the whole operation provided by the small group of highly motivated, dedicated and loyal RCAF officers was the key to the amazing success which was achieved in the early days of ONUC.

Despite the differences in operational techniques, flying standards and discipline, only one incident arose requiring the removal of a senior officer from command of a squadron. He was quietly sent home and I was severely chastised for being so "politically insensitive"! (He was not a Canadian.) With the direct link our long-range radios provided to our base in Trenton and to the Canadian capital, Ottawa, we were able to get support results that astounded the top-heavy UN Headquarters in Leopoldville and New York. Our Pakistani, Norwegian, Swedish, and Indian staff members working alongside the small Canadian staff deserve much credit also for helping to make this complex operation realistic and workable. To all of us, the 426 Squadron North Stars brought spares, Thanksgiving turkeys, and otherwise unavailable potables to supplement UN quartermaster provisions.

Retrospective

In retrospect, it is easy to see why the granting of independence to the Congo in 1960 was bound to fail. Belgian colonial policy had prevented native Congolese from holding positions of any authority above a basic level. Even in the indigenous 25,000-man Force Publique/ANC, the most senior Congolese rank was Sergeant – all the officers were Belgian. Thus, at their departure, the Force was virtually leaderless. Many members promoted themselves and Sese Seko (born Joseph-Désiré) Mobuto, a sergeant under the Belgians, made himself a colonel at the time of his coup d'état, a couple of months after they had left.

What education system there was taught only some reading, writing and mathematics – basic skills to peoples identified by the missionaries and other church officials as having potential. By 1960, only one indigenous Congolese, reportedly, was a university graduate. Some trade schools were run by the Belgian military, as were apprenticeship courses by the airline and mining interests.

However, no basic intellectual or educated infrastructure existed to take on the business of government. In this sad country, these problems still exist today.[13]

Conclusion

Much of the success during the early days of ONUC was due to the wholehearted manner in which the Canadian government responded to the UN request. It provided, with few strings attached, the best Canada had to offer. At the forefront of this largesse was the skill and universally admired professionalism of the Canadian soldier and airman. RCAF Air Transport Command's ability to get the job done was once again in evidence. The USAF of the day lauded our operation as the best military air transport in the world, not excepting its own military air transport system.

The RCAF was a key factor in ONUC's early success. Through name changes and force realignments, the core function continues to uphold the RCAF's tradition of, "Excelling at every task it ever undertook". *Sic Itur Ad Astra!*

---

13   In the original version of this chapter, as a paper in the *Canadian Aviation Historical Society Journal*, the author included further discussion of the historical, geographic, and social context of the Congo. This has been removed, as Kevin Spooner's contribution to this volume (Chapter 3) provides an overview of these issues and frames the broad historical context.

Chapter 2

Peacekeepers in Combat: Fighter Jets and Bombers in the Congo, 1961–1963

A. Walter Dorn

The United Nations Operation in the Congo (ONUC) was the largest, most complex, and most expensive UN peacekeeping mission of the Cold War. It was also the most robust operation, utilizing air power in an unprecedented and, in fact, unrepeated fashion among UN peace operations. It was, for example, the only UN peace operation to date to utilize bomber aircraft.[1]

The mission began as an effort to restore law and order in the Congo, a vast and newly independent country that had just elected its first democratic government. ONUC's military operations were first devoted to quelling the uprising of the riotous Congolese National Army (Armée Nationale Congolaise (ANC)). UN Secretary-General Dag Hammarskjöld did not at first allow ONUC to interfere in the internal and complex issue of the secession of the Katanga province. After Hammarskjöld's fatal plane crash while seeking a Katangan settlement in September 1961, the UN Security Council and the new Secretary-General, U Thant, adopted a firmer, more proactive stance, effectively siding with the Congolese central government to halt Katanga's secession. This effort then involved a myriad political and military Cold War intrigues, major US support, a murdered prime minister, and an operational mandate more forceful than had ever been put in place in UN peacekeeping. Katanga's resistance, especially in the air, necessitated the creation of the first "UN Air Force".[2] There followed the unique story of an aerial arms race.

---

1   In the Korean War, the "UN Command" utilized numerous combat aircraft but this was not a mission operated by the United Nations or its Secretary-General. The mission was authorized by the UN Security Council to engage in an enforcement action (war fighting), but it was under US command with minimal direction from New York.

2   The United Nations had used aircraft in pervious peace missions, but not for combat. In the UN's unarmed observer missions and its first peacekeeping force, it benefited from "dual use" aircraft for transport and reconnaissance. For instance, the United Nations Emergency Force (UNEF) deployed a few helicopters, DC-3 and Otter aircraft, later replaced by Caribou light cargo aircraft. By 1965, two light aircraft were deemed sufficient for air observation. Source: UN Doc. A/C.5/1049, "Survey of UNEF: Report of the Secretary General", 13 December 1965, quoted in Higgins, R. *United Nations Peacekeeping, 1946–1962: Documents and Commentary*, 1, *The Middle-East* (Oxford: Oxford University Press, 1970), 317.

Phase I: Deployment to Restore Order

During the first phase of the Congo Operation, from July 1960 to February 1961, ONUC's principal function was to restore order throughout a vast country that had fallen into widespread lawlessness and chaos. This tragic state arose immediately following the Congo's independence from Belgium on 30 June 1960 when the ANC mutinied against both its Belgian officers and the Congo's first democratically elected government. This triggered tribal uprisings against the central government. The national force that should have quelled these rebellions, notably the 25,000-strong ANC, began to plunder European property and even beat and kill many Belgians who had remained in the Congo, as well as their fellow Congolese.[3]

During this phase of the operation, the United States provided strategic airlifts to transport an unprecedented number of UN troops into the Congo. The US Military Air Transport Service, using about 50 C-124s, moved 9,000 UN troops, in about two weeks,[4] to positions across a country approximately the size of Western Europe. ONUC gradually re-established a semblance of law and order, and once the UN mission demonstrated an ability to protect civilians (including Belgian citizens) the Belgian troops began to depart. After ONUC's massive deployment was accomplished, air transport remained vital as almost all supplies had to be transported by air to ONUC troops dispersed across the vast country.[5]

During the first few months, UN troops were engaged in policing and training rather than fighting. As a result, the aerial contribution was limited to troop transport and supply – for a firsthand account from the individual responsible for UN air operations during this period of ONUC's operations see Chapter 1 in this volume. ONUC units succeeded in disarming many of the rebellious ANC troops,[6] which helped restore a degree of law and order. At this early juncture, ONUC's mandate forbade it from interfering in internal aspects of Congolese politics; thus, it did not undertake operations to force Katanga to end its secession. In fact, Secretary-General Hammarskjöld refused to comply with Prime Minister Patrice Lumumba's demands that ONUC enter Katanga, subdue that province's rebel forces, and compel Katanga's leaders to submit to the Congo's central government. On 9 August 1960 Security Council Resolution 146 mentioned Katanga for the first

---

3   Von Horn, C. *Soldiering for Peace* (New York: David McKay Company, 1967), 152.

4   Rikhye, I.J. *Military Adviser to the Secretary-General: UN Peacekeeping and the Congo Crisis* (New York: St. Martin's Press, 1993), 193.

5   To head the air logistics, the United Nations appointed Canadian Group Captain Bill Carr, later chief of the Canadian Forces Air Command. See Rikhye, *Military Adviser to the Secretary-General*, 97; see also Carr, W.K. "The RCAF in the Congo, 1960: Among the Most Challenging Assignments Undertaken by Canadian Forces in the Peace Keeping Role", *Canadian Aviation Historical Society Journal* 43(1) (Spring, 2005), 4–11, 31; Chapter 1 in this volume.

6   Von Horn, *Soldiering for Peace*, 172.

time, allowing UN forces to enter Katanga, but not to "intervene in or influence the outcome of any internal conflict".[7] Further complicating matters, the Congolese leadership fell into disarray. Joseph Kasavubu managed to eject Lumumba from power. However, the international mood of "non-interventionism" did not change until after Lumumba's murder on 17 January 1961 at the hands of his enemies in Katanga.

Phase II: The Fight for Katanga

The second phase commenced with Security Council authorization to take "all appropriate measures" to prevent the occurrence of civil war in the Congo, including "the use of force, if necessary, in the last resort".[8] This resolution was used to justify UN military operations to end the Katangan secession. Ironically, Prime Minister Lumumba's death triggered the fulfillment of his demands that the United Nations forcefully support his country's campaign against the secession. Also looming large was the threat of intervention by the Soviet Union, which was emboldened and angered after Lumumba's murder, and Moscow's offer to provide the Congolese government with personnel and materiel to suppress the secession. These developments combined to mobilize Western powers to request the United Nations to fulfill that role.

Katanga's leader, Moïse Tshombé, professed anti-Communism and was backed by powerful Belgian and other Western interests, especially the company Union Minière du Haute Katanga. Also Tshombé controlled Katanga's gendarmerie and a large cadre of mercenaries. The resolve of his secessionists hardened after some 1,500 of the central government's troops reached north Katanga in January 1961. Until that initiation of hostilities, the neutral zone negotiated by the United Nations with Tshombé on 17 October 1960 had held up but "it all came apart as pro-Lumumba troops captured Manono" in north Katanga.[9] After Manono, the situation deteriorated rapidly and negotiations broke down.

On 28 August 1961, the United Nations launched Operation Rumpunch to arrest and deport mercenaries in Katanga. Then, in September, the Indian-led UN forces in Katanga launched Operation Morthor ("morthor" is the Hindi word for "smash"), to further round up foreign mercenaries and political advisers and to arrest Katangese officials. The "arrest" operation, which violated Hammarskjöld's explicit directions to ONUC, quickly escalated into open warfare.

Almost immediately, air power in Katanga was brought in as a game-changer – but not by the United Nations. At this early stage of the conflict, the Aviation Katangaise (Avikat), also known as Force Aérienne Katangaise (FAK), held air superiority, though it consisted of only three Fouga Magister jet trainers.

---

7   Ibid, 225.
8   UN Security Council Resolution 161 of 21 February 1961.
9   Rikhye, *Military Adviser to the Secretary-General*, 182.

Remarkably, these aircraft were brought to Katanga in February aboard a Boeing Stratocruiser by the Seven Seas Charter Company, later identified as a US Central Intelligence Agency (CIA) contractor and possibly a front company. After UN officials observed the unloading of the aircraft, the mission grounded the company's entire fleet of planes, which the United Nations had earlier contracted to carry food. President John F. Kennedy decried the jet delivery and alleged in correspondence with President Kwame Nkrumah of Ghana that the transaction had taken place before the US government could stop it.[10]

In any case, the KAF fleet was quickly reduced in effectiveness: one Fouga Magister was lost when its pilot tried to fly under (rather than over) a power line; and UN forces captured another when they seized the airfield at Elisabethville, the Katangan capital, on 28 August 1961. This left the FAK with only one plane, but this single aircraft attained world renown during the hostilities of September by paralyzing UN supply efforts, which were mostly conducted by air transport aircraft. The single jet, flown by a Belgian mercenary from the Kolwezi airfield, also strafed UN positions, including the UN Headquarters in Katanga, and helped isolate a company of Irish troops who were forced to surrender to Katangan forces. Furthermore, the Fouga jet destroyed several UN-chartered aircraft at Katangan airports, including Elisabethville, the Katangan capital.[11] A US State Department official, Wayne Fredericks, commented: "I have always believed in air power, but I never thought I'd see the day when one plane would stop the United States and the whole United Nations".[12]

---

10   "UN Grounds 6 Planes; Report Says Congo Charter Carrier Aided Katanga", *New York Times*, 22 February 1961, 2. The CIA connection is described in Mahoney, R.D. *JFK: Ordeal in Africa* (New York: Oxford University Press, 1983), 81. Apparently, the agency at first saw the pro-West Katanga regime as a buttress against the pro-communist Lumumba. Further references on early CIA support to Katanga are given in Spooner, K. *Canada, the Congo Crisis and UN Peacekeeping, 1960–64* (Vancouver: UBC Press, 2009), 249 n.8. In late 1962, when the US was solidly against the secession of the Katanga province, the CIA put together an air combat unit to fight the Katangese, with pilots from Cuba (anti-Castro exiles), but it did not engage in combat in the Congo until 1964 during the Mulele revolt. See Hellström, L. *The Instant Air Force: The Creation of the CIA Air Unit in the Congo, 1962* (Saarbrüucken: VDM Verlag, 2008), 34, 39.

11   Johnson, R.C. "Heart of Darkness: The Tragedy of the Congo, 1960–67", *Chandelle: A Journal of Aviation History*, 2, no. 3 (October 1997), 11. Available at: http://worldatwar.net/chandelle/v2/v2n3/congo.html [accessed 7 May 2014]. In fact, Katanga had begun aerial attacks in this conflict on 30 January 1961 by using two converted commercial aircraft to bomb Manono and attack the UN garrison there, though the attack was not aimed at the United Nations but rather at the pro-Lumumba Congolese army that was forcefully entering Katanga. See Weissman, S.R. *American Foreign Policy in the Congo* (London: Cornell University Press, 1974), 183.

12   Quoted in Klevberg, H. "Logistical and Combat Air Power in ONUC", in Carsten F. Ronnfeldt and Per Erik Solli, eds, *Use of Air Power in Peace Operations* (Oslo: Norwegian Institute of International Affairs, 1997), 36–7.

Deadlock prevailed throughout 1961, and the indecisive outcome of the UN's August and September 1961 ground initiatives in Katanga (Operations Rumpunch and Morthor) spurred Hammarskjöld to try to negotiate a ceasefire with Tshombé. As the Secretary-General was flying to meet with the Katangan leader at the border town of Ndola, Northern Rhodesia, his plane crashed on the night of 17 September 1961, killing all onboard. Complicating the rescue effort, the plane had largely maintained radio silence and flew a circuitous route mostly at night in order to reduce the possibility of an attack by the "Lone Ranger" Fouga Magister. The Katangan jet had shot bullets into UN aircraft only days before. And Hammarskjöld's aircraft had been damaged by ground fire but was quickly repaired before take-off. The cause of the UN plane crash was never determined with certainty, though a UN commission concluded that it was probably due to pilot error during the approach to Ndola.[13]

With Hammarskjöld's death, the battle for Katanga entered a new phase. The new Secretary-General, U Thant, did not share Hammarskjöld's belief that the United Nations should not interfere in Congolese internal politics. Moreover, the general escalation of events spurred the Security Council to pass Resolution 169 on 24 November 1961, strongly deprecating the secessionist activities of Katanga and authorizing ONUC to use "the requisite measure of force" to remove foreign mercenaries and "to take all necessary measures to prevent the entry or return of such elements".[14]

Meanwhile, the United States, fearful of communist encroachment on the continent, was resolved in the Congo to keep the Soviet Union *out*, the United Nations *in*, and Belgian interference *down* in the former colony.[15] The Americans also wanted to stop the country from falling apart, viewing secession of mineral-rich Katanga as a threat to the economic vitality of the new country. In the background, decolonization was one of the great movements of the era and the United States was keen to show newly independent countries that it supported integral, viable new states. The disintegration of the Congo was a major concern, as was Soviet intervention. Therefore, international (United Nations) intervention

---

13   The plane crashed 15 km/9 mi into its approach to Ndola airport. There are many theories and alleged conspiracies behind the crash. See Kalb, M.G. *The Congo Cables: The Cold War in Africa – From Eisenhower to Kennedy* (New York: Macmillan, 1982), 298. See also Devlin, L.R. *Chief of Station, Congo: A Memoir of 1960–67* (New York: Public Affairs, 2007), 167–8.

14   United Nations, "Resolution [169] Adopted by the Security Council at its 982nd Meeting on 24 November 1961", Document S/5002, 24 November 1961. See also Findlay, T.C. *The Blue Helmets' First War? Use of Force by the UN in the Congo, 1960–64* (Canadian Peacekeeping Press, 1999), 247.

15   This type of formulation is borrowed from Lord Hastings Ismay, the first North Atlantic Treaty Organization (NATO) Secretary General, who described NATO's purpose as: "to keep the Russians out; the Americans in; and the Germans down". See Reynolds, D. *The Origins of the Cold War in Europe: International Perspective* (New Haven, CT: Yale University Press, 1994), 13.

in Katanga was deemed necessary, even if it meant intervention into the internal affairs of a new state (although at the request of that state). Thus the United States, which had previously refused Hammarskjöld's requests to ferry troops within the Congo and had only brought troops to the Congo from abroad, now provided four transport planes without conditions. President Kennedy even offered to provide eight fighter jets if no other member nations were willing to do so.[16] The US Joint Chiefs of Staff suggested these jets could "seek out and destroy, either on the ground or in the air, the Fouga Magister jets".[17] However, Thant sought to avoid direct superpower involvement in combat. Having promises of fighter jets from other nations, the American offer was turned down.[18] Instead, the United States provided over 20 large transport planes to ferry reinforcements and anti-aircraft guns into Katanga.

Before his death, Hammarskjöld had managed to obtain from several UN member states promises of combat aircraft, which were desperately needed for the field mission. In October 1961, Sweden provided five J-29 Tunnan ("The Flying Barrel") fighter jets – one of which is shown in Figure 2.1. Ethiopia sent four F-86 Sabre jets, and India backed the mission with four Indian B(I)58 Canberra light bombers. These aircraft became what mission personnel dubbed the first "UN Air Force".[19]

The UN's aerial assets soon joined the fray. In December, they attacked a military train east of Kolwezi and Katangan airfields at Jadotville and Kolwezi.[20] The United Nations created havoc among Katangan forces in much the same way that the armed Fouga Magister had earlier done to the UN mission. Charanjit Singh, one of the Indian UN pilots, described his attack on a Katangan camp in Elisabethville on 8 December 1961 in a cavalier fashion:

> … attacked an army police camp 2 km NE of old runway. Some vehicles were parked outside what looked like a headquarters building. I fired a full burst on those and saw them going up in smoke and flames. As I pulled out of the dive, I saw hundreds of men running out in utter panic. As I flashed past them, I gathered an image of men running in all directions, some in undies, others in halfpants, some in uniforms. I saw some enter a billet. Attacked the HQ building

---

16   Kalb, *The Congo Cables*, 303.

17   Quoted in Kalb, *The Congo Cables*, 303. Apparently, the Katangan authorities had managed to obtain other Fouga Magister jets by this time.

18   Besides, the United States did not want to be directly involved in combat, in part to avoid giving the Russians a stronger reason to deploy.

19   Memorandum, "Command and Control – Fighter Operations Group", 13 October 1961, in UN Archives, New York, DAG-13/1.6.5.8.3.0:1 6600/F-OPS Policy. October 1961–March 1963.

20   Rikhye, *Military Adviser to the Secretary-General*, 294. Descriptions of the Kolwezi attack by Indian Air Force officers are provided in Singh, P. "Canberras in the Congo", Bharat Rakshak website. Available at: http://www.bharat-rakshak.com/IAF/ History/1960s/Congo01.html [accessed 7 May 2014].

Figure 2.1   Rockets are uncrated before being deployed on Saab J-29 jets
Source: UN Photo 72379.

and vehicles again. Saw a vehicle turn over. At the end of four attacks, the whole thing looked like the Tilpat [air-to-ground practice firing range near Delhi] show.[21]

The net result of the UN buildup and its December 1961 offensive was that Katanga's "air superiority" was temporarily ended.[22] The fate of the infamous jet trainer became an object of much speculation. The UN pilots claimed to have destroyed it on the ground in an air attack on the Kolwezi airfield, but they actually hit a carefully crafted dummy. It was then believed that the Katangan Fouga had crashed while its South African mercenary pilot had parachuted to safety,[23] but this too was found to be false.

But even the UN's new aerial hardware was deemed insufficient for the robust mandate. The UN field mission pressed headquarters to obtain bombs for the Indian Canberra jets. "We need those bombs", Secretary-General U Thant would insist to the British government.[24] After weeks of stalling, the government of Prime Minister Harold Macmillan finally agreed on 7 December 1961 to supply 24 1,000-lb bombs. But the offer came with the condition that they could only be used "against aircraft on the ground or [against] airstrips and airfields".[25] Even still, Macmillan worried that his government might fall over its handling of the Congo crisis, given the fierce support in some Conservative quarters for the anti-communist Katanga regime.[26] In the end, the United States transported bombs directly from India.[27]

Realizing what an enormous role a single Fouga jet had played in the success of Katangan operations in September 1961, Tshombé began purchasing new aircraft and hiring foreign mercenary pilots of various nationalities to fly them. Indeed, throughout 1962, UN Air Command desperately tried to monitor the Katangan aerial buildup through both aerial surveillance of Katangan airfields and intelligence gathered by ONUC's Military Information Branch (MIB).[28] In an

---

21   Singh, C. "Congo Diary", *Air Power Journal*, 2, no. 3, 2005 (July–September), 36. Available at: http://www.isn.ethz.ch/Digital-Library/Publications/Detail/?ots591=0c54e3b3-1e9c-be1e-2c24-a6a8c7060233&lng=en&id=119942 [accessed 6 May 2014]. The Air Contact Team controls offensive air support missions in the battle zone. On 9 December 1961, ground forces put a bullet through the cockpit of Singh's aircraft, missing his head by only .5 m.

22   UN Archive, ONUC files, Annex A to MIL INFO 852, Leopoldville, 30 May 1962, Headquarters ONUC, "Katangese Air Capability, An Appreciation", 6, paragraph 22 in UN Archives, New York, S0829-1-14. (DAG 13/1.6.5.8.4.0).

23   Ibid.

24   Kalb, *The Congo Cables*, 315.

25   Kalb, *The Congo Cables*, 315, quoting from Harold Macmillan, *Pointing the Way* (New York: Harper & Row, 1972), 449.

26   Kalb, *The Congo Cables*, 315. The opposition Labour Party wanted the bombs delivered right away, while the Conservative Party wanted the United Kingdom to stop supporting ONUC altogether.

27   "Bomb Supply on Way", *New York Times*, 2 January 1962, 2.

28   Dorn, A.W. and Bell, D.J.H. "Intelligence in Peacekeeping: The UN Operation in the Congo 1960–64", *International Peacekeeping*, 2(1) (Spring, 1995), 11–33.

attempt to procure immediate intelligence on Katanga's air capability, a desperate ONUC on 9 March 1962 noted that aircrews from UN military air units and from its charter companies were making "important observations during their flights and stops at various airfields in the Congo".[29] The mission began mandatory debriefings of aircrews after landing. The mission also sought to create an air reconnaissance unit capable of meeting both long-term reconaissance and immediate operational requirements. One memo dated 10 March 1962 stated "it becomes imperative that the air recce unit should be allotted with both C-47s and jet recce aircraft such as S-29s or photo-recce Canberras".[30] ONUC's Chief of Military Intelligence requested three C-47 aircraft "to check the Katangan air movements through systematic visual reconnaissance of their airfields".[31] On 6 June 1962 the ONUC Force Commander cabled Ralph Bunche, the Under-Secretary-General at UN Headquarters responsible for peacekeeping operations that:

> ONUC suffers from a grave lack of reconnaissance facilities. As a result even the photographs available may contain much more information which it is NOT possible to get because of inadequate facilities in equipment and personnel for interpretation.[32]

In 1962, Sweden provided two J-29Cs, the photo-reconnaissance versions of the J-29 jet aircraft that proved of great worth.[33] The mission consequently added personnel designated as air intelligence officers. At the same time, the threat of re-emerging Katangan aerial capabilities was real. ONUC concluded in May 1962:

> [M]ercenaries, fighting for money and receiving higher salaries as FAK pilots than even Generals receive in UN service, are ruthless, cunning, non-conventional, clever and inventive. They have war experience, and they know where, when and how to hit ... there is no alternative but to consider FAK as a dangerous enemy in the air.[34]

---

29 Appendix 2 to Annex A, To MIL INT 121, 9 March 1962, "Debriefing of Aircrews", paragraphs 1–2, in UN Archives, New York, S0829-1-10 (DAG 13/1.6.5.8.4.0).

30 Memorandum, MIL INT 121, 10 March 1962 to Chief Fighter Operations Officer, "Air Reconnaissance", 1, paragraph 3 in UN Archives, New York, S0829-1-10 (DAG 13/1.6.5.8.4.0).

31 Memorandum, MIL INT 4/A/5, 16 March 1962 to Air Commander (through Chief of Staff), "Aerial Reconnaissance", in UN Archives, New York, S0829-1-10 (DAG 13/1.6.5.8.4.0).

32 MIL INFO 852, 6 June 1962, To Dr Ralph Bunche, United Nations – New York, From Force Commander – ONUC, Leopoldville, "Katangese Air Capability", 1, paragraph 4, in UN Archives, New York, S0829-1-14 (DAG 13/1.6.5.8.4.0).

33 "J 29 Tunnan", see http://historywarsweapons.com/page/15

34 Annex A to MIL INFO 852, Leopoldville, 30 May 1962, Headquarters ONUC, "Katangese Air Capability, An Appreciation", 9, paragraph 32, in UN Archives, New York, S0829-1-14 (DAG 13/1.6.5.8.4.0).

ONUC had success uncovering the extent of Tshombé's aircraft acquisitions through intelligence gathered by the MIB. Defectors and informants interviewed by the MIB revealed a wealth of information about Katangan aircraft both in Katanga and neighbouring countries. Lieutenant-General Kebbede Guebre (Ethiopia), the ONUC Force Commander, cabled Bunche at UN Headquarters on 24 August 1962, referencing a report that Katanga-owned jet fighters were hidden in Angola and/or Rhodesia. Kebbede requested Bunche to "check with Australia [about] the possibility of Australian trained jet [mercenary] pilots being available to Tshombe".[35] In another cable to Bunche dated 27 September 1962, he stated that:

> a fully reliable source reported ... that twelve Harvard aircraft have recently left South Africa, bound for Katanga ... equipped with guns and French rockets ... [and that] an unspecified number of P-51 Mustangs may have left South Africa recently ... intended for Katanga.[36]

Clearly, the United Nations perceived itself in an aerial arms race with the Katanga government. It was trying to persuade its member states to provide aircraft while the Katanga government was purchasing them clandestinely wherever possible.

General Kebbede again cabled Under-Secretary-General Bunche on 1 October 1962, comparing the air capabilities of the two protagonists. Katanga (FAK) was now estimated to have twelve Harvard single-propeller aircraft, eight or nine Fouga Magister trainer jets, four Vampire jet fighters and a large number of P-51 Mustang single-propeller fighters (being delivered).[37] The UN mission possessed six Canberra jet fighter–bombers, four Saab J-29B fighter–bombers, and four Sabre F-86 jet fighters.[38] At the time, the UN Air Division possessed no bombs – a serious deficiency, as it was considered the weapon needed to neutralize air bases and enemy forces on the ground.[39] Great Britain was still dithering on UN pleas for bombs for its Canberra aircraft. ONUC concluded once again that air resources were inadequate to meet the FAK threat. Due to serviceability problems, only about 60 to 70 percent of ONUC aircraft would be available for operations,

35 Outgoing Code Cable to Dr Bunche from General Kebbede, 24 August 1962, No. ONUC 5838, paragraph 1, in UN Archives, New York, S0829-1-14 (DAG 13/1.6.5.8.4.0).

36 Outgoing Code Cable to Dr Bunche from Gardiner, Leopoldville, 27 September 1962, No. G-1307, in UN Archives, New York, S0829-1-14 (DAG 13/1.6.5.8.4.0).

37 AEQ/6600/5/1/F.OPS (Classified "Secret"), 1 October 1962 to Dr Bunche, UN HQ from Lt. General Kebbede, Force Commander, ONUC, "Subject: Katangese Air Force (FAK) vis-à-vis UN Air Division", 1, paragraph 3, in UN Archives, New York, S0829-1-14 (DAG 13/1.6.5.8.4.0).

38 To Dr Bunche, UN HQ from Lt General Kebbede [Force Commander, ONUC], "Subject: Katangese Air Force (FAK) vis-à-vis UN Air Division", 2, paragraph 6, in AEQ/6600/5/1/F.OPS (Classified "Secret"), 1 October 1962, UN Archives, New York, S0829-1-14 (DAG 13/1.6.5.8.4.0).

39 Ibid, 2–3, paragraph 7c.

which would make it impossible to keep even a section of fighters on readiness and thus impossible to simultaneously defend even one airfield, conduct offensive sweeps, and escort transport aircraft. Moreover, since ONUC was entirely dependent on supplies delivered by air, of which 95 percent were lifted by civil chartered companies, a Katangan air threat would ground essential supply planes in the absence of UN fighter escorts.[40]

In the same October 1962 report to Bunche, General Kebbede recommended immediate steps be taken to reinforce the UN Air Division. The first recommendation was for the acquisition of two S-29E photo-reconnaissance aircraft and a complete photo-interpretation unit to monitor developments and activities at Katangan air bases. The second was to increase two UN fighter squadrons to eight fighters each (for a total of 16 fighters). The third was the addition of two additional Canberra aircraft. Also recommended was the acquisition of anti-aircraft defences for UN air bases and radar for Elisabethville, as well as heavy-calibre and napalm bombs for the Canberra bombers and additional communications equipment.[41] These recommendations were considered to be the bare minimum necessary for the operation.

Things became even worse when Ethiopia abruptly withdrew its Sabre aircraft after losing one in an accident. Furthermore, India experienced an urgent need to repatriate its Canberra bombers to fight in a border war with China.[42] On the positive side, Sweden promised more Saab jets and Norway offered an anti-aircraft battery. New air surveillance radars were deployed at Kamina and Elisabethville.[43]

A few days following Kebbede's UN requests, a cable from Robert Gardiner, the UN representative in the Congo, to Bunche reported that a South African aircraft company had offered Katanga 40 Harvard aircraft, each equipped with 40 rockets, for US$27,000 each. The planes were thought to be transported into Katanga through Angola, a Portuguese colony. Moreover, intelligence reported that the same company had previously sold 17 aircraft to Katanga.[44] On 17 October, Gardiner cabled Bunche that aerial photography had confirmed the presence of six Harvard aircraft at Katanga's Kolwezi–Kengere airfield.[45]

The UN mission was clamouring to increase its air force, particularly its fighter strength, despite UN Headquarters' concerns about costs, having overcome earlier inhibitions on combat. Intelligence evidence mounted regarding the acquisition of new aircraft by Katanga. The growing strength of Katanga's air force relative

---

40   Ibid, 2–3, paragraph 7.

41   Ibid, 3–4, paragraph 9.

42   Hellström, *The Instant Air Force*, 23.

43   Dorn and Bell, "Intelligence in Peacekeeping", 21.

44   Outgoing code cable, priority 743, to Dr Bunche from Mr Gardiner, 5 October 1962, No. G-1355, in UN Archives, New York, S0829-1-14 (DAG 13/1.6.5.8.4.0).

45   Outgoing cable, secret, to Bunche from Gardiner, 17 October 1962, No. G-1438, 1, in UN Archives, New York, S0829-1-14 (DAG 13/1.6.5.8.4.0).

to ONUC's had immediate military and strategic consequences. The ANC were frequently bombed and harassed by Katangan aircraft.[46] The UN Commander's assessment was that:

> Due to ONUC's limited strength of four fighters, we have to confine our action to Recce the area in question as often as possible during daylight and attack any Katangese aircraft flying in that area. We are *not* attempting to destroy any aircraft found in the airfield in the vicinity of that area because if we do locate one or two aircraft and destroy them, we feel that FAK will react against [our] Kamina Base and also disperse their aircraft from Kolwezi to other airfields, thereby making our task of locating and destroying these aircraft on the ground very difficult. Please advise dates by which additional four Swedish fighters, as promised, will be available and if any additional aircraft expected from other nations.[47]

The UN Commander's strategy was to wait until the new aircraft gave ONUC a fighter force capable of destroying the bulk of Katanga's air force on the ground in one overwhelming surprise attack. Another cable from Kebbede to Bunche on the same day (24 November 1962) stated that:

> on request from the ANC, air recce missions over Kongolo area are being provided by UN fighters … Missions will be confined to recce and destroying any Katangese aircraft if found *flying* over that area. Instructions have been issued NO repeat NO ground targets to be attacked.[48]

The ONUC Commander did not want to give the Katangese any reason to disperse or hide their aircraft but rather wanted them to feel that they were safe and secure when on the ground at their major airfields.

Meanwhile, efforts in New York had gained traction. Sweden sent two Saab photo-reconnaissance aircraft in November 1962, greatly facilitating the gathering of air intelligence, which permitted a revised estimate of Katanga's air capability. Doubts about FAK's endurance were reinforced because many of the aircraft appeared to be unserviceable, and stockpiles of ammunition as well as petroleum, oil, and lubricants could only be found at a few airfields. Furthermore, aerial photos showed that previous reports of underground shelter construction at some airfields were incorrect, and that underground shelters at the Kolewezi–Kengere

---

46 Outgoing code cable, "most immediate", from General Kebbede to Dr Ralph Bunche, 24 November 1962, No. ONUC 7926, in UN Archives, New York, S0829-1-14 (DAG 13/1.6.5.8.4.0).

47 Ibid. (24 November 1962). Emphasis added.

48 Outgoing code cable, most immediate, to Dr Ralph Bunche from General Kebbede, 24 November 1962, No. ONUC 7926, in UN Archives, New York, S0829-1-14 (DAG 13/1.6.5.8.4.0). Italics added.

airfield were vulnerable. Concerns about possible anti-aircraft batteries at some Katangan airfields were also shown to be misplaced.[49] This new appraisal of FAK's capability coincided with the arrival of ONUC's new fighters, and the bolstering of defences by a Norwegian anti-aircraft battery accompanied by 380 men.[50] The ONUC Commander's "wait until ready" strategy was near the point of fruition. The "UN Air Force" was poised to strike jointly with UN ground troops under a plan for an operation appropriately named "Grand Slam". However, a massive airlift capability was required to deploy the UN troops simultaneously. Though the United Nations by now had 65 transport aircraft, the largest were propeller-driven DC-4s and ONUC could not move its forces without major support from the US Air Force. The details of the UN requirements were passed to the US Department of Defense by Brigadier-General Indar Jit Rikhye, Thant's military attaché, now stationed in the Congo. A few days later, the United States responded that the United Nations could count on US support.[51]

The United States was again, as a year earlier, considering fighter support in addition to logistics and transport. In November, President Kennedy offered fighter planes without American pilots. Following that, the Pentagon went further, recommending a Composite Air Strike Unit to "destroy or neutralize" Katanga's air capability.[52] But the Joint Chiefs recommended the "direct commitment of US forces" only under dire circumstances. Kennedy asked his UN Ambassador, Adlai Stevenson, to determine if the United Nations desired the US-piloted jet planes. In a meeting on 16 December, Thant expressed confidence that the UN mission would be able to resolve the situation without the US fighter jets. Thant wanted to keep the veneer of UN impartiality, while trying to avoid a direct superpower clash in the Congo. He planned to enforce Security Council-mandated sanctions with forceful UN action from the air and on the ground. The Americans argued for an "overwhelming show of strength from the air". Thant said that if the situation remained deadlocked in the spring of 1963, he would consider the US offer.[53] This was not necessary, however, since the final fight over Katanga began just a few days later.

On Christmas Eve 1962, Katangese gendarmes shot at a UN observation helicopter, fatally wounding an Indian crewmember and forcing the aircraft to land. The crew was seized and beaten.[54] Elsewhere, Katangan forces began firing continuously at UN positions, fatally wounding several soldiers. The United

---

49   Annex C, HQ ONUC, MIL INFO 741, 22 February 1963, "Report of Visit to Kolwezi and Jadotville Airfields 25–29 January 63", 9, in UN Archives, New York, S0829-1-10 (DAG 13/1.6.5.8.4.0).

50   Kalb, *Congo Cables*, 360.

51   Rikhye, *Military Adviser to the Secretary-General*, 302.

52   Kalb, *The Congo Cables*, 362.

53   Kalb, *The Congo Cables*, 365.

54   Code Cable [Robert] Gardner [Officer in Charge, ONUC] to Bunche, 24 December 1962, UN Archives, S-0875-0006-06-00001.

Nations sought a ceasefire, even escorting Tshombé himself to a point near the fighting. Katanga's leader had to agree that the firing was coming only from Katangan positions and the United Nations was not engaging in combat. After four days of ceasefire efforts, the UN commander in Elisabethville, Major-General Prem Chand of India, finally persuaded Thant to approve an offensive operation, designed to be decisive.[55] The convincing argument came from radio intercepts that had revealed Katanga's commander had ordered his air force to bomb Elisabethville airport during the night of 29 December.[56]

Equipped with air transports and the newly acquired jet fighters, the United Nations launched Operation Grand Slam. The mission's Air Division struck Katangan air assets with confidence, achieving a certain level of surprise. Early on 29 December 1962 ONUC's J-29 fighters attacked the Kolwezi–Kengere airfield. They relied entirely upon their 20-mm cannons since the cloud ceiling was too low to use their 13.5-mm rockets.[57] Three UN aircraft were hit by ground fire. One plane suffered two bullets through its canopy which, fortunately, failed to hit the pilot. The UN attacks continued through the day and expanded to other Katangan airfields. On 29 December 17 fighter and three reconnaissance sorties were carried out by UN aircraft resulting in six Katangan aircraft destroyed on the ground and possibly one in the air. Five petrol dumps were set on fire at the Kolwezi–Kengere airfield, where the administrative building was also destroyed.[58] Active patrolling of the skies by the Swedish J-29s effectively cut the air bridge between Katanga and its allies in Portuguese West Africa and Southern Africa, precluding the introduction of new aircraft.[59] From 28 December 1962 to 4 January 1963 a total of 76 sorties were carried out by UN aircraft against Katanga's airfields and aircraft.[60]

As a result of these coordinated attacks, most of Tshombé's aircraft in Katanga were destroyed on the ground. The ONUC Commander's strategy had succeeded against very little resistance. One ONUC summary of the attacks concluded triumphantly that the "Katang[an] Air Force as such is no longer in existence". Out of the estimated dozen combat aircraft in the force (Harvards, Fouga Magisters, and Vampires), only one or possibly two Harvards were not confirmed destroyed.

---

55   Kalb, *The Congo Cables*, 366–7.

56   Dorn and Bell, *Intelligence in Peacekeeping*, 20.

57   Appendix I, Annex C, HQ ONUC, MIL INFO 741, 22 February 1963, "Summary of Air Activity 28 December–4 January 63", 1, in UN Archives, New York, S0829-1-10 (DAG 13/1.6.5.8.4.0).

58   Appendix I, Annex C, HQ ONUC, MIL INFO 741, 22 February 1963, "Summary of Air Activity 28 December–4 January 63", 1–2, in UN Archives, New York, S0829-1-10 (DAG 13/1.6.5.8.4.0).

59   Johnson, "Heart of Darkness: The Tragedy of the Congo, 1960–67", 11.

60   Appendix I, Annex C, HQ ONUC, MIL INFO 741, 22 February 1963, "Summary of Air Activity 28 December–4 January 63", 2, paragraph 7, in UN Archives, New York, S0829-1-10 (DAG 13/1.6.5.8.4.0).

Moreover, all vital air installations at the Kolwezi airfield had been demolished. Evaluating the threat, the summary concluded confidently:

> It is unlikely that any further offensive activity can be expected by Katangan aircraft in the near future. Should they, however, try to undertake any such action, the only [Katangan] course would be hit and run raids by individual aircraft from airfields outside KATANGA.[61]

During Operation Grand Slam, seven UN fighter aircraft and one reconnaissance plane were hit by ground fire but no pilots were injured.[62] In addition to kinetic action against Katangan air assets, UN fighter aircraft also provided close air support to UN ground forces.[63] Also UN forces entered a key mineral facility near Jadotville unopposed, despite threats of sabotage from mercenaries. Though defeated militarily, Tshombé sought to cut deals, but the United Nations demanded that he surrender his remaining military strongholds, given that he had broken many agreements before. Tshombé finally capitulated on 15 January 1963, renouncing for good his secession.

An ONUC intelligence team subsequently learned that the Katangan air force still had some 15 aircraft hidden in Angolan airfields;[64] this was later confirmed by Angolan authorities in a radio broadcast on 9 February 1963. According to Belgian mercenaries interrogated in Kolwezi by the UN intelligence team, these aircraft were placed there for use "in the next fight for Katanga's secession".[65] Moreover, when the December hostilities had begun, the Katangan air buildup had still been under way and at least some of Katanga's leaders had believed that they could seriously challenge the United Nations. This was expressed to the UN intelligence team in the following manner:

> If you had only given us four more weeks so that we could have got the Mustangs ready, you would have experienced the same disastrous surprise one early morning at your Kamina Base as we experienced at Kengere [Kolwezi] on 29 December.[66]

Clearly ONUC's victory had come just in time; indeed, it might have been a very close call for the United Nations since the Mustang aircraft purchased by Tshombé

---

61   Ibid, 3. Capitalization in the original.

62   U Thant, *View from the UN* (New York: Doubleday, 1978), 145.

63   Rikhye, *Military Adviser to the Secretary-General*, 310.

64   Annex C, HQ ONUC, MIL INFO 741, 22 February 1963, "Report of Visit to Kolwezi and Jadotville Airfields 25–29 January 1963", 8, in UN Archives, New York, S0829-1-10 (DAG 13/1.6.5.8.4.0).

65   Ibid.

66   Ibid.

were expected to arrive in Katanga in January 1963.[67] (That month the United Nations received additional Sabre jets from Italy, The Philippines, and Iran,[68] although these jets did not need to engage in combat.)

The United Nations confirmed that the Katangan air buildup in 1961–1962 had been accomplished with the knowledge and assistance of the governments of Angola, South Africa, and Rhodesia. A UN study concluded: "the need for an efficient *air intelligence service* appears to have been confirmed even for a 'peaceful' operation such as that of the UN in the Congo".[69]

The experience of robust, kinetic air power in the Congo had raised some ethical dilemmas that required tough decision-making by the UN Secretary-General. The day before the surprise attack on the airfield, Thant had cabled General Kebbede to forbid the use of napalm, which could be spread by the Indian Canberras and the Swedish Saab 29s. Thant had stated:

> We recognize that tactically napalm type bombs might have some special utility. But we are certain that the disadvantages, particularly as regards world opinion, outweigh the advantages. Therefore, it has been decided that they cannot be used.[70]

This order came several years before the United States used napalm in Vietnam with such a negative impact on world opinion.

The minimization of UN and civilian fatalities was also extremely important for the United Nations. After the surprise attack, the United Nations could confirm that no UN personnel were killed or injured as a result of the air attacks on Kolwezi, Kamatanda, and Ngule airfields.[71] Likewise there were no confirmed reports of civilian casualties during Operation Grand Slam. Thus a potential media relations disaster for ONUC was avoided, while the mission was accomplished. However, the number of Katangese gendarmes, civilians and mercenaries killed is not known.

---

67   Ibid, 7.

68   Cooper, T. "Congo, Part 1: 1960–63", Air Combat Information Group. Available at: http://www.acig.org/artman/publish/article_182.shtml [accessed 7 May 2014].

69   Annex C, HQ ONUC, MIL INFO 741, 22 February 1963, "Report of Visit to Kolwezi and Jadotville Airfields 25–29 January 1963", 9, in UN Archives, New York, S0829-1-10 (DAG 13/1.6.5.8.4.0). Italics added.

70   Incoming Code Cable, Priority, to Kebbede Leopoldville, from Secretary-General New York, 28 December, 1962, Number 9153, in UN Archives, New York, S0829-1-13 (DAG 13/1.6.5.8.4.0).

71   Annex C, HQ ONUC, MIL INFO 741, 22 February 1963, "Report of Visit to Kolwezi and Jadotville Airfields 25–29 January 63", 8, in UN Archives, New York, S0829-1-10 (DAG 13/1.6.5.8.4.0).

Some Lessons with Examples

The Congo mission in the early 1960s was a pioneering multidimensional mission that offered significant though long forgotten lessons on the benefits and challenges of air power in various roles.

Aerial Reconnaissance: Strengths and Weaknesses

While the importance of air reconnaissance was shown in the mission, the limitations were also illustrated. In a major example, a UN aircraft was sent on 13 November 1961 to observe the situation at the Kindu Airport after radio communications had been lost. The pilot did not observe or report anything abnormal or alarming. On the ground, however, the situation was anything but normal. There was a stand-off at the airport, with the rebel Congolese forces demanding possession of two Italian aircraft that had just flown in, as well as 14 Ferret Scout cars of the Malaysian Special Force. The Congolese forces surrounded the airport and, over the next eight hours, the Malaysian forces dug into defensive positions. The Malaysians, for their part, demanded the return of the 13 Italian airmen who had been seized by rebel Congolese forces. The rebel forces had erroneously mistaken the Italians for despised Belgian military personnel.

After three days, a lieutenant colonel from the UN Air Force arrived at the airport to determine the situation on the ground. The UN mission quickly reinforced its presence with two additional rifle companies and Canberra bombers flying overhead. It made plans for a ground and air attack on Congolese rebel forces in three locations. The Indian bombers made three sorties but did not need to engage. The Congolese rebels withdrew in the face of such military power. Unfortunately, it was too late for the Italian airmen who had been taken hostage. As reported by Belgian civilians, all 13 airmen suffered a gruesome death.[72] This dire situation on the ground was not apparent in the quick reconnaissance flyby on the first day. Apparently, the pilot saw the UN flags flying, the armoured vehicles in good condition, and "deemed the situation on the ground normal"![73]

Another situation also showed the limitations of air observation. In 1962, a Swedish transport aircraft was shot down by gunmen in the bush.[74] To begin

---

72   The 13 Italian airmen were shot dead. Some were then "butchered and sold as meat at the market in Kindu". Siebel, N.H., and Lam, M. "Congo Kindu Airport". Available at: http://www.the-blue-helmets.ca/documents/Congo%20Kindu%20Airport.pdf [accessed 7 May 2014].

73   Anonymous. "2Lt. N.H. Siebel PGB and Captain Maurice Lam PGB in the Congo". Available at: http://www.the-blue-helmets.ca/documents/Congo%20Kindu%20Airport.pdf [accessed 7 May 2014].

74   The UN investigation by the Board of Inquiry is described in UN Document S/5053/Add.12 of 8 October 1962. The ONUC Dakota was engaged in reconnaissance operations when it was shot down near Kabongo. After evacuating the burning aircraft

the search and rescue for survivors, the site of the crash was determined. A UN helicopter was to land close to the wreckage. An Indian Canberra, piloted by Squadron Leader Peter Wilson, provided cover for this operation. He reconnoitered the area and detected no hostile elements in the bush, and so radioed the "all clear" message. As the helicopter was landing, however, an estimated 50 people broke from cover and ran towards it. Wilson saw white Europeans in front but behind were Africans who were either following or chasing. The Indian Air Force website describes what happened:

> Wilson did not want to fire, as it was not clear if the Africans were hostile, and they were anyway too close to the Swedes; but to warn them off he made several low passes over them; low enough so that they threw themselves to the ground as he passed over. The helicopter pilot called Wilson on the R/T, "IAF Canberra please stop, you are frightening these people!" The Africans turned out to be friendly local Congolese, who had helped and looked after the Swedish survivors, rather than hostile Katangan rebels.[75]

Air Combat: The Risks of Using Force

Prior to ONUC, all UN peacekeeping missions were either unarmed or used force only in self-defence. Though the Security Council did not explicitly invoke the UN Charter's Chapter VII (Enforcement) in the Congo, ONUC was the first UN peacekeeping operation to put into effect a Security Council call for all "necessary" measures. The mission found itself in a de facto war between ONUC and Katanga. ONUC's Rules of Engagement (ROEs) were frequently amended due to changes in the circumstances in the Congo and in ONUC's mission. Indeed, ONUC's ROE were affected not only by the three successive Security Council resolutions but also by at least 10 different operational directives, as described by the academic Trevor Findlay.[76] Addressing the specifics and impact of each of these transitions is beyond the scope of this chapter. However, some points merit attention both in general terms and due to their relevance to the use of air power by a UN mission.

In ONUC's early stages Secretary-General Hammarskjöld refused to interfere in Congolese internal politics and saw ONUC's mission only in terms of restoring order and promoting stability. He would not take sides in the issue of Katanga's secession and refused to authorize military force to prevent it, denying the mission's early demands for air combat power. Even after Security Council Resolution 161

---

and extracting a dead colleague, the crew came under fire from ANC troops nearby who thought the flyers were from the Katangan forces, but when they realized the crew were UN personnel they provided full cooperation.

75   Kumar, K. Sree. "Encounters with Veterans: Air Commodore P.M. Wilson", *Indian Air Force History*. Available at: http://www.bharat-rakshak.com/IAF/History/1960s/Wilson01.html

76   Findlay, T.C. *The Use of Force in UN Peace Operations* (Solna: SIPRI and Oxford: Oxford University Press, 2002).

of 21 February 1961 authorized the UN operation to take "all appropriate measures to prevent the occurrence of civil war in the Congo, including ... the use of force, if necessary, in the last resort", Hammarskjöld's instructions to ONUC in September 1961 included numerous limitations on the use of military measures.[77]

Several of Hammarskjöld's qualifications were subsequently ignored and even broken by the Special Representative in Katanga, Conor Cruise O'Brien, especially in the launching of Operation Morthor in September 1961. As Findlay put it: "It involved significant use of force, caused hundreds of casualties, and exponentially increased the dissent that had plagued the UN operation in the Congo from its inception".[78] However, it did not involve the use of UN air power for combat. Though Hammarskjöld cancelled the continuation of Morthor, he sought promises from several UN member states to provide aircraft for "defensive" purposes, notably against Katanga's air assets, which had wreaked havoc on UN forces during the operation. After Hammarskjöld's death, Secretary-General Thant was more ready to use force. Though a Buddhist pacifist in personal theology,[79] Thant believed that ONUC's mandate to prevent civil war implied armed force, necessitated combat air power, to suppress armed secessionist activities.

The arrival of aircraft from Sweden, Ethiopia, and India led to the creation of ONUC's air wing, which required explicit direction. Thant authorized ONUC to protect UN troops from Katangan actions that endangered the lives of peacekeepers, including Katangan efforts of "actually attacking them or by moving directly against them with hostile intent".[80] November 1961 marked the first time the United Nations issued ROE for the use of combat air power. The instruction to engage in pre-emptive defensive action in the case of hostile intent added a new dimension to traditional self-defence rules of ground forces in peacekeeping operations. Subsequently, more detailed instructions on ONUC's use of air power were promulgated, placing strong command and control limitations:

> The UN jet Air Force will not be used in a supporting role without the authority of the Air Commodore under instructions from Dr. Linnér [Special Representatives of the Secretary-General] and the Force Commander ... The possible use of this Air Force will not be conveyed to the ANC in the form of a threat or otherwise except on authority from Dr. Linnér and the Force Commander ... [aerial action] is to be taken as a last resort and should be limited to those measures clearly necessary to the defense of ONUC troops

77   Findlay, *The Use of Force*, 74.

78   Findlay, *The Use of Force*, 75.

79   Dorn, A.W. "U Thant: Buddhism in Action", in Kent Kille (ed.), *The UN Secretary-General and Moral Authority: Ethics and Religion in International Leadership* (Washington, DC: Georgetown University Press, 2007), 143–86. Available at: http://www.walterdorn.net/pdf/UThant-BuddhismInAction_Dorn_SG-MoralAuthority_2007.pdf

80   Outgoing Code Cable, no. 8155, 20 November 1961, from the Acting Secretary-General to Linner, Yacob, Leopoldville, UN Archival document DAG1/2.2.1, #10.

and other personnel. No action of this kind is to be ordered, however, without prior warning in ample time, to the authorities concerned. Due care should be exercised to avoid casualties among non-involved civilians.[81]

On 5 December 1961, with the launching of the UN's military operation, Thant authorized "all counter-action – ground and aerial – deemed necessary to restore complete freedom of movement in the area".[82] Leaflets were dropped by air, telling the Katangese that the United Nations was a force of peace. The two-way air war commenced 5 December 1961 with the Katangan bombing of Elisabethville airfield. The next day, ONUC's first airstrike occurred when Indian Canberra bombers attacked the Kolwezi airstrip. The Jadotville airstrip and other Katangan installations were also attacked and by 8 December, ONUC commanded the skies over Katanga.

There followed a year-long shaky truce during which time Tshombé steadily built a new and highly credible air force. During the buildup, Thant explained, in March 1962, why the United Nations could not use force to end Katanga's secession: "The UN has been authorized to use force only in three situations: one, to prevent civil war; two, to arrest foreign mercenaries; and three, to retaliate when attacked".[83] By October 1962, the ANC and the ONUC were again suffering from direct attacks by Katanga's aircraft. As such aggression was tantamount to civil war, violating the 21 February 1961 Council resolution, ONUC ordered Katanga to ground all military aircraft and declared it would shoot down aircraft engaged in offensive operations.[84] This was, in effect, a UN-mandated no-fly zone.

The escalation of events in December 1962, including the shooting down of an ONUC helicopter, led to UN warnings to Tshombé that unless firing against UN forces ceased, the mission would take "all necessary action in self-defense and to restore order".[85] Tshombé's refusals to order his troops to stop firing, and radio intercepts that revealed that Katanga's commander had ordered his air force to bomb the Elisabethville airport during the night of 29 December, compelled Thant to acquiesce to requests from the Special Representative Gardiner and Force Commander Prem Chand to commence military operations. On 27 December 1962 the UN air wing was issued Fighter Operations Order 16 to retaliate (that is, "shoot

---

81   Fighter Operations Order No. 6, AHQ/66PP/1/F-OPS, undated, UN Archival document no. DAG13/1.6.5.0.0, #7; cited in Findlay, *The Blue Helmets' First War?*, 117.

82   Findlay, *The Blue Helmets' First War?*, 118.

83   UN Press Services, Note No. 2548, "Note to Correspondents: Press Conference by the Acting Secretary-General at UN Headquarters on Tuesday, 27 March 1962"; also cited in Findlay, *The Blue Helmets' First War?*, 124.

84   Findlay, *The Blue Helmets' First War?*, 127; referring to *Yearbook of the United Nations: Special Edition*, UN Fiftieth Anniversary 1945–1995, 39.

85   Lefever E.W., and Joshua, W. *United Nations Peacekeeping in the Congo, 1960–1964: An Analysis of Political, Executive, and Military Control, 1–4* (Washington, D.C.: Brookings Institution for the US Arms Control and Disarmament Agency (ACDA), 30 June 1966), 2, 108; as cited in Findlay, *The Use of Force in UN Peace Operations*, 129.

down") any Katangan aircraft that attacks "any target, whether belonging to UN or NOT". Furthermore, any Katanga aircraft "carrying visible offensive weapons, such as bombs or rockets", should be shot down.[86] This strong aerial ROE could only be justified by the extreme circumstances that existed in late December 1962, as the fighting had started by Katanga and had continued one-sided for days. The success of the United Nations in destroying the Katangan air assets came as an aerial arms race was growing fierce.

There were objections to the escalation of force, among both UN diplomats and service members on the ground. In 1961, Swedish pilots refused some requests for close air support to ground troops, reasoning that the risk of civilian casualties was too high. In November 1962, the Swedish air commander refused a direct order to shoot down Katangan aircraft. The UN Air Commander (an Indian) resigned in complaint and the American air attaché in New York lamented that concerted efforts by UN headquarters "can be nullified by the actions of one officer".[87] Nevertheless the decision was supported by the Swedish government, which wanted its aircraft to be used for purely defensive purposes only. This shows how national caveats can be as troublesome for UN missions as they are for NATO and other multinational missions, even half a century later in Afghanistan. Furthermore, finding the balance between defence and offence is difficult in any military mission, including those of the United Nations.

Conclusion

ONUC was a pioneering mission. It was the first UN mission to engage in combat against rebels and mercenaries, and the first mission to implement a no-fly zone and an arms embargo – including by detaining aircraft and crews that were bringing arms and military supplies into the Congo. Most significantly, it was the first peacekeeping mission to use combat to carry out the decisions of the Security Council. Air power played a large role in the operational mandate. Coordinated air-to-ground attacks were used for the first time in the history of peacekeeping (and ONUC was one of the few UN missions to do this before the twenty-first century). Had ONUC's air contingent failed to destroy Katanga's considerable air assets at the outset of Operation Grand Slam, Katanga's air capability would have made it impossible for the United Nations to resupply its ground troops, and ONUC would likely have failed in its mission. Many of the UN forces might even have been cut off and possibly forced to surrender, as the UN's Irish troops had been forced to do a year earlier. In that instance, Katanga's single Fouga jet had helped prevent their resupply and reinforcement.

---

86   ONUC LEO (ONUC Headquarters Leopoldville), "Fighter OPS ORDER NO 16", 27 December 1962, AHQ/6600/1/F-OPS, UN Archives DAG13/1.6.5.0.0, #9 (series 0787), paragraph 4; also cited in Findlay, *The Blue Helmets' First War?*, 130.

87   Hellström, *The Instant Air Force*, 24, citing the J.F. Kennedy presidential archives.

The aim of this chapter has been to examine the role of kinetic air power in ONUC, a forerunner of modern multidimensional missions, and to draw some lessons from this experience. It was demonstrated that initially air transport, mostly provided by the United States,[88] was crucial in bringing troops to the Congo and, later, to transport them to Katanga. Because of the absence of an air fighter contingent early in the mission, the whole endeavour was jeopardized. Just prior to his death, Secretary-General Hammarskjöld procured fighter aircraft – including from his native Sweden – as a deterrent to Katanga. One of the last acts of his legacy, as a result, was to create the "UN's first Air Force". When air operations began against Katangan forces, detailed surveillance from the air was key, especially surveillance of Katanga's airfields. What enabled the UN Air Force to prevail was a viable air strategy. Notably the ONUC commander ordered his forces to forego attacks on Katanga's airfields, thus giving the Katangese a false sense of security, until such time as the United Nations had adequate air assets to destroy almost all Katanga's planes on the ground in an overwhelming surprise assault.

This mission also demonstrated that air power, while enabling the United Nations to project force at a relatively safe distance, can be quite politically sensitive in ways that force on the ground is not. Air power in the Congo had a strong offensive element, rather than the usual self-defence-only rules provided to peacekeepers. Thant's decision to forbid the use of napalm not only resulted in less bloodshed but also likely averted a public relations disaster for the United Nations. By this time, ONUC had already attracted enormous criticism from member states and from the international media for its decision to side with the Congolese central government. Had the United Nations used napalm, the world might have viewed pictures of burn casualties from a UN-perpetrated atrocity.

Sadly, while air power played a crucial role in this UN operation, it was not without drawbacks and collateral damage. UN aircraft allegedly bombed a hospital at Shinkolobwe, northwest of Elisabethville,[89] and the Lido Hotel in Elisabethville.[90] Also an aircraft narrowly missed a building where, unbeknownst to its UN pilots, 150–200 women and children had taken shelter. "It was only due to poor aiming that the bombs did not hit the building" containing the civilians, wrote the Canadian Consul General in Leopoldville. He also wrote that two Canberra

---

88   The US Air Force flew 2,310 missions, moving 63,884 personnel and 37,202 tons of material from 33 countries during three and a half years of service. See Van Nederveen, Captain Gilles K. "USAF Airlift into the Heart of Darkness, the Congo 1960–1978: Implications for Modern Air Mobility Planners", Research Paper 2001–04, Air University, Maxwell Air Force Base, AL, September 2001.

89   Ericson, Lars (ed.). *Solidarity and Defense: Sweden's Armed Forces in International Peacekeeping Operations during the 19th and 20th Centuries* (Stockholm: Svenska Militärhistoriska Kommissioner/Swedish Military Commission, 1995), 76, as cited in Findlay, *The Blue Helmets' First War?*, 155.

90   "Four UN Jets Bomb Elisabethville Hotel". *The New York Times*, 14 December 1961.

aircraft were on their way to bomb Tshombé's residence before Air Commander H.A. Morrisson (from Canada) in Leopoldville managed to stop the attack.[91] Despite these close calls, far more collateral damage was done by ONUC's ground troops, including the alleged striking of another Katangan hospital by mortar fire.[92]

The Congo mission highlighted many organizational difficulties for the United Nations. As a mission of unprecedented complexity, cobbled together in a rush, it experienced difficulties with command and control, intelligence (at least initially), and the application of force. When the United Nations returned to the Congo four decades later, it faced many of the same problems. But by the time the UN Mission in the Congo (MONUC) was created in 1999, the lessons of ONUC had been forgotten and the UN's historical knowledge buried in its archives. The lessons and historical actions need to be explored, described, and revealed, especially as the United Nations re-engages in robust peacekeeping in Central Africa (see Chapter 14 on the Congo in the 2000s). After four years the modern mission began to employ attack helicopters, though no fighter jets or bombers, to deal with the Congo's "wild east", especially the Kivu provinces bordering Rwanda and Uganda. Fortunately, the southern province of Katanga is relatively peaceful and has been integrated into a united Congo, thanks in part to the robust actions of UN peacekeepers in the 1960s.

Acknowledgments

The author thanks Robert Pauk, former peacekeeper, for his help in researching and drafting the paper on which this chapter is based. The author also gratefully acknowledges the Canadian Forces Aerospace Warfare Centre for the funds which made his research assistance possible. An earlier version of the paper was published in the *Journal of Military History* (2013) under the title "The UN's First 'Air Force': Peacekeepers in Combat, Congo 1960–64". The publisher-granted right to publish this shortened and revised version is much appreciated. Finally, the author thanks Leif Hellström for his expert commentary on the paper, based on his own in-depth investigations.[93]

---

91 Canadian Consulate General Leopoldville to Under-Secretary of State for External Affairs Ottawa. "Conduct of UN troops During December Fighting in Elisabethville", Cable of 16 January 1962, Library and Archives Canada, RG25, 5224, file 6386-C-40 part 18, copy kindly provided by Kevin Spooner and referenced in Spooner, *Canada, the Congo Crisis and UN Peacekeeping*, 179.

92 Dennis Neeld, "UN Bombs Katanga Positions After Night of Heavy Action", *The Washington Post*, 15 December 1961.

93 Hellström, L. *Fredsflygarna: FN-flyget i Kongo 1960–1964* (Freddy Stenboms Förlag, Svenska, 2003) [in Swedish].

Chapter 3

A Fine Line: Use of Force, the Cold War, and Canada's Air Support for the UN Organization in the Congo

Kevin A. Spooner

The Congo Crisis of 1960, erupting as it did mere days after that nation's independence and at a time of heightened Cold War tensions, seemed to utterly threaten not only the stability of the new state but also the delicate balance of the East/West rivalry in Africa. For certain, both the West and the Communist bloc were already actively jockeying to garner new friends amongst the decolonized and decolonizing of the continent, but there was considerable fear in the international community that quiet, behind-the-scenes, diplomatic and even covert activity could erupt into open conflict, with the two sides of the ideological divide supporting opposing Congolese political factions. That fear, not at all unreasonable, prompted the international community to respond to the crisis with a UN peacekeeping force – the United Nations Organization in the Congo, but known typically by its French name, the Opération des Nations Unies au Congo (ONUC) – in an attempt both to assist the Congolese government to restore law and order and to insulate the country against outside, direct interference. Assessments of ONUC's success are mixed, but one finding can be made with certainty: the mission severely tested the peacekeeping and financial capacity of the United Nations, especially as it began to exercise an increasing degree of force to carry out its mandate.[1] And, with the Cold War as an ominous backdrop, the implications of a more forceful ONUC were never far from the minds of those engaged in decisions related to the operation.

The use of aircraft in support of UN operations was not new when peacekeepers were deployed to the Congo, but the large scale and the diversity of aircraft used was certainly impressive for the time. In Chapter 2 in this volume, A. Walter Dorn has thoroughly addressed many operational and international aspects of air support in ONUC, while in Chapter 1, William K. Carr provides a detailed review of the mission from his personal experiences in establishing UN

---

1 Carole J.L. Collins provides a particularly critical perspective, even suggesting ONUC "served as an unconscious midwife to the arrival of the Cold War in Africa". See Collins, C. "The Cold War Comes to Africa: Cordier and the 1960 Congo Crisis", *Journal of International Affairs* 47(1) (1993), 244–5.

air operations. This chapter, by comparison, is focused much more directly on the Canadian political and policy dimensions of the Royal Canadian Air Force (RCAF) contributions to ONUC. Canadian foreign policy in the Congo Crisis proved very complex; at various times between 1960 and 1964, the governments of John Diefenbaker and then Lester Pearson were forced to weigh dozens of factors when formulating responses to what proved to be a very fluid and volatile international situation. Still, two significant themes already identified were consistently evident and shaped policy options and decisions throughout the period: the underlying impact of the Cold War and the increased level of force exercised by ONUC to fulfill its evolving mandate. To see how these two themes were relevant to the RCAF and ONUC, this chapter examines the political considerations that influenced decisions on the Canadian contribution of North Star aircraft for airlift to the Congo, a Canadian attempt to provide Caribou aircraft to ONUC, and the provision of command personnel for ONUC's air operations. It also reviews Canadian responses to UN requests that arrived at times when conditions in the Congo were particularly troubled or as ONUC contemplated exercising a greater degree of force to address secession in the Congolese province of Katanga.

The North Star Airlift

When the Congo Crisis erupted in July 1960, the international community responded with a surprising degree of alacrity – not at all typical of the diplomatic dithering so often seen in other situations. In Canada, National Defence and External Affairs were equally quick to recognize the United Nations might approach Canada for assistance. Immediately, concern arose over what shape such assistance might take. The existence of a standby battalion, previously earmarked for UN service, raised expectations in the press and parliament that Canadian soldiers might be dispatched. However, the Diefenbaker government recognized the inherent dangers of sending white, North Atlantic Treaty Organization (NATO)-aligned, combat-ready troops into a Congolese imbroglio that threatened to become a Cold War proxy conflict. At the United Nations, Secretary-General Dag Hammarskjöld shared these concerns; after a conversation with Hammarskjöld, Canadian Permanent Representative to the United Nations Charles Ritchie advised Ottawa that there was no question of sending a Canadian contingent and certainly not the standby battalion. Instead, External Affairs recommended that Cabinet consider providing food and supplies, and the necessary air transportation, to get these provisions to the Congo. The UN had already informally requested the use of two Canadian aircraft serving in the United Nations Emergency Force (UNEF), for the purpose of ferrying supplies and personnel to the Congo. The RCAF saw no objection to such service and so it was suggested Canada provide the aircraft if the United Nations formally requested their use.[2] Subsequently, in the House of Commons, Prime Minister Diefenbaker announced that Canada would respond favourably to UN requests for technical advisors, food, or transportation. These were recognized as "the most useful contribution" Canada could make.[3]

By 21 July, the Canadian Cabinet had agreed to provide the United Nations with North Stars[4] and crew to airlift supplies from Pisa, Italy, to the Congo. Pisa had been designated a staging area for the gathering of material in support of ONUC and the four Canadian aircraft were already in the area, having just delivered Canada's gift of food to the Congo. The United Nations proposed the airlift be shared between Italy and Canada, in a 30-day arrangement that would see costs of the operation reimbursed by the United Nations. The Canadian Assistant Trade Commissioner in Leopoldville also wired Ottawa, requesting the government consider authorizing the aircraft to land at points within the Congo, in addition to Leopoldville.[5]

The North Stars quickly became the object of controversy, related directly to the government's concern that Canada's ONUC contribution be perceived entirely as non-combat and neutral. Cabinet understood the planes would be used only to transport supplies and equipment, to and within the Congo. Then Hammarskjöld approached Ritchie with a request to use the planes to transport troops from the capital, Leopoldville, to secessionist Katanga, in advance of a planned withdrawal of Belgian armed forces. Keen to maintain the appearance of neutrality, or at least objectivity, the government was not eager to facilitate a plan that directly involved its NATO ally, particularly if this involved shifting troops to Katanga. Prime Minister Diefenbaker turned down the request. Secretary of State for External Affairs Howard Green was notified and Ritchie in New York was told to make this policy clear to the UN Secretariat.[6]

There was considerable concern in Ottawa when it was learned the planes had already been used to transport forces to numerous locations within the Congo, facilitating the withdrawal of Belgian paratroopers. UN officials had urged the senior RCAF officer in the Congo to contact Air Transport Command Headquarters in Trenton to obtain permission to use the Canadian aircraft to deploy Moroccan and Tunisian troops.[7] In New York, Hammarskjöld's Executive Assistant assured

---

2 Robertson to Green, 13 July 1960; G.S. Murray, 13 July 1960; Telegram 911 PERMISNY to External: Disorders in Congo, 13 July 1960, RG25, 5208, File 6386-40 part 6, Library and Archives Canada (LAC).

3 Canada, House of Commons Debates, 14 July 1960, 6273-4.

4 *Editor's note*: The North Star is the Canadian version of the four-engined DC-4 transport aircraft.

5 Cabinet Conclusions, 19 July 1960, RG2, 2747, LAC; Cabinet Documents 228/60, 236/60, 238/60, RG2, 5937, LAC; Cabinet Conclusions, 21 July 1960, RG2, 2747.

6 Memo UN Div to DL (1): Congo: Development of UN Force, 22 July 1960, RG25, 5208, File 6386-40 part 7, LAC.

7 Fred Gaffen suggests that it was Lieutenant-Colonel J.A. Berthiaume who had "managed with the connivance of some members of the RCAF to circumvent this restriction

Ritchie the whole incident had been a "crash operation". Nevertheless, the Canadian Representative asked that the planes be used only for purposes explicitly identified by Ottawa and insisted that all future requests of a *political* nature be forwarded through the Permanent Mission.[8] This position was reinforced after ministerial consultations between Howard Green and George Pearkes at National Defence. External Affairs and Department of National Defence (DND) officials were told to restrict use of the aircraft to the transport of supplies and equipment from Pisa to Leopoldville; more to the point, they were advised, "The use of these aircraft for the transportation of troops is not authorized by Cabinet and is to cease forthwith". Pearkes exactingly interpreted these instructions, suggesting they even prohibited the return transportation of anything from Leopoldville back to Pisa. Noting the United Nations was to reimburse Canada for the airlift costs, officials at External Affairs were concerned the instructions were too restrictive and, after other nations came forward to provide internal airlift, lobbied to ease conditions. The entire episode demonstrated the government's determination to participate in ONUC only in a non-combat capacity; any use of the North Stars that even appeared to compromise this principle was quickly curtailed.[9]

The aircraft had first arrived in Leopoldville with their cargo of food aid on 21 July, and within three days, more than 60 crewmembers had arrived, filling every bed in the official residence of the Canadian Trade Commissioner and of a local company's guesthouse. Once the Diefenbaker government committed to send personnel from the Royal Canadian Corps of Signals to provide communications for ONUC – another significant Canadian contribution that lasted throughout the peacekeeping mission – the North Stars were reassigned to transport the men and equipment. From Trenton, they embarked on a 40-hr, 6,320-mi trip to the Congo, with stops in Gander, Lajes, Dakar, and Accra.[10] They were assisted by the United States Air Force (USAF) which used C-124 Globemasters to transport

---

and airlift the Tunisian contingent to Kasai province and its capital Luluabourg". Gaffen, F. *In the Eye of the Storm: A History of Canadian Peacekeeping* (Toronto: Deneau and Wayne, 1987), 233–4.

8   Memo for Minister: Congo: use of RCAF aircraft, 21 July 1960; Memo for Minister: Congo: Use of RCAF aircraft, 23 July 1960, 5219, File 6386-C-40 part 1, LAC. Telegram PERMISNY to External: Congo: Use of RCAF Aircraft, 23 July 1960, RG25, 5209, File 6386-40 part 8, LAC.

9   Memo SSEA [Secretary of State for External Affairs] to USSEA [Under-Secretary of State for External Affairs]: Congo – modification of the use of the RCAF North Stars, 23 July 1960; Letter Pearkes to Green, 25 July 1960; Letter Green to Pearkes, 27 July 1960; Memo for Minister: Congo: Use of RCAF Aircraft, 30 July 1960, RG25, 5219, File 6386-C-40 part 1, LAC. Memo UN Div to USSEA: Use of RCAF Aircraft on Pisa-Leopoldville Airlift, 27 July 1960, RG25, 5209, File 6386-40 part 8, LAC. Letter Green to Pearkes, 4 August 1960, RG25, 5219, File 6386-C-40 part 2, LAC.

10   War Diaries CDN HQ UNEF in Congo and No. 57 Canadian Signal Squadron, August 1960, RG24, 18482, LAC; "UN Force Receives Side Arms", *Montreal Gazette*, 5 August 1960, 1.

vehicles and equipment too heavy for RCAF aircraft; in addition, the USAF flew 117 peacekeepers to the Congo. As historian J.L. Granatstein has observed, the Canadian military's reliance on US planes, in this instance, serves as a stark reminder that peacekeeping was not as "independent" as it was often assumed to be.[11]

The RCAF initially envisaged their contribution to ONUC as an Air Transport Unit (ATU) consisting of two key elements: four Caribou aircraft to be employed in support of Canadian forces, and the routine North Star airlift between Pisa and Leopoldville. The latter was considered a temporary commitment, initially undertaken for 30 days, while the Caribou were seen as the key long-term commitment. Ironically, as we will see, the Caribou portion of the ATU never materialized for political reasons; on the other hand, the arrangements governing the "temporary" North Star airlift were repeatedly renewed every 90 days in the months and years ahead.[12]

Decisions to renew the Pisa–Leopoldville airlift, however, were by no means automatic. As early as October 1960, Douglas Harkness – who had replaced Pearkes as Minister of National Defence – was already keen to review the RCAF commitment. Following the initial deployment, Ottawa had agreed to the first UN request for a 90-day extension of RCAF participation in the airlift. Flights in support of the Canadian contingent had simply been integrated into this airlift. The agreement with the United Nations was scheduled to expire on 9 December. While Green at External considered the airlift a means to assist the United Nations without further "commitment of Canadian personnel and equipment in the Congo itself", Harkness at Defence was not entirely convinced. The Chief of the Air Staff inquired at the United Nations whether it was possible to reduce the airlift by transporting more supplies by sea. The Secretariat quickly responded with an urgent request to continue the existing airlift, with an assurance that a "constant check" would be maintained to determine if or when it would be possible to reduce or discontinue flights. Given the limited transportation infrastructure throughout the Congo, air support was considered especially critical. In late November Green reminded Harkness that a decision was required and the airlift agreement was extended for another 90 days.[13]

---

11 March notes that ONUC "involved almost every long-range transport aircraft that the RCAF had available at the time. Only essential transport runs were maintained throughout North America and Europe". March, W. "The Royal Canadian Air Force and Peacekeeping", in *Peacekeeping 1815 to Today*, ed. Serge Bernier (Ottawa: International Commission of Military History, 1995), 471. See also Wainhouse, D. *International Peacekeeping at the Crossroads: National Support – Experience and Prospects* (Baltimore, MD: Johns Hopkins University Press, 1973), 284; Granatstein, J. "Canada and Peacekeeping: Image and Reality", *The Canadian Forum* 54(263) (1974), 16.

12 Organisation Order 8.13, 2 August 1960, RG24, 3022, file 895-8/115, LAC.

13 Memo for the Minister: Congo – Use of RCAF North Star Aircraft, 24 October 1960; Letter Green to Harkness, 24 October 1960; Letter Harkness to Green, 27 October 1960, RG25, 5220, File 6386-C-40 part 5, LAC. Letter Vaughan (UN) to Speedie, 10

For the next two years, extensions of the airlift became routine, with mutual agreement from both External Affairs and National Defence, partly because the government was reluctant to curtail an essential form of logistical support for the peacekeeping mission. Indeed, it feared ending the airlift might suggest a "declining interest" in ONUC or intent to "scale down Canadian participation in the Force" at a time when Congolese political conditions were still unsettled.[14]

Then, in July 1962, there was a significant about-face. Just the month before, the government had agreed to yet another 90-day extension. Now, National Defence told External Affairs the agreement would not be renewed again in September. The Diefenbaker government was confronting serious economic difficulties that had already led to the devaluation of the Canadian dollar in May; after a subsequent currency exchange crisis in June, Cabinet approved emergency measures, including significant reductions in government spending. National Defence justified its decision to end the airlift on the grounds that the government's austerity measures required the review of "all extraneous commitments in order to effect every economy possible". The airlift was to be replaced with bimonthly, non-stop Yukon flights in direct support of Canadian peacekeepers in the Congo. An important Canadian contribution to the UN operation was about to come to an end.[15]

The political implications of this decision were immediately apparent to External Affairs, where officials acted quickly to get the decision reversed. They questioned National Defence's argument that canceling the airlift would result in financial savings for the government as a whole, given most of the expenses involved were recoverable from the United Nations. Officials further argued that:

> [t]he announcement that Canada is curtailing its assistance to ONUC at such a critical juncture in the Congo would throw unfavourable light on the Canadian attitude toward the UN without bringing us any substantial advantage in terms of the austerity programme.[16]

November 1960; Telegram PERMISNY to External: Congo: Use of RCAF North Star Aircraft, 7 November 1960; Memo for Minister: Congo – Use of RCAF North Star Aircraft, 16 November 1960, RG25, 5220, File 6386-C-40 part 6, LAC. Letter Green to Harkness, 21 November 1960, RG25, 5221, File 6386-C-40 part 7, LAC.

14   Despatch SSEA to PERMISNY: UN Airlift–Replacement of North Star Aircraft, 23 February 1961; Memo for Minister: Congo – Use of RCAF North Star Aircraft, 20 February 1961; Letter Harkness to Green, 24 February 1961, RG25, 5222, File 6386-C-40 part 11, LAC.

15   Telegram External to PERMISNY: Congo – Extension of RCAF airlift, 4 June 1962, RG24, 7169, File 2-5081-6 part 16, LAC. Letter CCOS to USSEA, 13 July 1962, RG25, 5224, File 6386-C-40 part 20, LAC. The monetary exchange crises Diefenbaker faced are addressed by Smith, D. *Rogue Tory: The Life and Legend of John G. Diefenbaker* (Toronto: Macfarlane, Walter and Ross, 1995), 437–9, 442–5.

16   Memo for Minister: Canadian Airlift to the Congo, 17 July 1962, RG25, 5224, File 6386-C-40 part 20, LAC.

Howard Green instructed his Under-Secretary, Norman Robertson, to discuss the matter with the Chairman, Chiefs of Staff, Air Marshal Frank R. Miller.

At National Defence, they believed ending the airlift was an administrative decision, so there was utter dismay when External expressed its intention to raise the matter in Cabinet, if Defence proceeded with its plans. Air Marshal Miller wrote to the Chief of the Air Staff, Air Marshall Hugh Campbell, noting, "It is apparent that if we are to get approval on this we will be up against [External] Affairs in Cabinet". He asked, "Have we got enough ammunition to win?" National Defence persevered, maneuvering to resolve the matter at the administrative level. In mid-August, Air Commodore (A/C) Leonard Birchall simply told the Defence Liaison Division at External that 426 Transport Squadron had been disbanded as part of the government's austerity program; the RCAF just did not have the aircraft to continue the Congo airlift. With a looming September deadline approaching, Birchall advised notifying the United Nations so it could make alternate arrangements because there was no longer enough time to arrange for the flights to be resumed, even if Cabinet did consider the issue. The Chairman, Chiefs of Staff, later suggested there might be some flexibility with the September date but reaffirmed it was best to "notify the UN as requested and reconsider the matter when the inevitable 'protest' follows". At this point, Howard Green actually appealed to the Prime Minister to see if Diefenbaker would ask Harkness to reconsider. Diefenbaker said he would not object if Green asked Harkness to review the decision, but he would not direct the Minister of National Defence to alter it. Ultimately, Green chose not to make any further representations to Harkness.[17] The recurring North Star airlift commitment did indeed come to an end in the fall of 1962.

Caribou Aircraft

In the early days of the crisis, as Canada considered and dispatched assistance to ONUC, the Diefenbaker government found itself in a difficult position because of its commitment to send four Caribou aircraft to support Canadian peacekeepers serving in the Congo. On 1 August 1960 the government announced in the House of Commons its intention to purchase these planes from the de Havilland Aircraft Company. Because of their ability to take off and land within short distances, they were considered ideally suited to conditions in the Congo. Arrangements to purchase the aircraft were quickly completed, with delivery of the first operational

---

17   Note CCOS to CAS, July 1962; Letter USSEA to CCOS, 20 July 1962, File 73/1223, part 463, Directorate of History and Heritage National Defence [DHH]. Memo DL (1) to Ross Campbell: Provision of RCAF Air Transport for UN Operations, 15 August 1962; Memo for Prime Minister: Canadian Airlift to Congo, 16 August 1962; Outgoing Message External to PERMISNY: Congo – RCAF Airlift, 17 August 1962; Memo SSEA to Ross Campbell: Canadian Airlift to the Congo, 1 September 1962, RG25, 5224, File 6386-C-40 part 20, LAC.

aircraft by 15 August. Air and ground crews would be trained by the time the aircraft arrived.

Once all the Canadian forces, mostly signallers, were airlifted to the Congo and the North Stars had returned to duties on the Pisa–Leopoldville external airlift, the new Caribou were expected to provide internal air support for the Canadian forces. After the needs of Canadian forces were met, the planes would be made available for other UN duties. Officials at External, however, expected it to be difficult to persuade the United Nations to accept the Caribous if they were to be used in direct support of Canadian peacekeepers but not placed under the operational control of ONUC's Commander. Such an interpretation of the Caribous' role would have required the Canadian government to negotiate a direct bilateral agreement with the Congo government, something considered politically impractical. Minister Pearkes believed a compromise was possible: the RCAF Caribou unit could be placed under the operational control of the UN Commander, with priority given to Canadian force requirements.[18]

By mid-September, the Caribou problem was still not resolved; it actually became more complicated. When Canadian officials offered the Caribou to the United Nations, the Secretary-General was neither in New York nor Leopoldville. His official representative in the Congo, Ralph Bunche, initially reacted favourably, given the UN's very real need for air transport. Subsequently, Hammarskjöld made it clear that he considered it politically inadvisable to increase the number of Canadians serving in ONUC. In Leopoldville, Bunche was contacted by the Secretariat to clarify this difference in opinion. He confirmed the practical advantages of the Canadian offer but added that ONUC's Supreme Commander, General Carl von Horn, rejected the Canadian proposal that the Caribou be used primarily to support the Canadian Signals Unit or that priority should be given to their requirements. Von Horn wanted the Caribou assigned to the ONUC Air Transport Unit, under his command. In the end, Bunche said he "understood" Hammarskjöld's view of the political implications of accepting the Caribou. In effect, the United Nations had decided not to accept Canada's offer and the Permanent Mission concluded only a direct approach to the Secretary-General might reverse this position. Stories of a "mixup" began to appear in the Canadian press. One report, while noting that no one was willing to make an official comment on the situation, surmised that the government had ordered the aircraft before finding out if the United Nations wanted them. Moreover, it correctly traced the root of the problem to the government's decision to limit the use of the aircraft to supplying only Canadian forces.[19]

---

18   Memo: Congo Operations – RCAF Participation [Chief of Air Staff to MND], 9 August 1960; Memo: Canadian Contribution to UN Forces in the Congo Republic, Synopsis No. 13, 12 August 1960, RG24, 21484, File 2137.3 part 1, LAC. Robertson to Green, 9 August 1960; Pearkes to Green, 11 August 1960, RG25, 5219, File 6386-C-40 part 2, LAC.

19   Telegram PERMISNY to External: Congo – UN Use of Caribou Aircraft, 18 August 1960, RG24, 21484, File 2137.3 part 1, LAC. "Mixup Stalls RCAF Craft For

Even though the confusion over the Caribou had the potential to become a public embarrassment, External Affairs decided not to press the Canadian position in New York after it learned that Hammarskjöld "responded negatively in very firm terms" to the compromise proposal suggested by the government. The Secretary-General had recently become the target of a virulent and nasty Soviet campaign of criticism. They had been especially critical of his decision to include Canadian signallers in ONUC, and he believed a proposal to send a Canadian air unit would leave him in an "untenable position". Hammarskjöld suggested that the Caribou might still be used if Canada was prepared to make them available on a "lend-lease" basis, so that aircrews from other UN units could staff them. When a Canadian officer, Colonel Albert Mendelsohn, returned to Canada from the Congo in September 1960 to give a preliminary report, he argued that, in spite of the Secretary-General's concerns, there was an urgent need for the Caribou and that this need was fully recognized by von Horn. The only thing standing in the way was Hammarskjöld's desire not "to upset the Russians".[20] The political realities of Canada's position in the Cold War had a real impact on the nature and composition of Canada's contribution to ONUC's air operations.

Command Personnel

Canadians served within most branches of ONUC Headquarters, for example as Chief Operations Officer, Chief Signals Officer, and Chief Air Officer (see Chapter 1 in this volume). In fact, for the duration of ONUC there were almost always more Canadians serving as officers at headquarters than was the case for any other nationality. At least one scholar has attributed this disproportionate presence to "their language capability, peacekeeping experience, generally good

---

Congo", *Montreal Gazette*, 9 September 1960, 1. The Canadian government was guilty of "placing the diplomatic cart before the horse". Precedent established that no nation had the right to insist on the inclusion of its forces in a peacekeeping mission. While the Canadian government did not quite go this far, it was certainly premature to announce the acquisition of the Caribou for service in ONUC prior to consultations with the UN Secretariat on the appropriateness and suitability of such a contribution. On the development of this precedent, see David Wainhouse, *International Peacekeeping at the Crossroads*, 558.

20 Defence Liaison to Cadieux, 30 August 1960; Murray to Robertson, 14 September 1960, RG25, 5219, File 6386-C-40 part 4, LAC. Telegram PERMISNY to External: Congo Caribou Aircraft, 2 September 1960, RG24, 21485, File 2137.3 part 2, LAC. Aide Memoire: Notes on Congo, Meeting with Col. Mendelsohn, 12 September 1960, File 112.1.009 (D21) 4, DHH. Note for File by Labouisse, 29 August 1960, File S-0209-0003-16, United Nations Operations in the Congo (ONUC) – Records on Foreign Countries, Canada volume 1 – ONUC military assistance, United Nations Archives [UNA]. Telegram Cordier to SG and von Horn, dated 14 August 1960, File S-0791-0045-10, UN Operation in the Congo, Force Commander, 3310/2/Ops – Canadian Troops – Personnel, UNA.

political acceptability, professionalism, and familiarity with both Commonwealth and U.S. military procedures".[21]

The RCAF provided valuable assistance in the early days of ONUC and the United Nations had been especially impressed with the services of A/C F.S. Carpenter, present in the Congo when the first peacekeepers arrived. After Carpenter's return to Canada, the Secretary-General asked if he could be sent back to Leopoldville, accompanied by five RCAF staff officers, to form an air staff at von Horn's headquarters. Group Captain W.K. Carr was dispatched in Carpenter's place, along with 10 other personnel to serve at Force Headquarters and as RCAF communications technicians and operators (see Chapter 1 in this volume). As von Horn prepared his proposed establishment of the United Nations Air Transport Force, he had specifically requested a Canadian to fill the position of air commander, or at the very least senior air staff officer. Indeed, he also wanted Canadians as the chief operations officer, engineering officer, and supply officer. It is significant that von Horn anticipated Hammarskjöld would think he was relying too heavily on Canadians – recall the Secretary-General's political difficulties in New York over Canada's participation; ONUC's commander actually couched his request with a plan to reduce the number of Canadians at UN Air Transport Headquarters by one-third, over a period of three months (and overall RCAF strength did fall from 58 personnel in August to 15 by December).[22]

In July 1961, A/C H.A. Morrison, considered one of the air force's "most experienced officers in the air transport field", had been chosen as the latest ONUC Air Commander.[23] Later that year, however, the Chief of the Air Staff issued instructions to develop a case to get the RCAF out of providing an officer to serve in this position. The timing of this decision, coinciding as it did with the addition of jet fighters and light bombers to ONUC's air services, suggests National Defence was uncomfortable having a Canadian oversee operations that went beyond transportation of supplies and personnel. In the midst of the second round of serious fighting in Katanga, Harkness wrote to Green to say once A/C Morrison completed his tour in the Congo he would not be replaced, justifying his decision largely on the grounds that ONUC's military action in Katanga, including both offensive and defensive operations, would require an enlarged staff drawn from countries other than Canada. The

---

21   Beauregard, J.P.R.E. "UN Operations in the Congo, 1960–1964", *Canadian Defence Quarterly* (August 1989): 27; Wainhouse, *International Peacekeeping at the Crossroads*, 308; Memo: Canadian Contribution to UN Forces in the Congo Republic, Synopsis No. 13, 12 August 1960, RG24, 21484, File 2137.3 part 1, LAC.

22   Spooner, K. *Canada, the Congo Crisis, and UN Peacekeeping, 1960–64* (Vancouver: University of British Columbia Press, 2009), 87.

23   Letter CCOS to USSEA: UN Request for Assistance – Congo and attached biography of H.A. Morrison, 30 May 1961, RG25, 5223, File 6386-C-40 part 14, LAC.

country supplying the largest elements of the force, Harkness argued, should also provide the commander.[24]

The UN Division at External Affairs expressed considerable concern at this decision. General E.L.M. Burns' command of the UNEF was used by External as a ready example of how the United Nations did not consistently follow the principle of appointing commanders from the largest troop-contributing states. Various other arguments were rallied to the cause, but above all, the political implications of not replacing Morrison were noted:

> We should not wish to expose ourselves to a charge of backing away from the United Nations operation at a time when our support was needed most. There is no doubt in my mind that if we do not replace Morrison the news about our refusal will spread.[25]

When Green wrote Harkness to ask for the decision to be reconsidered, the Minister of Defence was unmoved. Green was told to "inform the UN authorities promptly of our desire to withdraw Air Commodore Morrison by the end of this year". Harkness was not entirely uncompromising: he was willing to give the United Nations an additional two weeks of service in order to find a replacement. Green decided not to press National Defence any further and issued instructions to inform New York. The Secretariat was disinclined to accept "no" for an answer, however. They contacted External Affairs and pleaded that the UN command "had become accustomed to dealing with RCAF officers on air matters and that the smoothest cooperation had been possible because the RCAF officers 'understood the United Nations'". They were so disturbed in New York that U Thant, the UN's Acting Secretary-General, directly appealed to Diefenbaker to replace Morrison. This resulted in further consultations between External Affairs and National Defence; Morrison's term was extended by an additional three months, after which time it was made clear National Defence would neither renew Morrison's term again nor provide a substitute.[26] Notably, after all the serious hostilities had been brought to an end and as ONUC entered its final months, National Defence

---

24   Memo Chief of Air Staff to Vice Chief of Air Staff, 24 October 1961, RG24, 3022, File 895-8/115, LAC. Letter Harkness to Green, 7 December 1961, RG25, 5223, File 6386-C-40 part 17, LAC.

25   Memo UN Div to USSEA: Congo Replacement for A/C Morrison, 13 December 1961, RG25, 5223, File 6386-C-40 part 17, LAC.

26   Letter Green to Harkness, 18 December 1961; Letter Harkness to Green, 20 December 1961, RG25, 5223, File 6386-C-40 part 17, LAC. Memo UN Div to USSEA: Canadian Military Assistance to the UN in the Congo, 3 January 1962, RG25, 5214, File 6386-40 part 29, LAC. Memo SSEA to Ignatieff, Campbell DL (1), Euro Div and attached telegram, 5 January 1962; Memo Robinson to USSEA: Replacement for A/C Morrison, 11 January 1962; Memo for Prime Minister: Congo – Replacement for A/C Morrison as UN Air Commander, 8 January 1962; Memo from Prime Minister to Robinson, 10 January 1962, RG25, 5224, File 6386-C-40 part 18, LAC.

responded favourably to a renewed UN effort to once again appoint a Canadian to this position, with the promotion of someone serving in the Congo to the rank of Group Captain in order to serve as both Air Commander and Coordinator Air Transport Operations.[27]

Muscular Peacekeeping and Canadian Concerns

The debate over the replacement of A/C Morrison was indicative of the official Canadian attitude towards ONUC's use of force, as the peacekeeping mission laboured to achieve its mandate. Canadian authorities were never entirely comfortable with the form of muscular peacekeeping that ultimately evolved in the Congo, though by the time hostilities came to a head in ending the Katanga secession in early 1963, they were reluctantly resigned to the idea that some degree of force would be necessary to resolve the crisis and to secure conditions that would permit ONUC's eventual withdrawal. But even as this premise was accepted, Ottawa maintained a cautious and quite hesitant view towards permitting Canadians to serve in ONUC in periods of heightened tension and in capacities that directly contributed to the peacekeeping operation's ability to exercise greater force. This was equally true with respect to Canadians serving as signallers, in ONUC headquarters, and as part of the RCAF contribution.

An early indication of this cautious attitude can be seen in January 1961, when Canada turned down a UN request for 27 RCAF technical personnel, some three months after the UN had initially asked. This was the first significant ONUC request the government chose to decline. Initially, details from the United Nations were unclear and when DND prompted External Affairs for clarification of the UN's precise needs, the Chairman, Chiefs of Staff, advised the Under-Secretary:

> The organization of the RCAF is such that they are much more able to contribute a complete unit such as a squadron, rather than to weaken several units by supplying a piecemeal group as requested by the United Nations.[28]

By mid-November, details had been obtained, planning was undertaken, and the RCAF approved a plan to provide the necessary personnel to operate a telecommunications network for ONUC's three main air transport bases in Leopoldville, Stanleyville, and Kamina. The Chief of the Air Staff abandoned his earlier reservations because Canada had since been asked to fill the position of Air Commander in ONUC, and he did not want either the flexibility or safety

---

27　Telegram PERMISNY to External: ONUC – Request for Assistance Air Commander, 6 November 1963; Letter CCOS to USSEA: ONUC – Air Commander, 6 December 1963, RG25, 10648, File 21-14-6-ONUC-5 part 1, LAC.

28　Letter CCOS to USSEA: RCAF Personnel for the Congo, 14 October 1960, RG25, 5220, File 6386-C-40 part 5, LAC.

of the air operations to be compromised because of inadequate communications. Cabinet, however, postponed a decision on the request because of the "disturbed" political situation in the Congo. Following discussions with Group Captain Carr, the Chief of the Air Staff asked Harkness to raise the matter in Cabinet again. The Minister suggested a further delay of two weeks. When that interval passed and political conditions in the Congo had still not improved, DND finally asked External Affairs to advise the Secretary-General that it would not be possible to meet his request.[29] Precarious political conditions in the Congo could clearly be a decisive and significant factor when assessing UN requests.

At times, relations between the United Nations and the Congolese authorities became terribly strained. In such moments, Canada also proved reluctant to assist ONUC if the result could be increased tension or even violence between ONUC and the Congolese armed forces. For instance, in April 1961 the United Nations approached Canada for assistance in airlifting Indian troops from Dar es Salaam to Kamina. The United States had transported 2,300 Indian peacekeepers by sea to Tanganyika but backed out of an earlier commitment to airlift half these troops onwards to Katanga. The Secretary-General's Military Adviser, Major General Indar Jit Rikhye, then turned to Canada with an *informal* enquiry for assistance, not wanting to put the Canadian government in the awkward position of having to turn down an official request. External Affairs, after recognizing political difficulties with the UN's appeal for help, was lukewarm towards it and simply asked that DND just give it sympathetic consideration. In Leopoldville, Consul General Michel Gauvin urged Ottawa to decline the request in light of Congolese opposition to the arrival of additional Indian peacekeepers. The US, he noted, was criticized for airlifting the first 1,000 Indians. He advised:

> If without letting down UN too badly and if it is possible to discourage their request I would think it wise to do so especially since nature of [Canadian] contribution to ONUC has been such up to now that we have been able to avoid being involved in controversial issues between [the] Congolese and ONUC.[30]

The United Nations made other arrangements to transport the troops before a final decision could be reached, and the enquiry was suspended.[31]

---

29 Memo for Minister (DND): United Nations Request for Assistance in the Congo, 18 November 1960, RG25, 5220, File 6386-C-40 part 6, LAC. Cabinet Conclusions, 25 November 1960, RG2, 2747, LAC; Spooner, *Canada, the Congo Crisis, and UN Peacekeeping, 1960–64*, 127–8.

30 Telegram Leopoldville to External: UN Request for RCAF Planes, 13 April 1961, RG25, 5222, File 6386-C-40 part 13, LAC.

31 Memo for Minister: United Nations Request for Air Lift Assistance, 11 April 1961; Letter Green to Harkness, 11 April 1961; Telegram PERMISNY to External: Congo: Enquiry re RCAF assistance in airlift Indian troops from Dar to Kamina, 14 April 1961; Spooner, *Canada, the Congo Crisis, and UN Peacekeeping, 1960–64*, 154–5.

When serious fighting broke out between ONUC and armed elements in Katanga in the fall of 1961, in operations Rumpunch and Morthor, Canada was again compelled to consider UN requests for additional assistance at a moment when peacekeepers were engaged in open hostilities. The Canadian government was clearly ill at ease with developments in Katanga and was hardly enthusiastic when new UN appeals for help arrived. On 20 September, the Secretariat urgently requested transport aircraft, aircrews, maintenance personnel, and spare parts for airlifts within the Congo for three to five weeks. ONUC relied, to a considerable extent, on charter airlines for internal transport of supplies and personnel. During Operation Morthor, Katangese jet fighters damaged or destroyed a number of these charter planes, so most airlines withdrew their services, reducing available charter aircraft from thirty to three. The aircraft requested were to resupply forces stationed throughout the Congo; Sweden and Ethiopia had already offered jet fighters to escort the transport aircraft. By the end of five weeks, ONUC expected the threat from the Katangese jets to be resolved and planned to revert to chartered transport. Officials warned Howard Green that there could be "armed resistance and renewed hostilities" if the United Nations moved to arrest mercenaries in Katanga. Cabinet considered the request and Green acknowledged that the "decision was a difficult one". Although the aircraft would be at risk of attack, especially if an existing ceasefire ended, Cabinet agreed on 23 September to send two C-119s for one month, together with the necessary crews to permit their operation 24 hours a day; the planes and personnel left the next day.[32] In acceding to the request, Cabinet identified a number of important factors: the need to support Canadian and other peacekeepers deployed throughout the Congo, the significance of UN success in Katanga for the organization's future effectiveness, and public opinion.

Two weeks later, a second request arrived from the Secretariat. ONUC now required eight control tower officers and two maintenance ground communication technicians to aid in the operation of the Swedish and Ethiopian jet fighters and Indian light bombers recently attached to ONUC. Because of the policy implications of this request, further information was sought from New York. Ottawa learned that ONUC intended to use the fighters and bombers if hostilities resumed, both to defend its transport aircraft and to "render unuseable" the runways available to Katanga's jets. Should the ceasefire be breached, ONUC's Commander, General Sean MacEoin, planned to move all jets to Kamina to operate from within Katanga. External Affairs was very concerned about the implications of Canadian involvement in this aspect of ONUC's operations. Robertson wrote:

---

32   Memo to Minister: Congo: UN Request for Assistance, 22 September 1961; Memo for Minister: Congo – UN Request for Canadian Transport Planes and Crews, 22 September 1961, RG25, 5223, File 6386-C-40 part 16, LAC. Cabinet Conclusions, 23 September 1961, RG2, 6177, LAC. Memo Chief of Air Staff to Minister National Defence, 25 September 1961, RG24, 21485, File 2137.3 part 5, LAC.

There is, of course, a possibility that if we agree to the present UN request, we could be placed later on in an awkward position if the UN engages in warlike operations in the Congo, and particularly in Katanga.[33]

The Under-Secretary was especially worried that such action might be taken in circumstances that would prove troubling to Canada, but Howard Green did ultimately ask the Minister of National Defence, Douglas Harkness, to give "sympathetic consideration" to the request. The personnel involved, it was argued, would still be considered non-combatant and the aircraft would provide protection for members of both the RCAF and 57th Signals Unit stationed in the Congo. Harkness advised Green on 25 October that there was "an acute shortage" of suitable personnel required to meet the UN's request, so it could be met only by sacrificing the operational efficiency of RCAF units in Canada. He asked External Affairs to inform the Secretariat "Canada would prefer not to accept this commitment".[34]

Disappointed and deeply concerned by the negative reply, UN Under-Secretary Bunche personally approached Canadian Ambassador Ritchie and asked if Canada would reconsider its decision. The American and Ethiopian missions also expressed concern. The United States went so far as to threaten not to provide the necessary communications equipment unless Canadians agreed to operate it, even as the need for this equipment became acute when Katangese planes carried out bombing raids in Kasai. In a meeting of the Secretary-General's Advisory Committee, Bunche revealed that ONUC had warned the Katangese authorities that any further offensive action would be countered, with the destruction of "all planes involved either in air or on ground". But, the United Nations would not be able to carry out this threat without the American equipment and Canadian personnel. Green wrote Harkness asking him to reconsider his decision. The Minister observed:

[I]t would appear that Canada would be the object of widespread criticisms by Afro–Asian countries, particularly those who are members of the Congo Advisory Committee, if it is felt during the forthcoming developments that the capacity of the UN to resist aggression is seriously impaired because of our inability to provide the communications personnel needed for the servicing of the UN aircraft.[35]

---

33   Memo for Minister: Congo: UN Request for Assistance, 17 October 1961, RG25, 5223, File 6386-C-40 part 16, LAC.

34   Memo for Minister: UN Operations in Katanga, 18 October 1961; Letter Green to Harkness, 17 October 1961; Letter Harkness to Green, 25 October 1961, RG25, 5223, File 6386-C-40 part 16, LAC.

35   Letter Green to Harkness, 2 November 1961, RG25, 5223, File 6386-C-40 part 16, LAC.

Before Harkness received Green's appeal, the Minister of Defence raised the matter in Cabinet on his own initiative, and the earlier decision was reversed. Cabinet also granted a 30-day extension on the loan of the two C-119s but cautioned, "there was no intention of continuing this arrangement indefinitely".[36] In the end, Canada may have provided critical assistance for an important episode with ONUC, but the deliberations related to these decisions demonstrated considerable concern, angst, and serious reservations.

While Ottawa hardly needed a demonstration of just how dangerous and unpredictable the situation in the Congo could be, the legitimacy of the Canadian government's concerns was made all too apparent when Congolese forces seized a Yukon turboprop when it landed in Leopoldville on 20 November 1961. The plane was released only after A/C Morrison appealed directly to Congolese Prime Minister Cyrille Adoula and Joseph Mobutu, the chief of staff of the Congolese armed forces at that time. Worried that additional aircraft might be detained, National Defence suspended all Yukon flights to the Congo, a decision subsequently endorsed by Cabinet. It was late December before the matter was reviewed. At that time, the Chairman, Chiefs of Staff, asked Robertson to seek assurances from the United Nations that any RCAF aircraft flying within or into the Congo in support of ONUC would not "be subject to seizure or impoundment". External Affairs learned from Leopoldville and New York that the Yukon incident was an isolated case of mistaken identity. The Congolese were confused by the unfamiliar design of the plane and because it bore only RCAF insignia, not UN markings. To reassure Ottawa, the United Nations enacted measures to ensure Congolese authorities were given adequate notice prior to the arrival of each flight.[37] On the page opposite, Figure 3.1 shows the Yukon aircraft.

It has been suggested that incidents such as that with the Yukon, happened "frequently enough" to cause Ottawa to become less "eager" to provide ONUC with assistance generally and to meet a particular request in November 1961 for help in establishing a security service. While the threat of violence towards Canadian peacekeepers was always a concern and a factor weighed by the Government when it assessed UN requests, political and even administrative concerns were

---

36   Memo Ross Campbell to DL (1): Congo: UN Request for Assistance, 3 November 1961; Letter Harkness to Green, 6 November 1961, RG25, 5223, File 6386-C-40 part 16, LAC. Cabinet Conclusions, 23 October 1961, RG2, 6177. Doc. 117, Telegram From the Department of State to the Embassy in the Congo; Doc. 120, National Security Action Memorandum No. 97, *Foreign Relations of the United States, 1961–1963*, XX. Memo Bowitz to Bunche, 30 October 1961, United Nations Archives (UNA), S-0209-0003-13, United Nations Operations in the Congo (ONUC) – records on foreign countries, Canada – correspondence, cables, etc.

37   Memo for Minister: Incidents Involving Canadians in the Congo, 28 November 1961; Letter CCOS to USSEA, 18 December 1961, RG25, 5223, File 6386-C-40 part 17, LAC. Cabinet Conclusions, 20 November 1961, RG2, 6177, LAC. Telegram PERMISNY to External: Congo: RCAF Airlift, 24 January 1962; Numbered Letter Leopoldville to USSEA: Congo: RCAF Airlift, 18 January 1962, RG25, 5224, File 6386-C-40 part 18, LAC.

**Figure 3.1   A Canadian Yukon aircraft at Leopoldville airport being
inspected by Congolese and UN military officers**

often the more significant factors when it was decided to turn down requests or
scale back Canadian involvement in ONUC. By early 1963, UN requests for
various additional personnel for ONUC itself were increasingly scrutinized,
especially by National Defence. The Secretariat asked Canada to provide four
training and administrative officers for service with two Congolese National
Army (ANC) battalions, helicopter pilots, ground crew and movement control
personnel. Following consultation with the naval and air forces, the Chairman,
Chiefs of Staff, turned down the request for helicopter personnel because it would
seriously prejudice other commitments. External Affairs was not surprised by this
and decided it was best not to pressure National Defence in order to preserve intra-
departmental goodwill for future and more important UN appeals for assistance.
The Chief of the General Staff, Lieutenant-General Geoffrey Walsh, was clearly
frustrated by "piecemeal" requests which were said to be making it "almost
impossible to do any career planning for the officers concerned" and because they
were having "an adverse effect on the proper general administration of the Army".[38]

_____

38   Granatstein, J. and Bercuson, D. *War and Peacekeeping* (Toronto: Key Porter,
1991), 221. Letter USSEA to CCOS: UN Request for Assistance, 18 January 1963; Memo
DL(1) Div to Ross Campbell: Congo: UN Request for Assistance, 6 February 1963; Letter
CCOS to USSEA: Congo – UN Request for Assistance, 12 February 1963; Telegram
PERMISNY to External: Congo: UN Request for Assistance, 27 February 1963,
RG 25, 5224, File 6386-C-40. part 21, LAC. Letter CCOS to USSEA: UN Request for
Assistance – Congo Helicopter Pilots and Ground Crew, 7 March 1963; Outgoing Message
External to PERMISNY: Congo Ops, 23 March 1963, RG25, 5224, File 6386-C-40 part 22,

To conclude, the RCAF made significant contributions to ONUC throughout its operations in the Congo. Particularly important were the services provided by officers in the command and coordination of UN air operations and the essential airlift from Pisa to Leopoldville. It is important to recognize, however, the historical contexts and political circumstances that often dictated and shaped the nature of Canada's peacekeeping contribution. In the earliest days of the crisis, the government embraced the opportunity to provide air transport as a means to play down expectations Canada would send combat forces – something seen as politically inadvisable by both the Diefenbaker government and the United Nations. The politics of the Cold War were an ever-present determinant of policy in these years. They were evident most notably in the UN's decision to decline Canada's offer of Caribou aircraft and crews, but they were also at play when decisions were made regarding staffing at ONUC Headquarters. The increasingly offensive or muscular nature of ONUC's activities were not especially welcomed in Canada and they served as a backdrop for increasing reticence to maintain or bolster Canada's contributions to air operations in the Congo. The decision to end A/C Morrison's appointment in April 1962 came at a critical juncture in this respect and External Affairs was especially disturbed by how his departure would likely be perceived.

A fine line connects the practical decisions related to the precise contributions a country is prepared to make to international peacekeeping with the domestic and international political considerations and contexts that shape those decisions. In the case of ONUC, Canada provided essential support to various elements of the UN's air operations in the Congo, but the willingness behind, capacity to provide and political suitability for this effort appeared tenuous at times. The influence of the Cold War, given Canada's position as a Western-aligned nation, and specific concerns about ONUC's use of force also represented "a fine line" of sorts – a line Canada crossed with difficulty with respect to the Cold War and a line to be crossed only with extreme caution and care with respect to muscular peacekeeping.

LAC. Letter CCOS to USSEA: UN Request for Assistance – Congo, 22 March 1963, RG24, 21487, File 2137.3 part 10, LAC; Spooner, *Canada, the Congo Crisis, and UN Peacekeeping, 1960–64,* 206.

PART II
Airlift: Lifeline for UN Missions

From its earliest peacekeeping experience, the United Nations has used airlift to deploy, employ, and sustain its missions, especially in difficult conflict zones in remote locations. To move military forces and their equipment, including the weapons and ammunition, from around the world in a timely manner, air transport remains essential. Locally, ground transport is usually too slow or even impossible because roads were primitive or impassable, if they exist at all in remote war zones.

The United Nations gets its air transport from nations and from contractors. It does not have its own aircraft. Part II looks at how national contributions are made. Chapter 4 presents a case study of the Canadian aircraft provided to the UN mission in Kashmir, which was mandated to oversee a shaky ceasefire between Indian and Pakistani forces in that divided territory. Matthew Trudgen shows how the Canadian airlift traversed the mighty Himalayas to support these UN observer missions. The national dilemmas and decision-making illustrated are typical of many nations involved in many UN missions, past and present.

The emergency humanitarian operation in Haiti provides excellent examples of aerial coordination with national forces, especially the US Air Force. After the devastating January 2010 earthquake, the international community poured humanitarian aid into the long-suffering country. The local infrastructure in Haiti, however, could not support the world's generosity. The Haitian government gave permission to the US Air Force to run the country's main airport in Port-au-Prince. Robert C. Owen was a keen observer of that effort and he shares his insights on the vital US–UN cooperation in Haiti in Chapter 5.

The UN fleet of aircraft deployed in conflict areas is not solely arranged by and for peacekeeping missions. The United Nations Humanitarian Air Service (UNHAS), run by the World Food Programme based in Rome, charters over 50 aircraft to service the humanitarian community, not only for United Nations and governmental agencies but also to assist non-governmental partners. A. Walter Dorn and Ryan W. Cross provided a pioneering academic paper on this little-known but life-saving UN service in Chapter 6. The UNHAS airlift and airdrops serve as a living lifeline for millions.

Chapter 4

Above the Rooftop of the World: Canadian Air Operations in Kashmir and Along the India–Pakistan Border

Matthew Trudgen

From 1949 to 1995, the Canadian government worked to assist the efforts of United Nations in bringing peace to the Indian subcontinent. This was first done through the provision of military observers for the United Nations Military Observer Group India–Pakistan (UNMOGIP). A Canadian officer, Brigadier Harry H. Angle, also served as the first Chief Military Observer of UNMOGIP until his tragic death in a plane crash in July 1950.[1] However, beginning in the mid-1960s, Canada's role began to evolve.

In 1964, the Canadian government dispatched one CC-108 Caribou along with three officers and five ground crew from No. 102 Composite Squadron to support UNMOGIP. This unit would eventually be renamed 424 Squadron and would be later re-equipped with a CC-138 Twin Otter. Furthermore, as part of Canada's commitment to the newly created United Nations India–Pakistan Observation Mission (UNIPOM), which was formed after the second Indo–Pakistani War, in 1965, the Canadian military provided not only many of the missions' observers and its commanding officer, Major General Bruce Macdonald, but also UNIPOM's air component of two Caribous and three CC-123 Otters, as well as their crews and support personnel. This Canadian air contingent was also assigned the task of supporting UNMOGIP while it was in the area, carrying out the twin tasks of air transport and air observation. Although most of these aircraft would be withdrawn with the end of UNIPOM's mission in 1966, a single Canadian transport plane remained in the area to support UNMOGIP until 1975. In addition, CC-130 Hercules aircraft would continue to assist UNMOGIP until 1995, even after the last Canadian Army observers were pulled out in 1979.

---

1 Though rumour circulated of sabotage, the plane seems to have gone down during a severe storm with the autopilot engaged. See Granatstein, J.L. "Canada: Peacekeeper – A Survey of Canada's Participation in Peacekeeping Operations", in *Peacekeeping: International Challenge and Canadian Response*, by David Cox, J.L. Granatstein and Alistair Taylor (Toronto: Canadian Institute of International Affairs, 1968), 101–02; Maloney, S. *Canada and UN Peacekeeping: Cold War by Other Means, 1945–1970* (Toronto: Vanwell Publishing, 2002), 26.

This chapter will use this experience as a case study to understand the contribution of the Royal Canadian Air Force (RCAF) to UN operations during the Cold War. I first outline the background to the crisis between India and Pakistan that led to the dispatch of UN military observers to the region in the late-1940s and the reasons why Canada decided to contribute personnel to this force. I also describe how Canada's contribution to peace observer missions on the Indian subcontinent evolved in the mid-1960s due to developments in Ottawa and because of the outbreak of the second Indo-Pakistani War. Then I discuss the Canadian air operations as part of UNMOGIP in the late 1960s and the early 1970s using oral history interviews and other primary source material provided by former Canadian Forces personnel. I conclude with an examination of what lessons can be learned by the Canadian government and the RCAF from this experience.

Background to the Mission

The origins of the UN's involvement in the affairs of the Indian subcontinent began during the period of partition and independence. Much of the troubled relationship between India and Pakistan had its origins in the fate of the predominantly Muslim Princely State of Jammu and Kashmir. This state, which arguably should have become part of Pakistan, instead became part of India. Kashmir's Hindu Maharaja, Hari Singh, decided to accede to India in order to get Indian troops to protect his kingdom and his ruling dynasty from irregulars from Pakistan. However, it is unclear whether the Pakistani irregulars who invaded Kashmir were Pathan tribesmen out for loot or trying to liberate their Muslim brothers from an oppressive regime. Moreover, how much the Pakistani government and military had to do with these events is still the subject of debate.[2] The result of all these factors was a sustained limited conflict between India and Pakistan that was confined to Kashmir and which ended in stalemate. Eventually, through the work of the United Nations Commission for India and Pakistan (UNCIP), a ceasefire was agreed. It was then determined that UN military observers under the control of UNCIP would monitor this agreement and report on the compliance of the Indians and Pakistanis. When the UNCIP was dissolved in 1950, this observer force became known as UNMOGIP.[3]

---

2   Extract from Minutes of Meeting of Heads of Divisions, 6 January 1948, reproduced in Mackenzie, H. (ed.), *Documents on Canadian External Relations* 14, 1948 (Ottawa: Department of Foreign Affairs and International Trade, 1994), 232–3. For more information on these events see Schofield, V. *Kashmir in Conflict: India, Pakistan and the Unending War* (London: I.B. Tauris, 2010), 49–72.

3   Granatstein, "Canada: Peacekeeper", 101–3; Maloney, Canada and UN Peacekeeping, 25–26. Canadian Forces, "Details/Information for Canadian Forces (CF) Operation *United Nations Military Observer Group in India and Pakistan*". Available at:

This leads to the question: Why did the Canadian government contribute personnel to this mission? It is first important to emphasize that this decision was made at a time when peacekeeping was not seen as an important role for Canada in the world. When Brooke Claxton, post-war Minister of National Defence, referred this matter to the Cabinet, Canadian ministers were "allergic" and asked two questions: "Why is Canada one of the countries invited to appoint observers?" and "What other countries have accepted the invitation?"[4] It was likely that there was some resistance to the mission from the Canadian military due to the problems caused by the rapid post-war demobilization. By 1947 defence spending had fallen to C\$200 million; by 1948 the strength of the entire Canadian armed forces was only 34,000 personnel.[5]

But there were a number of factors that worked to ensure that Canada would contribute to this mission, including that it was on the United Nations Security Council at this time.[6] Furthermore, Canada's Ambassador to the UN, General (Retired) Andrew McNaughton, was then serving as its president and had played a role in trying to mediate this conflict.[7] However, the most important factor was the positions of Prime Minister Louis St. Laurent and Lester Pearson, then Secretary of State for External Affairs. The Cabinet had "decided to leave this matter" to them and they agreed that Canada should send four observers to assist the United Nations in this region.[8] It was likely that such a commitment was seen to be line with the greater Canadian interest in international affairs represented by St. Laurent's

---

http://www.cmp-cpm.forces.gc.ca/dhh-dhp/od-bdo/di-ri-eng.asp?IntlOpId=292&CdnOp Id=352 [accessed 1 June 2011].

4   Memorandum from Acting Under-Secretary of State for External Affairs to Secretary of State for External Affairs, 15 January 1949, reproduced in Mackenzie, H. (ed.) *Documents on Canadian External Relations*, Vol. 15, 1949 (Ottawa: Department of Foreign Affairs and International Trade, 1995), 302–3. Canada was eligible to send observers partly because Canadian military personnel were able to speak English, but also because it had no colonial possessions in Asia and it was not a country directly involved in the Kashmir dispute.

5   Bercusson, D. *True Patriot: The Life of Brooke Claxton, 1898–1960* (Toronto: University of Toronto Press, 1993), 158, 169, 177–80.

6   Granatstein, J.L. "Peacekeeping: Did Canada Make a Difference? And What Difference did Peacekeeping Make to Canada?", in *Making a Difference? Canada's Foreign Policy in a Changing World Order*, ed. John English and Norman Hillmer (Toronto: Lester Publishing, 1992), 225.

7   Reford, R.W. "UNIPOM: Success of a Mission", *International Journal* 27(3) (1972), 406. Sean Maloney has argued that the need for bases in Pakistan and Afghanistan under American–British–Canadian War Planning also encouraged Canada to support this mission. Maloney, *Canada and UN Peacekeeping*, 26–9.

8   Memorandum from Acting Under-Secretary of State for External Affairs to Secretary of State for External Affairs, 15 January 1949, reproduced in Mackenzie, *Documents on Canadian External Relations*, 302–3.

Gray Lecture in February 1947.[9] Moreover, because the mission involved a dispute between two Commonwealth countries, St. Laurent and Pearson concluded that it was important to prevent a serious crisis from breaking out between members of this organization. The situation in Kashmir also came to be seen in Ottawa as a real threat to world stability. Indeed, Canadian officials would increasingly value the fact that Canada's military observers provided an accurate picture of what was happening on the ground.[10] These debriefings were even distributed to Washington and London under the "CAN/US/UK Only" level of classification.[11]

Although Canada would later increase its number of observers to eight, and allow Brigadier Angle to serve as the commander of this mission until his death in a plane crash on the mission, along with some of his staff, in July 1950 there was very limited interest in this mission in the House of Commons and UNMOGIP was given little public attention.[12] This reality was partially due to the perception that UNMOGIP would be better off if it had a low public profile, given the sensitive nature of this conflict. But a more important factor was that the Kashmir dispute was seen in Ottawa to be a "delicate and embarrassing question in terms of Commonwealth relations". It therefore became Canadian practice to "not to mention UNMOGIP except when necessary".[13] Nonetheless, in 1964 and 1965, Canada's role in the international effort to address this situation would change dramatically.

Evolution of Canada's Role

The first and most obvious reason for Canada's participation was the sharp rise of tensions in the area that culminated in the war between India and Pakistan in 1965. However, an increased interest in peacekeeping operations already existed in Ottawa from the early days of Pearson's government. One example was the attention given to the idea of a UN "standby" peacekeeping force by the Prime Minister in this

---

9   St. Laurent, L. *The Foundations of Canadian Policy in World Affairs* (Toronto: University of Toronto Press, 1947).

10   Background Paper, 5 April 1965, RG 25 Department of External Affairs, Vol. 10121, File 21-13-UNMOGIP Military Actions – Armistice – United Nations Military Observer Group in India and Pakistan Vol. 1, LAC.

11   Information Report, 17 1964, RG 25 Department of External Affairs, Vol. 10121, File 21-13-UNMOGIP-1 Military Actions – Armistice – United Nations Military Observer Group in India and Pakistan – Canadian Contingent Part 1, LAC.

12   Lourie, S. "The United Nations Military Observer Group in India and Pakistan", *International Organization* 9(1) (1955), 24; Dawson, P. *The Peacekeepers of Kashmir: The UN Military Observer Group in India and Pakistan* (Bombay: Popular Prakashan PVT, 1995), 37.

13   Background Paper, 5 April 1965, RG 25 Department of External Affairs, Vol. 10121, File 21-13-UNMOGIP Military Actions – Armistice – United Nations Military Observer Group in India and Pakistan Vol. 1, LAC.

period.[14] The Secretary of State for External Affairs, Paul Martin Sr, also argued that Canada's participation in these operations contributed to it being:

> accepted and welcomed as a participant in important ventures. Those who ask whether we have an independent identity before the world must consider all this evidence of decision, action and participation in international affairs.[15]

Consequently, peacekeeping received increased attention in the 1964 Defence White Paper.[16]

This interest was further shown by Canada's holding of "The Meeting of Military Experts to Consider the Technical Aspects of UN Peace Keeping Operations", in late 1964. This conference accomplished little in concrete terms, but the fact that the Canadian government went to the trouble of holding it, overcoming some Soviet opposition in the process, showed the increased attention to these operations in Ottawa. One of the participants of this conference, Major General Indar Jit Rikhye of the Indian Army, even suggested after seeing a demonstration of Canadian forces that "airlift for peace-keeping operations might be supplied by Canada".[17]

These factors therefore help to explain why, when the United Nations requested Canadian air support for UNMOGIP in August 1963, some Canadian ministers and officials were interested in providing a transport aircraft. It should be emphasized that this form of assistance was vital to UNMOGIP's operations because of the need to transport the military observers and their supplies to the base camps on both sides of the Line of Control in this disputed border region, which was extremely difficult to access due to its rugged terrain.[18]

The RCAF was less than enthusiastic about this potential assignment, as the Chief of the Air Staff (CAS), Air Marshal C.R. Dunlop, concluded that no suitable aircraft was available.[19] Despite this negative response, the United Nations continued to lobby Canada. UN officials argued that only Canada and the United

---

14   Pearson, L. "A New kind of Peace Forces", in *Canadian Foreign Policy Since 1945 Middle Power or Satellite?*, ed. J.L. Granatstein (Toronto: Copp Clark Publishing Company, 1973), 151–4; Pearson, L. "Lecture by the Right Honourable Lester B. Pearson, Prime Minister of Canada, in the Dag Hammarskjöld Memorial Series at Carleton University, Ottawa, 7 May 1964 (extracts)", in *Canadian Defence Policy Speeches and Documents 1964–1981*, ed. Larry R. Stewart (Kingston: Centre for International Relations Queen's University, 1982), 156–62.

15   Martin, P. "An Independent Canadian Foreign Policy", in *Paul Martin Speaks for Canada* (Toronto: McClelland and Stewart, 1967), 20.

16   Government of Canada, *White Paper on Defence* (Ottawa: Queen's Printer, 1964), 15, 24.

17   Girard, C. *Canada in World Affairs, 1963–1965* (Toronto: Canadian Institute of International Affairs, 1980), 320–27.

18   The "Line of Control" is the term given to the ceasefire line of 1949.

19   Memorandum to Chairman Chief of Staff, 30 August 1963, RG 24 Department of National Defence, Vol. 21489 File 2137.5 UN India and Pakistan. LAC Part 1, LAC.

States could provide an aircraft with a sufficiently well-trained crew for this task, and since the Americans could not take up this role, Canada needed to do it.[20] The Indian government and UNMOGIP's Chief Military Observer, General Robert Nimmo, further urged Canada to provide this capability.[21] In late November 1963, even the Americans showed some interest in this issue when Turner Cameron, the Director of Southwest Asian Affairs for the State Department put it to Canadian officials that a "reliable nation" needed to provide aircraft.[22]

In December 1963, after the United Nations had again requested Canada's help,[23] Martin wrote to the Minister of National Defence, Paul Hellyer. Martin argued that Canada had gained an "enviable reputation" for providing airlift to a variety of peacekeeping operations. He added that

> I think we should endeavour, so far as possible, to meet well-founded requests
> for internal air transport for UN peacekeeping operations as something of
> a Canadian specialty. I realize that this request may involve the purchase of
> another Caribou ... but I would hope that this aircraft might be regarded as
> giving the RCAF some extra flexibility to meet requests of this nature.[24]

Martin continued to press Hellyer in January 1964 when he re-emphasized UNMOGIP's need for the aircraft and asserted that he would support the acquisition of an additional Caribou to allow the RCAF to provide support for this mission.[25]

Ultimately, Martin's arguments paid dividends when Hellyer wrote to Dunlop that "I think it would be politic for us to agree to this request if it is possible for us to do so".[26] Hellyer informed Martin in early 1964 that while an aircraft was not available and the RCAF would not immediately be able to meet this need, the Air Force had put in a request to the Treasury Board to acquire an additional Caribou

---

20   Memorandum to the Cabinet Committee on External Affairs and Defence, 6 February 1964, RG 24 Department of National Defence, Vol. 21489, File 2137.5 UN India and Pakistan Part 1, LAC.

21   Tremblay to Ottawa, 10 January 1964, RG 24 Department of National Defence, Vol. 21489, File 2137.5 UN India and Pakistan Part 1, LAC.

22   Message from the Canadian Delegation to the UN, 30 November 1963, RG 24 Department of National Defence, Vol. 21597, File 2-5081-9 United Nations Emergency Force UNMOGIP, LAC.

23   Provision of Chartered Aircraft, 19 August 1965, RG 25 Department of External Affairs Vol. 10121, File 21-13-UNMOGIP-1 Military Actions – Armistice – United Nations Military Observer Group in India and Pakistan Part 1, LAC.

24   Martin to Hellyer, 5 December 1963, RG 24 Department of National Defence, Vol. 21489, File 2137.5 UN India and Pakistan Part 1, LAC.

25   Martin to Hellyer, 14 January 1964, RG 25 Department of External Affairs, Vol. 10121, File 21-13-UNMOGIP Military actions – Armistice – United Nations Military Observer Group in India and Pakistan Part 1.2, LAC.

26   Hellyer to the CAS, 10 December 1963, RG 24 Department of National Defence, Vol. 21489, File 2137.5 UN India and Pakistan Part 1, LAC.

aircraft. Hellyer then argued that he would support the dispatch of an aircraft to support UNMOGIP if this commitment was reviewed every six months and if it was authorized by the full Cabinet.[27] On 10 February 1964, the Treasury Board granted permission for this purchase and on 18 February the Cabinet approved Canada's participation in this mission.[28] Canada's contribution was in the form of one Caribou transport, three pilots and five ground crew.[29]

The interest in providing air support to UN peacekeeping missions would continue to influence Canadian policy in the aftermath of the second Indo-Pakistani War in 1965. The origins of this conflict were in the ongoing dispute between India and Pakistan over Kashmir, and the pressures from the Pakistani people on their government to rectify the situation. But a number of other factors also played a role, namely certain perceptions and misperceptions by the Pakistani leadership. At this point, one should note that poor Pakistani strategic decision-making is not a new phenomenon. These perceptions in Islamabad included the idea that the riots in Kashmir following the theft of a Muslim relic, a hair of the Prophet's beard known as the Moi Maquaddas, symbolized both growing Muslim sentiment in the province and wider discontent with Indian rule. Furthermore, Indian military weakness during the Sino–Indian War in 1962 and in border clashes over the Rann of Kutch, a dissolute region inhabited largely by flamingos and wild donkeys, as well as various internal difficulties in India that emerged after Prime Minister Jawaharlal Nehru's death in 1964, convinced Pakistan that an opportunity to resolve this situation was at hand. Their solution was Operation Gibraltar, which included the introduction of "guerrillas" into Kashmir, who would start a rebellion against Indian rule. This move would then be followed up by Operation Grand Slam, involving additional Pakistani forces. The problem was that the Muslim population of Kashmir did not revolt and Pakistani military operations quickly became bogged down. It was still worse for Pakistan that, unlike during the 1947 war, India expanded the conflict to Pakistan by striking towards the Pakistani city of Lahore. At this point, equipment and parts shortages caused by the heavy fighting and a British and American arms embargo, when combined with the fact that both countries' military operations had reached stalemate, led India and Pakistan to accept a UN ceasefire. However, this truce created the need for additional UN observers.[30] But since UNMOGIP

---

27   Hellyer to Martin, 18 January 1964, RG 24 Department of National Defence, Vol. 21489, File 2137.5 UN India and Pakistan Part 1, LAC.

28   Record of Cabinet Decision, 18 February 1964, RG 24 Department of National Defence, Vol. 21489, 2137:5 UN India and Pakistan Part 2, LAC.

29   It would take months of negotiations for Canada and the UN to agree on the terms under which this aircraft would operate, Memorandum to the Cabinet, 27 August 1965, RG 24 Department of National Defence, Vol. 21597, File 2-5081-9 United Nations Emergency Force UNMOGIP, LAC.

30   Schofield, *Kashmir in Conflict*, 103–10; Lamb, A. *Kashmir: A Disputed Legacy, 1846–1990* (Oxford: Oxford University Press, 1991), 247; Nawaz, S. *Crossed Swords:*

was confined to Kashmir and it was restricted by its terms of reference to having only a limited investigative role, the UN's Secretary General, U Thant, concluded that it would be better to create a separate mission, UNIPOM, that would have a more flexible mandate and would also be able to monitor the ceasefire along the rest of the India–Pakistan frontier.[31]

As for Canada, Pearson's government quickly moved to become involved in resolving the crisis. While Pearson's attempt to become a mediator was rejected by Thant, after some lobbying by Martin, Canada was able to take up extensive involvement in the mission.[32] The Canadian government not only secured the appointment of Major General Bruce Macdonald as UNIPOM's commander but also provided all the air transport assets for the mission. These aircraft, which included two Caribou and three Otters, as well as their crews and almost 100 maintenance personnel, were placed under the command of Macdonald's Air Adviser, RCAF Group Captain George Murray. Moreover, Canada provided twelve of UNIPOM's military observers.[33] Canada's contribution in all numbered 112 personnel and represented the core of this peace observer mission.[34] Aside from the seriousness of the situation on the Indian subcontinent and the threat it posed to world stability, Canada's strong commitment to UNIPOM was the result of several factors. These included Pearson's and Martin's renewed attention to peacekeeping

---

Pakistan, its Army and the Wars Within (Oxford: Oxford University Press, 2008), 194–5, 202–14, 220–22.

31   Martin to Ottawa, 23 September 1965, RG 24 Department of National Defence, Vol. 21597, File 2-5081-9 United Nations Emergency Force UNMOGIP Part 2, LAC; Message from External Affairs to New Delhi, 4 October 1965, RG 25 Department of External Affairs, Vol. 10121, File 21-13-UNMOGIP Military actions – Armistice – United Nations Military Observer Group in India and Pakistan Vol. 1, LAC; Dawson, *The Peacekeepers of Kashmir*, 237–8.

32   Maloney, *Canada and UN Peacekeeping*, 227; Granatstein, "Canada: Peacekeeper", 104.

33   Terms of Reference United Nations Air Transport Unit India–Pakistan, 28 September 1965, RG 24 Department of National Defence Vol. 21597, File 2-5081-9 United Nations Emergency Force UNMOGIP Part 2, LAC; Macdonald to Moncel, 26 November 1965, 2005/05 Alan James Papers, Box 29, File 4 India and Pakistan – MacDonald Papers, DHH; Outline of UNIPOM's Activities September 1965 – March 1966, Undated, 2005/05 Allan James Fond, Box 29, File 4 India and Pakistan – MacDonald Papers, DHH, 4.

34   Maloney, *Canada and UN Peacekeeping*, 227; Granatstein, "Canada: Peacekeeper", 104. Brazil, Burma, Ceylon, Ethiopia, Ireland, Nepal, the Netherlands, Nigeria and Venezuela contributed observers to this force. Some of the observers were borrowed from UNMOGIP and the United Nations Truce Supervision Organization. See, respectively, Part Three: India/Pakistan The UN Military Observer Group in India and Pakistan and the India–Pakistan Observation Mission, 2005/05 Allan James Fond, Box 29, File 14 India and Pakistan – United Nations Military Observer Group in India and Pakistan, DHH; Outline of UNIPOM's Activities September 1965 – March 1966, Undated, 2005/05 Allan James Fond, Box 29, File 4 India and Pakistan – MacDonald Papers, DHH, 3.

and the government's interest in providing air transport for these kinds of missions. Canada's strong commitment to the Commonwealth and its close relations with India and Pakistan also played a role in this decision.

Another reason that arose was the need to maintain Canada's international reputation. One memo to the Cabinet argued that Canada needed to participate in UNIPOM because of expectations created from its past support of peacekeeping and the leadership that the Canadian government had taken in this field. In addition, some ministers and officials understood that the desire to be involved in "crisis diplomacy" and to play more of a role in international affairs needed to be backed up. For example, although Pearson's efforts to serve as a mediator had been rejected, Martin argued that Canada's "willingness to support [the] SECGENS [Secretary General's] efforts for a ceasefire have undoubtedly encouraged expectations here [at the United Nations] of a favourable CDN [Canadian] response". Another draft memorandum even noted that there was a need to support "the leading role played by the Canadian Prime Minister in offering his services as a mediator to the two countries".[35] This point was removed from the final submission to the Cabinet, but it does give a sense of what the thinking was behind the scenes in Ottawa.[36] The result of all these factors was that Canada committed significant resources to this peace observer mission, which undoubtedly helped to preserve its reputation as a leader in the field of peacekeeping operations. But what should not be forgotten was that it was left up to Macdonald, his fellow observers and the RCAF contingent to make this very difficult mission work.

Canada's Participation in the UN India–Pakistan Observer Mission

Aside from the reality that the UN observers were "utterly dependent upon the good will and cooperation of both sides", and if either country wanted to fight the mission was powerless to stop it, UNIPOM had numerous other problems such as the lack of suitable vehicles.[37] Moreover, not only had the mission to accommodate officers from ten different countries but also many of these men

---

35   Martin to Ottawa, 23 September 1965, RG 24 Department of National Defence, Vol. 21597, File 2-5081-9 United Nations Emergency Force UNMOGIP Part 2, LAC; Memorandum to the Cabinet (DRAFT), 24 September 1965, RG 24 Department of National Defence Vol. 21597, File 2-5081-9 United Nations Emergency Force UNMOGIP Part 2, LAC.

36   Memorandum to the Cabinet, 24 September 1965, RG 24 Department of National Defence, Vol. 21597, File 2-5081-9 United Nations Emergency Force UNMOGIP Part 2, LAC.

37   Macdonald to Moncel, 6 October 1965, 2005/05 Alan James Papers, Box 29, File 4 India and Pakistan – MacDonald Papers, DHH; Macdonald to Family, 19 October 1965, 2005/05 Alan James Papers, Box 29, File 4 India and Pakistan – MacDonald Papers, DHH.

lacked the proper kit and even inoculations.[38] Macdonald also did not consider himself properly briefed for the political background of the issues surrounding the second Indo–Pakistani War. He later noted that the Under-Secretary-General for Special Political Affairs, Ralph Bunche, expected him "to know things because he knows them – not because he or anyone else has ever told me". Macdonald added that the United Nations "really couldn't have given me a tougher job with less preparation and less briefing".[39]

Other difficulties surfaced with the air component of the mission. There was disagreement between the UN and UNIPOM on how many aircraft were needed, as UNIPOM's officers wanted six Otters instead of the three assigned to them. Ultimately, these Otters were not available for financial reasons.[40] In addition, the RCAF Caribou originally assigned to UNMOGIP had been destroyed on the ground by the Pakistani Air Force (PAF) during an air strike on 7 September, when the aircraft had been parked at an Indian airfield at Srinagar. It had been destroyed despite the fact that it was located "where they [the PAF] knew it was parked", as Nimmo later angrily complained to the Chief of the Pakistani General Staff.[41]

Although this aircraft was replaced by an RCAF Caribou borrowed from the United Nations Truce Supervision Organization, another Caribou was damaged during a botched landing during operations and was unavailable for the rest of the mission.[42] To add to the difficulties, there were disagreements between General Macdonald and the commander of the air contingent, Group Captain Murray, over the conduct of air operations. Murray was greatly worried about the safety of his pilots and aircraft, whereas Macdonald was "concerned with carrying out

---

38   Outline of UNIPOM's Activities September 1965 – March 1966, Undated, 2005/05 Allan James Fond, Box 29, File 4, India and Pakistan – MacDonald Papers, DHH, 12; Reford, "UNIPOM: Success of a Mission", 411. Furthermore, Reford described the lack of basic supplies such as radios, which required the officers to use the communications systems of the Indian and Pakistani armies.

39   A Job for Soldiers, Undated, 2005/05 Alan James Papers, Box 29, File 5 India and Pakistan – MacDonald Papers, DHH.

40   Ibid.; Message from Canadian Delegation to the UN, 16 October 1965, RG 24 Department of National Defence, Vol. 21597, File 2-5081-9 United Nations Emergency Force UNMOGIP Part 2, LAC.

41   Dawson, *The Peacekeepers of Kashmir*, 76. Fortunately there were no casualties. The United Nations accepted full liability for the loss of the aircraft. See: PERMISDNY to EXTERNAL, 7 September 1965, RG 24 Department of National Defence, Vol. 21597, File 2-5081-9 United Nations Emergency Force UNMOGIP, LAC; Lansky to Trimble, 8 September 1965, Vol. 21597, File 2-5081-9 United Nations Emergency Force UNMOGIP Part 2, LAC.

42   A Job for Soldiers, Undated, 2005/05 Alan James Papers, Box 29, File 5 India and Pakistan – MacDonald Papers, DHH; Memorandum to the Minister, 7 September 1965, RG 25 Department of External Affairs, Vol. 10121 File 21-13-UNMOGIP-1 Military Actions – Armistice – United Nations Military Observer Group in India and Pakistan – Canadian Contingent Part 1, LAC.

my mission in the most efficient manner possible". In particular, Macdonald wanted the aircraft to fly as close to the ground as possible for better observation. Eventually, it was decided that the captain would always be "fully responsible for the operation of the aircraft, irrespective of the rank of the passengers". The flight safety regulations further outlined that in the forward areas, the aircraft would not be able to fly lower than 1,500 ft while they would have to fly above 5,000 ft if they were crossing the international border.[43] Nonetheless, the RCAF's contingent played a major role in the success of the mission.

The fact that the Canadian aircraft were deployed and made operational in a short period of time was of great importance in allowing the UN observers to accomplish their mission. The Caribous and Otters fulfilled several roles, including the aerial supply of isolated outposts and the movement of observers and VIPs, as well as flying numerous reconnaissance missions that were invaluable for allowing the military observers to inspect Indian and Pakistani activities in the desert regions along the border. Indeed, it would often "take eight or more hours of driving over difficult country to inspect an area which could be reached by air in a matter of minutes". Moreover, these aircraft served as a vital and secure means of communication between Macdonald and the UN outposts before the mission had been equipped with enough radios. The ability of the pilots to do what Macdonald termed "bush flying" was also of great help in coping with the primitive conditions of the area.[44] Thus, with the help of the RCAF presence, the UN observers were able to do their job well.

In fact, the presence of the observers when combined with the Tashkent Agreement, brokered by Soviet Prime Minister Alexei Kosygin in 1966, succeeded in getting both sides to withdraw to their former positions.[45] As a result, UNIPOM was disbanded in March 1966. One historian later noted that "the successful completion of its task within a short time and eventual disbandment of UNIPOM was a rather rare occurrence in the UN's experience".[46] Nevertheless, this success represented the high point of UN peacekeeping efforts on the subcontinent.

---

43   The clause about flights over the international boundary was included at the insistence of the Indian government. A Job for Soldiers, Undated, 2005/05 Alan James Papers, Box 29, File 5 India and Pakistan – MacDonald Papers, DHH; Macdonald to Moncel, 6 October 1965, 2005/05 Alan James Papers, Box 29, File 4 India and Pakistan – MacDonald Papers, DHH; Macdonald to Moncel, 26 November 1965, 2005/05 Alan James Papers, Box 29, File 4 India and Pakistan – MacDonald Papers, DHH.

44   Outline of UNIPOM's Activities September 1965 – March 1966, Undated, 2005/05 Allan James Fond, Box 29, File 4 India and Pakistan – MacDonald Papers, DHH, 13; A Job for Soldiers, Undated, 2005/05 Alan James Papers, Box 29, File 5 India and Pakistan – MacDonald Papers, DHH.

45   Schofield, *Kashmir in Conflict*, 111–12.

46   Shafqat Hussain Chauhdry. *United Nations India–Observation Mission (UNIPOM) 1965–66 Vol. II*. Ph.D. Dissertation 1979, 2005/05 Allan James Fond, Box 29, File 16 India and Pakistan – UNIPOM 1965–66 Vol. II, DHH.

The outbreak of the third Indo–Pakistani War in 1971 may have exposed the limitations of UNIPOM's achievement, but the conflict passed without another peacekeeping force being created. Furthermore, after the war, the Indians concluded that UNMOGIP's mandate had lapsed. Although Secretary-General Thant disagreed and UNMOGIP remained in the area, India has restricted UN operations on its side of the Line of Control ever since 1971.[47] Yet despite this fact Canada did not immediately end its contributions to this mission: 424 Squadron would remain in the area until its departure on 31 March 1975,[48] and the last of the Army Observers would only be withdrawn in early 1979. Canada would continue to supply a Hercules transport to move UNMOGIP's headquarters from Rawalpindi, Pakistan, to Srinagar, India, and back again every six months until 1995 when the United Nations decided to use trucks instead. Whether this decision was taken as the result of budget cuts in Ottawa or by a decision by the United Nations to reduce costs is unclear, but whatever the case, this move ended Canada's participation in UNMOGIP after almost half a century.[49]

Canadian Air Operations in UN Military Observer Group India–Pakistan

Having provided an overview of Canada's involvement with these UN peace observer missions, there is still the need to explore Canadian air operations in greater detail. This analysis will be done through the use of interviews with former Canadian Forces personnel who served in UNMOGIP, as well as other primary sources.[50] According to these individuals, the role of Canada's UNMOGIP air contingent was to fly the military observers and mail in and out of the UN base

---

47   The Indian government has continued to supply UNMOGIP with the same level of administrative support as before 1971. United Nations, *UNMOGIP Background*. Available at: http://www.un.org/en/peacekeeping/missions/unmogip/background.shtml [accessed 1 June 2011]; Part Three: India/Pakistan the UN Military Observer Group in India and Pakistan and the India–Pakistan Observation Mission, 2005/05 Allan James Fond, Box 29, File 14 India and Pakistan – United Nations Military Observer Group in India and Pakistan, DHH.

48   Untitled Press Release, 4 February 1975, 75/179, DHH.

49   Canadian Forces. *Details/Information for Canadian Forces (CF) Operation United Nations Military Observer Group in India and Pakistan*. Available at: http://www.cmp-cpm.forces.gc.ca/dhh-dhp/od-bdo/di-ri-eng.asp?IntlOpId=292&CdnOpId=352 [accessed 1 June 2011].

50   I conducted interviews with Merv Matiowsky in Trenton, Ontario, on 4 March 2011, and with Bernard "Ted" Green in Belleville, Ontario, on 6 June 2011. Mr Matiowsky was one of the RCAF's pilots in UNMOGIP from May 1970 to September 1971. Mr Green served as a member of the ground crew from June 1967 to May 1968. Moreover, I utilized the diary of Sergeant James H. Baker who served with UNMOGIP from June 1966 to June 1967. Finally, I consulted Dr Allan English on 15 May 2011 about his experiences with this mission.

camps where the observers were stationed.[51] Many of these flights were made so the observers could take their rest leave. In addition, in the early 1970s UNMOGIP's aircraft was responsible for transporting the mission's headquarters back and forth from Srinagar to Rawalpindi every six months.

While they were part of UNMOGIP, the Canadian airmen sought to provide the best support possible to the army observers. As one of the interviewees put it, it was well understood that the observers were relying on them "to come in with mail, to come in with supplies ... and the guys [the observers] wanted to get out for their R and R [rest and recuperation]. So we [the pilots] tried to be very dedicated that way". However, it should be emphasized that in UNMOGIP, unlike UNIPOM, the Canadians did not fly any reconnaissance missions. Instead, it was the job of the army observers to monitor the ceasefire.[52]

I also learned that, at least in the opinion of the former Canadian military personnel I interviewed, the Air Force did an excellent job in supporting the mission. It provided effective training that prepared them for what they were going to face in addition to allocating capable aircraft to UNMOGIP. Certainly, the transport aircraft that were used by Canadian personnel were not perfect. The Caribou, because it had piston engines, had a limited service ceiling and rate of climb, both of which were issues, given the mountainous terrain in the area. Although the Twin Otter had turboprop engines, it did not have the cargo capacity of the Caribou. This was a problem when the Twin Otter was used to move UNMOGIP's HQ.[53] Nonetheless, given that these were the aircraft available either in the inventory or off the shelf, they both served this mission well. Furthermore, there were no complaints about the quality of the logistic support that they received from Canada.[54]

The interviewees related other interesting details. One example was that the RCAF, like the Canadian Army, briefed and debriefed the officers who served on UNMOGIP to prepare them for the local conditions and, presumably, to learn lessons from this experience.[55] In addition, the Canadian Army and Air Force personnel worked extremely well together and, in general, the observers from all the countries got along with the exception of the Chilean contingent.[56] A word should also be said about the issue of corruption and UNMOGIP. It goes without saying that there was a lot of small-time corruption, namely the use of a bottle of whisky at the airport to get what you wanted through customs;[57] but there were some more serious cases. One retired Canadian officer, Dr Allan English, told

---

51   Matiowsky interview (see Note 50).

52   Ibid.; Merv Matiowsky, email message to author, 22 January 2012.

53   Matiowsky interview (see Note 50).

54   Ibid.

55   This practice was not extended to NCOs. Green Interview (see Note 50).

56   Mr Matiowsky did not know the reason for this situation. (Matiowsky Interview, see Note 50).

57   Ibid.

me a story of when he was part of the aircrew for a Hercules flight into Kashmir. There was supposed to be no cargo for the flight back to Rawalpindi, Pakistan, but he quickly noticed that logs were being loaded onto the Hercules. According to the local ground crew, this was being done on the orders of the UNMOGIP commander. He later learned that the timber was being smuggled into Pakistan to be used in the production of furniture.[58]

These interviews provided a detailed picture of the lives of Canadian personnel serving in UNMOGIP. On one hand, they were living really well, spending half the year in Kashmir, which was a particularly beautiful part of India. They had access to servants, known as bearers, and duty-free liquor and cigarettes, which led to an enjoyable atmosphere with parties every weekend.[59] One interviewee described the atmosphere as one where they "worked hard and played hard". Not surprisingly, he described the mission as a great adventure and even had volunteered to stay three months extra in country in order to help with the transition from the Caribou to the Twin Otter.[60] Another interviewee also told me a story about a fly-fishing trip in the Himalaya Mountains where he caught five trout in 15 minutes.[61]

Despite the pleasant aspects of this mission, these men did face some serious difficulties, including illness and disease. All personnel on the mission suffered from chronic diarrhoea, which had several colloquial names including "Gypo Gut", "Delli Belli", and "Pindi Trots".[62] Major General Macdonald even wrote to one of his fellow officers in Canada about "a type of projectile diarrhoea, reminiscent of Cape Canaveral on a busy day, [which] is something which has to be experienced to be believed".[63] In addition, there were more serious cases, as one of my interviewees was hospitalized with dysentery and other personnel required evacuation back to Canada.[64] Moreover, the ground crew had to deal with the extreme levels of heat and humidity while their pilot counterparts faced the hazards of flying in the Himalayas, particularly in the winter months.[65] There was also the isolation and loneliness of being in an alien culture. The writer of one diary I consulted stated that his time in UNMOGIP was "the longest year in history".[66] There were other difficulties related to the local population. For example, in June 1967, a number of Kashmiris in Srinagar rioted after they had

---

58   English Interview (see Note 50).
59   Matiowsky Interview (see Note 50).
60   Ibid.
61   Green Interview (see Note 50).
62   Ibid; Matiowsky Interview (see Note 50).
63   Macdonald to Parker, 14 February 1966, 2005/05 Allan James Fond, Box 29, File 6 India and Pakistan – MacDonald Papers, DHH.
64   Green Interview (see Note 50).
65   Ibid; Matiowsky Interview (see Note 50).
66   Sergeant James Baker wrote on his arrival home "Hooray!! HOME AT LAST! (Thank God!) THE LONGEST YEAR IN HISTORY (+LONELY)". Diary of Sgt. James H. Baker (see Note 50).

heard that the Israelis had bombed Mecca during the Six Day War. During the riot, they burned a couple of Christian churches and attacked the UN compound.[67]

Of course, being stationed between two heavily armed and aggressive nations was stressful as well. One of my interviewees stated that there was a brief "flare-up" between the Indians and Pakistanis that caused some concern in the mission.[68] It is important to remember that two of the aircraft, a Caribou and a Twin Otter, sent to support UNMOGIP were destroyed in the 1965 and 1971 wars. Finally, for those who noticed, there was a sense of futility. In particular, Matiowsky quickly realized that the United Nations simply lacked the resources to prevent the outbreak of conflict. To his mind, in a place like Kashmir with its valleys and mountains, there were too many places on the border where both armies could hide excess men and artillery from the UN observers. This factor, combined the inability of the observers to do snap inspections due to the need to get permission from the Indians or Pakistanis, meant that UNMOGIP's mission was fatally flawed.[69]

Conclusion

From 1948 to 1995 the Canadian government and armed forces worked to assist the UN's peace observer missions on the Indian subcontinent. This effort first emerged through the dispatch of a handful of Canadian Army observers to the area. However, in the mid-1960s, Canada's role changed first through the dispatch of one CC-108 Caribou aircraft, along with its crew and maintenance personnel, to support UNMOGIP. Then, in 1965, in response to the outbreak of the Second Indo–Pakistani War, Canada played a leading role in the formation of UNIPOM by providing its commanding officer and its air transport component of two Caribous and three CC-123 Otters. This decision was taken in response to a crisis that threatened world peace and the stability of the Commonwealth, but also reflected an increased interest in peacekeeping in Ottawa and the desire to ensure that Canada's reputation at the United Nations was maintained. While UNIPOM would be disbanded after the successful completion of its mission in 1966, Canada would continue to use its air assets to support UNMOGIP into the 1990s. Therefore, having examined the Canadian experience on these peace observer missions, one other issue remains: what can be learned?

One lesson is that the decisions in Ottawa whether or not to support these missions were heavily influenced by individual personalities. For example, Canada's dispatch of observers to serve in the region in the late-1940s was driven by St. Laurent and Pearson. Moreover, the allocation of the Caribou to assist UNMOGIP in 1964 and Canada's commitment to UNIPOM were largely the

---

67 Ibid.
68 Matiowsky Interview (see Note 50).
69 Ibid.

result of Martin's strong lobbying behind the scenes. He not only pushed Hellyer to supply the aircraft but also argued that Canada needed to contribute significant personnel and aircraft to UNIPOM.

Canada's involvement with these operations further illustrated that while having a positive reputation in some field of international endeavour is a good thing, it always must be remembered that this status does not come without its costs. Indeed, the fact that Martin perceived that Canada needed to allocate resources to UNIPOM just to maintain its position as a leader in the field of international peacekeeping is an important lesson that good reputations have burdens as well as benefits. This point further shows that for Canadian diplomacy to be at its best, it needs to be backed up by a well-equipped and trained military that can effectively fulfill the commitments made by Canadian officials.

This experience also demonstrates the problems of participating in small UN peace observer missions, namely that these operations will only do useful work when the parties involved want them to. As Macdonald stated, UNIPOM could do nothing if the Indians or the Pakistanis decided they wanted to fight. UNIPOM was ultimately successful, but without the political will to solve the underlying problems, it was only a "Band-Aid" solution, as was shown by the outbreak of the third Indo-Pakistani War in 1971. This reality was even more the case with UNMOGIP, as there were simply too many places for the Indians and the Pakistanis to hide weapons and soldiers in the region and the system of inspections in use was utterly inadequate.

Finally, for the RCAF, the lessons of UNMOGIP and UNIPOM are that despite the difficult conditions of the region, it did its job well. The Air Force provided quality training and capable equipment to support its personnel in the region. This experience was a good affirmation of the work that had been done to rebuild this force out of the wreckage of the postwar demobilization, which had resulted in the emergence of one of the world's best air forces. The Air Forces' ability to maintain this level of excellence for a period of time afterwards, despite reductions in its funding, is a tribute to the officers and men of the period. There were many issues with this mission, but the RCAF's contribution was not one of them.

Acknowledgments

I would like to thank Major Mat Joost for assisting me in accessing the Alan James Papers during my visit to the Directorate of History and Heritage (DHH) in Ottawa.

Chapter 5

Humanitarian Relief in Haiti, 2010: Honing the Partnership between the US Air Force and the UN

Robert C. Owen

The recent interaction between the United States Air Force (USAF) and UN organizations and personnel during the 2010 Haiti Earthquake relief effort points to a further opportunity to refine their ability to partner in future humanitarian relief (HR) operations. During the Haiti operation, UN and USAF personnel cooperated to a greater degree than they had in years, both in the field and at a key operational headquarters. The exceptional circumstances of the emergency mandated this close cooperation. Logistically, the early weeks of the Haiti relief constituted a High-intensity, Restricted-infrastructure (HIRI) airlift operation. With the main seaport inoperative, large quantities of relief supplies had to move through Port-au-Prince's Toussaint Louverture International Airport (International Civil Aviation Organization (ICAO) airport identifier: MTPP). In short order, the press of government, chartered, and private aircraft trying to get into the field overwhelmed its ground infrastructure and created a hazardous air safety environment. Too battered by the disaster to deal with the situation itself, the Haitian government placed control of access to the airport and ground operations on the main parking ramp in the hands of USAF organizations engaged in the relief effort. When faced by growing criticisms of its efforts to prioritize access to the airport by aircraft operated by dozens of governments and civil relief agencies, the USAF for the first time invited the UN World Food Programme's Humanitarian Air Services (UNHAS) into its regional headquarters to supervise the lift. The spirit and mixed results of this effort point to an opportunity and need to normalize USAF and UN cooperation in future HIRI–HR efforts.

Normalizing the USAF and United Nations relationship could come in the form of institutional, doctrinal, and/or personnel changes and improvements. Institutional changes would include altering and/or creating organizations to improve the mechanisms by which the United Nations and USAF relate in airlift matters. Doctrinal changes probably would involve formal changes in procedures, while human changes likely would involve specialized training and focused selection of liaison and staff personnel. At one extreme, the US Department of Defense (DOD) might direct its Air Force to create a dedicated organization focused on facilitating humanitarian relief coordination. At the other extreme, each organization simply might train its personnel to understand and work more effectively with those of the other.

To assess which of these three options or, more likely, what mix of them will be most effective, this analysis will have three parts. The first will be a description of the background of the UN–USAF relationship in the realm of HR airlift operations. The second part will provide a brief discussion of UN–USAF coordination during the Haiti airlift. The final part will assess the implications of that coordination and suggest directions for improving it in future HIRI emergencies. In keeping with the principle of economy of effort, these suggestions for new directions actually will be quite modest. In retrospect, UN–USAF cooperation during the Haiti crisis was effective, if somewhat delayed in coming into play. Consequently, the experience points to a need for some doctrinal refinement, some adjustments in organizational and individual training, and perhaps the creation of a contingency organization to be consolidated and activated in future crises.

The Relationship: Long but not Particularly Deep

The historical relationship between the USAF's airlift commands and the United Nations goes back to the very foundation of the international organization. Indeed, many of the delegates to the San Francisco Conference of 1945 traveled on USAF Air Transport Command aircraft. Through the 1960s, the two organizations cooperated frequently. Operation New Tape was the highlight of this interaction. Between 1961 and 1964, the USAF flew 2,128 missions in support of the UN peacekeeping mission in the Congo, ultimately carrying 63,798 passengers and 18,593 t of cargo.[1] Around the world, American and UN conflict resolution and humanitarian relief policies paralleled one another and the long-range air transport capabilities of the USAF's Military Air Transport Service were unique in their scale and availability. For a number of reasons, this relationship weakened during the 1970s. Administrative changes in the way USAF airlift was financed made it less affordable and available to non-Defense Department users, including US agencies and foreign governments. UN and US humanitarian assistance policies drifted apart, with the former focusing on human relief and the latter integrating that objective with the promotion of US national security. Nevertheless, UN personnel and American airmen frequently found themselves working side by side in peacekeeping and HR operations; usually in cooperation, but sometimes at cross purposes.

By the turn of the millennium the infrequent and sometimes rocky interaction between the United States and the United Nations had exacted a toll. A RAND study at the time summarized the relationship between the American military

---

1   Kennith Blan, Interview with Robert C. Owen, 14 March 2001, tape 1, side B, index 000–030; Eckwright, R.E. and Cantwell, G.T. *The Congo Airlift – 1960* (Historical Division, Office of Information, Headquarters, United States Air Forces in Europe, 1961).

and the community of civil relief organizations as characterized by a "mutual lack of familiarity" and "little understanding of each other's organization and procedures".[2] Throughout their report, the RAND researchers argued that opportunities for improved peacekeeping and humanitarian operations were lost because both sides of the relationship disliked and were suspicious of the other. American airmen saw the personnel of the United Nations and those of the general community of non-governmental and private volunteer organizations (NGOs and PVOs) as Byzantine in their disorganization and feckless or unfriendly politically. Civil relief personnel, including those in the United Nations, understood that the priority of American military personnel in peacekeeping and even humanitarian operations is the achievement of United States rather than international policy objectives. They also were uncomfortable or intimidated by displays of uniforms, weapons, hierarchical organization, and force protection measures.[3]

In keeping with their general discomfort with the employment of military forces in humanitarian relief, the United Nations and probably most other civil relief organizations endorse the so-called Oslo Guidelines. First sponsored by the United Nations in 1994, these guidelines provide that uniformed Military and Civil Defense Assets (MCDA):

> should be employed by humanitarian agencies as a last resort, that is, only in the absence of any other available civilian alternative to support urgent humanitarian needs in the time required. Any use of MCDA ... should be ... clearly limited in time and scale and present an exit strategy element that defines how the function it undertakes could ... be undertaken by civilian personnel.

Thus, though it could not prevent military forces from entering a conflict or disaster area without UN sponsorship, the organization would not invite, endorse, or align itself with them, if doing so undermined its humanitarian principles, endangered its neutrality, or threatened civil control.[4] By implication, then, UN policy generally

---

2 Byman, D., Lesser, I.O., Pirnie, B.R., Benard, C. and Waxman, M. "Strengthening the Partnership: Improving Military Coordination with Relief Agencies and Allies in Humanitarian Operations", *RAND Monograph Report*, MR-1185-AF, 2000, xvii, 114. Available at: http://www.rand.org/pubs/monograph_reports/MR1185.html [accessed 23 July 2012].

3 Ibid, 101–19.

4 The UN Office for the Coordination of Humanitarian Affairs (OCHA) published two documents spelling out disaster relief policies – the so-called "Oslo Guidelines". United Nations Office for the Coordination of Humanitarian Affairs, *Guidelines on the Use of Military and Civil Defence Assets to Support United Nations Humanitarian Activities in Complex Emergencies*, Geneva, March 2003. Available at: http://www.unhcr.org/refworld/ docid/3f13f73b4.html [accessed 24 July 2012]; United Nations Office for the Coordination of Humanitarian Affairs, *Guidelines on the Use of Foreign Military and Civil Defence Assets In Disaster Relief* ("Oslo Guidelines"), Geneva, November 2007. Available at: http:// www.unhcr.org/refworld/docid/47da87822.html [accessed 24 July 2012].

views the presence of uniformed military personnel, including those conducting airlift operations, as an undesirable though sometimes unavoidable feature of specific missions.

In addition to worries about neutrality and civil control, self-interest feeds the reluctance of the United Nations and other civil HR organizations to see American military assets flood into a disaster area, particularly if they are not under UN control. There are thousands of participants in the humanitarian relief industry; including 10,000–20,000 NGOs and PVOs, dozens of governments, international alliances, individual corporations, sincere or merely grandstanding politicians and celebrities, and others. Some of the NGOs and PVOs field relief programs nearly as large as those of the United Nations, with thousands of employees and large budgets, and some are as small as husband and wife missionary teams working on shoestrings. All are locked in continual quests for funding and other forms of support, usually in direct competition with at least some other organizations. Success in this competition depends on gaining access to funds and support, which result from effective field operations and self-promotion. High visibility disaster relief activities provide excellent opportunities for organizations to gain visibility and credibility with donors. As a consequence, even relief organizations that had no prior engagement in a place like Haiti will flock by the hundreds to do good and, at least secondarily, gain notice in the media frenzy.

The United Nations does pretty well in this competition. Its charter to provide humanitarian relief, prestige, global access, and specialized relief organizations usually place it at the top of the churning heap of competing organizations. Moreover, UN agencies and personnel are deployed worldwide. In the case of Haiti, a large UN contingent had been present in the country for years, its most recent incarnation being the UN Stabilization Mission in Haiti (MINUSTAH). This presence meant, first, that many UN workers died in the earthquake; and second, that UN leaders had direct access to surviving senior government officials. Thus, as is often the case, the United Nations enjoyed the position of "senior lodger" in the Haiti relief; occupying a position of more or less natural leadership among many of the smaller NGOs and PVOs in the area. That status, along with its experienced personnel and "UN"-emblazoned vehicles and aircraft made it highly visible and credible. Its only real competition for "being in charge" came from the large military contingents arriving in the area, as in the case of the United States and Haiti. The flags, tent cities, energetic soldiers, photogenic generals, and big aircraft of the military contingents drew the cameras away from the United Nations, except to show it as the recipient of the military's largesse. Given the impact such a diminished stature can have on future prestige, donations, and its long-term development plans, it is little wonder that the UN's general policy is to accept military support only reluctantly and to send it away as soon as possible.

In this goal of minimizing military involvement, United Nations and American policy are in complete accord. Department of Defense policies recognize that the military's "unmatched capabilities in logistics, command and control communications, and mobility are able to provide rapid and robust response".[5] But they also emphasize that those capabilities will only be committed at the request of the Department of State, which would have to pay for them, and that the DOD's response would be "subject to overriding military requirements".[6] The DOD also endorses the Oslo Guidelines explicitly, including the proviso for clear exit strategies to hand operations over to civilians as quickly as possible.[7]

The challenge of these operations for the US military usually does not lie in their scale. While the USAF historically participates in 20 or 30 relief operations a year, most are small. They involve only handfuls of airlift sorties and deployments of small medical, engineer, logistics, or other units for a few days or weeks. Only a few relief efforts, such as Operation Unified Response, the Haiti relief, are large. In support of Unified Response the USAF drew personnel and materiel resources from 53 of its wings to support US Joint Task Force – Haiti (JTF-H), activated by the US Southern Command (USSOUTHCOM) to handle on-scene operations. In total about 3,330 US military-controlled aircraft sorties delivered about 31,000 tons of cargo into Haiti under JTF-H control.[8] On its part alone, the USAF's primary airlift arm, the Air Mobility Command (AMC), put in about 2,680 military and commercial charter aircraft sorties to move 26,781 passengers and 14,135 tons of cargo into and out of the area.[9] During the first 5 weeks of the operation the total relief flow averaged around 82 large aircraft transiting MTPP per day; of which about 35 were commercial charters, 32 were international civil and military aircraft, and 15 were US military.[10] Fifteen daily missions was not a daunting number for a command having access to nearly 1,000 large military and Civil Reserve Airlift Fleet aircraft.

But even this relatively small event imposed disproportionate strain, because the AMC system routinely operates fully tasked by its day-to-day commitments. There is no component of the US airlift system sitting in reserve. Virtually every AMC and theater-assigned transport aircraft is committed every day providing logistics support for ongoing conflicts and contingencies, supporting the routine logistics

5   United States Joint Chiefs of Staff, "Foreign Humanitarian Assistance", *Joint Publication* 3–29, 17 March 2009, sections I–2 and III–12. Available at: http://www.dtic.mil/doctrine/new_pubs/jp3_29.pdf [accessed 23 July 2012].

6   United States Department of Defense, *DOD Directive* 4100.46, 4 December 1975.

7   JP 3-29 2009, sec. III–10.

8   Lt. Gen. Ken Keen, "JTF–Haiti Forming a Joint Task Force in a Crisis: Observations and Recommendations", Commander JTF-H Briefing, n.d. (probably mid-April, 2010): 11.

9   Wallwork, E.D. et al. *Operation Unified Response: Air Mobility Command's Response to the 2010 Haiti Earthquake Crisis*, Scott Air Force Base, IL: Air Mobility Command Office of History (2010), 97. Note, these are figures updated in an email between the author and AMC historian Mr Mark Morgan on 7 June 2011 and are slightly higher than those contained in the official AMC history.

10   Longmire, J., Rieger, A. and Whiting, M. "612 Air and Space Operations Center/ Air Mobility Division Haiti Flight Operations Coordination Center After Action Report", March 2010, A4; Wallwork et al., *Operation Unified Response*, 23–4.

of a global military system, conducting training, or undergoing maintenance. So despite its access to so many resources, Haiti was a zero-sum game for US airlift forces and obliged AMC to rob other missions and operational commands to find aircraft. Expressing the practical reality of this game, the Commander of AMC's Tanker Airlift Control Center, Major General Brooks Bash, recounted that while Haiti was but a "blip on the scope" of AMC's daily schedule, the command still had to borrow C-17s from the Air Education and Training Command and Pacific Air Forces to fill the gap.[11]

The costs of AMC operations also influence its HR activities and its availability to outside users. AMC airlift operations are financed in two ways. To conduct training, maintenance, and some exercise operations, AMC receives an annual budget allocation for Operations and Maintenance. But most operations for the movement of passengers and cargo are financed by the organizations supported by them, through payments to the DOD Transportation Working Capital Fund. This is a revolving fund within a broader DOD industrial funding system for logistics, transportation, and other activities. As would any civil carrier, AMC keeps "solvent" by charging users to recover the operational, maintenance, amortization, and personnel costs of the airplanes they charter.[12] Currently, non-US government organizations pay US$7,580 per flight hr for C-130s and US$20,421 for C-17s. These rates are comparable to the US$23–28,000 per hr charged for a Boeing 747 on the commercial market. For this reason, the United Nations and other relief organizations prefer to charter civil carriers to move their cargoes – they are cheaper and AMC aircraft usually are not available to them anyway.

Although not precisely germane to the present discussion, it is useful to understand that the United Nations provides most of its own humanitarian airlift requirements through long-term charters. Under normal circumstances, the UNHAS organization, a component of the UN World Food Programme, provides routine, regional passenger and cargo airlift for UN and other relief organizations worldwide. UNHAS operates over 50 aircraft under long-term charters, augmented by short-term contracts. Most of these aircraft are small single- and twin-engine aircraft, though UNHAS does charter larger aircraft for "strategic" missions and to move larger amounts of cargo in emergencies. Generally, however, UNHAS provides passenger and high-priority cargo movements to augment surface modes and to cover the distances from major airports to isolated humanitarian operations

---

11   Interview by Mark Morgan, AMC History Office, with Major General Brooks L. Bash and Brigadier General Bradley R. Pray, 13 April 2010, 11–12 (Wallwork et al., *Operation Unified Response*, 47). General Bash had published an article, "Air Power and Peacekeeping", in the spring 1995 edition of the *Air Power Journal* and was a graduate of the Air Force's School of Advanced Air Power Studies.

12   United States Air Force Air Mobility Command. "Financial Management: Transportation Working Capital Fund Budget Guidance and Procedures", *Air Mobility Command Instruction*, 65–602. Scott Air Force Base, IL: Air Mobility Command (2009), 6–10.

locations. In 2009 UNHAS transported 323,714 passengers and 12,412 tons of cargo in support of over 700 different agencies.[13] Not a lot in comparison to the capabilities of national air forces, perhaps, but vital in the support of long-term relief and development programs. See Chapter 6 for more on UNHAS.

These issues of conflicting cultures, competition for visibility and influence, and economics, largely account for the historic coolness between the United Nations and the various elements of the US military that come in contact with it. The United Nations shares the uneasiness and suspicion of most civil relief agencies towards military presence in humanitarian operations. Most military organizations, if not all military personnel involved, reciprocate. At the same time, the military understands that the United Nations usually is the first among equals in large humanitarian activities. But it is only first among *equals*; it does not run the show. Consequently, a key challenge for American and other military forces in each new humanitarian relief operation is to figure out the real balance of power among the many civil organizations present or arriving from all directions. Thus, US Joint Doctrine Publication 3–29, "Foreign Humanitarian Assistance", advises Joint Force Commanders in their initial planning to find out:[14]

- Who are the relevant governmental and nongovernmental actors in the operational area? What are their objectives? Are their objectives at odds or compatible with the Joint Force Commanders' objectives?
- Who are the key communicators (persons who hold the ear of the populace, for example, mayors, village elders, teachers) within the operational area?
- What relief agencies are in place, what are their roles and capabilities, and what resources do they have?

Finally, UN and military forces in most cases arrive at disaster locations independently and with little interdependence logistically or interest in interacting beyond, perhaps, information sharing and coordinating distribution efforts. Of course, in exceptional circumstances like the relief of Haiti, pragmatic concessions to this distant relationship can be necessary.

Operation Unified Response: A High-intensity, Limited-infrastructure Incident

For the most part, the US government committed to, organized, and executed Operation Unified Response in accordance with an explicit body of policy and doctrinal guidance. This body of guidance begins with congressional legislative acts and presidential directives. It filters down through DOD directives, Joint

---

13 World Food Programme. *World Food Programme Aviation Annual Report 2009* (Geneva: WFP Information Unit).

14 United States Joint Chiefs of Staff (2009), sec. III–7.

Doctrine Publications, and handbooks and guides for various participants.[15] This body of literature generally fit the circumstances of Haiti well. Though large in scale and particularly tragic in the casualties it produced, the Haiti earthquake of 12 January 2010 presented the American military with an almost routine problem of responding, conducting rescue operations, mitigating secondary social and health effects, and generally giving the Haiti government time to reorganize itself. The United States and most developed countries in the world had participated in such activities many times before. But there was one wrinkle to the norm – the infrastructure available to support transport operations was restricted to an exceptional degree in relation to need. Haiti's only developed deepwater port was heavily damaged and inoperable, and its national airport, Toussaint Louverture, was inadequate to handle the flood of aircraft about to descend on it. It would be this challenge of conducting high intensity operations into a restricted airfield infrastructure that would push the United Nations and the USAF into an unusually close working relationship and, thereby, point to a need and opportunity to normalize that closeness.

The various components of the US government involved in crisis relief responded to the news of the Haiti disaster with practiced choreography. President Barack Obama immediately pledged massive support and dispatched a personal representative to survey the situation. After meeting with the National Security Council, he directed the Department of State to take its accustomed lead of the relief effort. Also as normal, the task of coordinating the interagency response fell on the United States Agency for International Development (USAID). Explaining the importance of the US role, the president announced a few days later that:

> our nation has a unique capacity to reach out quickly and broadly to deliver assistance that can save lives. That responsibility obviously is magnified when the devastation that's been suffered is so near to us.[16]

In support of the Department of State, the Defense Department issued a warning order on 13 January 2010 to the Combatant Commands that would have direct roles in the relief effort. USSOUTHCOM received overall military lead, since Haiti was within its geographic area of operation. The United States Transportation Command (USTRANSCOM) was responsible for organizing air and sea lines of communications and ports in support of USSOUTHCOM and the overall relief effort. The Commander of USTRANSCOM, General Duncan J. McNabb, gave

---

15   The most relevant volumes are: "Civil Support" (Joint Publication 3–28); "Foreign Humanitarian Assistance" (Joint Publication 3–29); "Civil–Military Operations" (Joint Publication 3–57); and "Air Mobility Operations" (Air Force Doctrine Document 3–17). But many other joint and service documents contain guidance relevant to air mobility operations in support of humanitarian relief.

16   Thompson, M. "The U.S. Military in Haiti: A Compassionate Invasion", *Time*, 16 January 2010.

verbal guidance to his component commands to get moving, and followed up with an execute order early on the 14th.[17] The need for quick action was becoming more apparent by the hour, as the world became aware of the extent of the devastation. Compounding the problem, the government of Haiti (GoH) and many relief organizations normally present in the country, including the United Nations Stabilization Mission in Haiti (MINUSTAH), had suffered many casualties and were struggling to reorganize and recover their morale, even as they began rescue and relief activities.

Even before formal orders came down USAF operational commands took initial steps to mitigate suffering and to posture themselves for the big push. USSOUTHCOM's air component (AFSOUTH) is the Twelfth Air Force based at Davis–Monthan Air Force Base in Tucson, Arizona. Lieutenant General Glenn F. Spears, AFSOUTH Commander, asked AMC for airlift planning and operations experts to beef up the Air Mobility Division of his Air and Space Operations Center (ASOC). He also asked for a senior Director of Mobility Forces to provide him with expert advice and coordinating authority. AMC sent out Brigadier General Robert K. Millmann Jr, who was the Air Force Reserve mobilization assistant to AMC's Eighteenth Air Force Commander and had had directed airlift operations during several previous disaster relief efforts.[18]

AMC operations began on 13 January, when a KC-135R tanker refueled two Air Force Special Operations Command (AFSOC) MC-130 aircraft carrying a team of tactical air traffic controllers to Toussaint Louverture Airport to re-establish air traffic control. Under the direction of Chief Master Sergeant Antonio D. Travis, the AFSOC team set up a card table at the edge of the runway and began controlling the traffic pattern with portable radios only 28 minutes after arriving.[19] The next day, an AMC C-17A delivered an urban rescue team into Toussaint, the first tangible US aid to arrive. While all of these activities were underway, AMC operations personnel were putting aircrews into rest, requesting loans of seven C-17s from Pacific Air Forces and six from the Air Education and Training Command; and organizing Homestead Air Reserve Base, Florida, and Charleston Air Force Base, North Carolina, as the primary Aerial Ports of Debarkation for Unified Response. Posturing AMC for a major operation into Port-au-Prince was a formidable task, since the command was "pretty much maxing out" already with the movement of "surge" forces to Afghanistan.[20]

---

17   Wallwork et al., *Operation Unified Response*, 26.

18   Major General Richard K. Millermann, Jr, Interview by Robert C. Owen, 19 July 2011, tape 1, side A, index 030.

19   Major David Small, "Airman Named to Time Magazine's '100 Most influential people' list for Haiti airfield Efforts", Air Force National Media Outreach Office, 30 April 2010. Available at: http://www.defense.gov/news/newsarticle.aspx?id=58970 [accessed 7 May 2014].

20   Ball, L., "Team Hickam, PACAF Aircraft Depart to Assist Haiti Relief Effort", Pacific Air Forces News Release, 21 January 2010. Available at: http://www.15wing.af.mil/

AMC's preparations allowed it to begin moving ground support elements into Toussaint on the heels of USTRANSCOM's warning order. A 13-person Joint Assessment Team from the 621st Contingency Response Wing arrived later in the morning of the 13th to begin assessing the condition of the airport and its readiness to begin receiving heavy aircraft. Eight hours later, the first 21 members and 44 tons of cargo from the 621st Wing's 818th Contingency Response Group (CRG) arrived from McGuire Air Force Base (AFB), New Jersey, to begin organizing aircraft parking and unloading operations for AMC aircraft and any other planes coming in. Eventually, the CRG's contingent would grow to over 200 members.

The CRG personnel discovered a situation in immediate need of the kind of organization they were trained to impose on contingency airfields in combat and non-combat situations. The Port-au-Prince ramp was crowded already with other-nation military and civil relief aircraft and a chaos of vehicles and crowds of people wandering around. The small parking ramp was saturated with aircraft, and more were coming in. Many of the aircraft were filled with piece cargo, and there were no organized teams to unload them. Determined to bring some order to the hubbub, the CRG's personnel began setting up camp, while their leaders discussed control arrangements with airport authorities. By the next day, the 818th CRG was in control of ramp and unloading operations, the AFSOC air traffic specialists were providing positive control of arriving aircraft, the US Army's 688th Rapid Port Opening Element was arriving to move cargo from the Toussaint ramp to a USAID-controlled distribution point nearer the city, and USAF Security Forces were patrolling the airport and its perimeter.[21] By that time, AFSOUTH had certified that under CRG and AFSOC control, Toussaint could handle 90 flights per day, compared to the 25 handled under normal circumstances. Very visibly, the Americans had taken control of airlift relief at MTPP.

Also beginning on 14 January 2010, AFSOUTH and the GoH took actions to gain control of the flow of aircraft into Haiti. Knowing that his own headquarters at AFSOUTH did not possess the capabilities needed, General Spears asked the First Air Force, the air component of United States Northern Command (AFNORTH), for help. AFNORTH had two mobility-related resources of immediate value to the building airlift. Colonel Warren Hurst, its Deputy Director of Mobility Forces, was involved already, coordinating between AFSOUTH, SOUTHCOM, and AMC on mobility issues, setting up an Aerial Port of Debarkation at Homestead AFB, Florida. Within the Air Mobility Division of its Air and Space Operations Center, AFNORTH also possessed the only standing Regional Air Mobility Control Center

---

news/story.asp?id=123186711 [accessed 17 June 2010]; Wallwork et al., *Operation Unified Response*, 25.

21   Johnson, Tech. Sgt Denise, "Air Force Provides Rapid Response to Disaster in Haiti", Joint Base McGuire–Dix–Lakehurst News Service, Public Affairs, Joint Base McGuire–Dix–Lakehurst, NJ, 15 January 2010. Available at: http://www.jointbasemdl. af.mil/news/story.asp?id=123185701 [accessed 10 June 2011]; Wallwork et al., *Operation Unified Response*, 15–21.

(RAMCC) in the USAF. On behalf of their Joint Combatant Commands, overseas Air Force components had established temporary RAMCCs to supervise relief operations into Bosnia in the mid-1990s and in Iraq and Afghanistan in the 2000s. But given its more or less continual obligation to respond to natural disasters in the United States, AFNORTH organized the 601st RAMCC on a permanent basis in 2007. Thus, when the earth heaved in Haiti in 2010, the 601st possessed the trained personnel, procedures, and communications capabilities needed to receive access requests from the dozens of operators wanting into Toussaint and assigning them arrival slot times.

The availability of the 601st RAMCC allowed Haitian President René Préval to authorize the US DOD to prioritize fixed-wing flight arrivals and departures at MTPP to facilitate the distribution of relief supplies as quickly as possible.[22] At the same time President Préval made it clear that Haitian sovereignty over its airspace remained intact. After signing the memorandum of understanding releasing control to the Americans, he verbally stipulated to American leaders that Haiti would resume control if he sensed that the airflow and/or airfield were being managed improperly, or if Haiti's desires were being ignored, or if individuals he identified, particularly the First Lady of Haiti, were being denied unrestricted access to the airport.[23] In keeping with these stipulations, AFNORTH redesignated the 601st RAMCC as the Provisional Haiti Flight Operations Coordination Center (HFOCC). The name change was useful; first because it clarified that the Center was coordinating and not controlling anyone's aircraft directly, and because it made the organization's role clearer to non-USAF operators.

The RAMCC's mechanisms for controlling the flow of aircraft into Port-au-Prince were apportionments and prioritization. For the HFOCC, apportionment meant "the percentage or number of contingency ramp slots allocated in advance to a specific category or agency". Initially the HFOCC Chief, Lieutenant Colonel Bradley G. Graff, planned to allocate 50 percent of all slots to American military and civil aircraft, and 50 percent to all other categories. But these percentages were only general guides, subject to the more precise task of prioritization; establishing "a current and specific ranking of what relief supplies are needed in the disaster".[24] In the first days of the emergency, the HFOCC staff broadly prioritized slot-time requests in accordance with a list it had developed from experience with previous emergencies. In short order, however, the standing list was superseded by priority lists arriving from SOUTHCOM, the United Nations, the World Food Programme, and USAID.

---

22   601st Air and Space Operations Center (2010), *Haiti Flight Operations Coordination Center After Action Report – March 2010*, Air Mobility Division (Florida: 601st Air and Space Operations Center, Tyndall AFB), Appendix 1.

23   601st Air and Space Operations Center (2010), 23.

24   601st Air and Space Operations Center (2010), 15; Wallwork et al., *Operation Unified Response*, 5.

Prioritization, nevertheless, remained a challenge for Colonel Graff and the HFOCC staff, until the declining pace of operations in late January mooted the issue. Most importantly, as Director of Mobility Forces, General Millmann later reported, the lists of SOUTHCOM, the United Nations, the World Food Programme, and USAID usually "did not line up".[25] The tyranny of time also obliged the HFOCC staff to grant slot times to requesters as they called in, without the luxury of waiting for later callers who might have more immediately important loads to deliver. The staffers did not want to resort to first-come-first-served allocations, but their reality was that the requests came in on that basis and they could not hold approvals in escrow until they could build a completely rationalized flow plan. Moreover, ad hoc demands for priority access came from many aid organizations, DOD, and GoH. These demands were troublesome since, once slots were assigned to users on the basis of the primary priority list, it was almost impossible to shift them to satisfy later requests coming in from such authoritative organizations. The imperative to pump as many aircraft as possible through the MTPP main ramp only intensified the pressure on the HFOCC. Literally, a delay in granting a slot time or mishandling a request could mean suffering and death for Haitians already on the edge of survival.[26]

The deployment of 4,000 soldiers of the 2nd Brigade Combat Team of the US 82nd Airborne Division greatly complicated the prioritization task and provided a major point of misunderstanding over the HFOCC's management of the relief airlift flow. Partly as a more or less automatic response in such situations, and partly out of specific fears that Port-au-Prince would descend into chaos, the US government began sending airborne units to Haiti on 14 January 2010. From the start, the Chairman of the Joint Chiefs of Staff, Admiral Michael G. Mullen, made it clear that the 82nd's movement was his number one priority. Responding accordingly, USTRANSCOM directed the 618th Tanker Airlift Control Center at Scott AFB, Illinois, to wedge the necessary 91 C-17 sorties into the schedule. According to the Vice Commander of the Tanker Airlift Control Center, Brigadier General Randy A. Kee:

> that was a mountain of stuff ... that became, for us, the number one priority. We still had only so many slots ... [so] we had to wedge priorities ... giving priority to the priority.

For the HFOCC, the movement of the 82nd simply changed the make-up of the American share of the overall airlift flow. Otherwise, it simply continued to meter the flow of American aircraft in accordance with the 50 percent allocation it had given them already. So, whatever the merit of many changes at the time and after that, the deployment of an entire brigade was unnecessary

---

25   Millermann: tape 1, side A, index 149–64.

26   Hertz, A. "WikiLeaks Haiti: The Earthquake Cables", *The Nation*, 15 June 2011. Available at: http://www.thenation.com/article/161459/wikileaks-haiti-earthquake-cables [accessed 15 June 2011].

or nefarious, there was no change in the apportionment of slots to the United States. The diversion of slots to the US Army undoubtedly delayed the initial deliveries of relief supplies by US aircraft. But, in coordination with the GoH, the United States felt that such security was a necessary precaution, given the circumstances.[27]

While aware that they were not fully qualified to "take the list of all the flights and put it in order of most important to least important", the HFOCC team also knew that the task was theirs to perform.[28] So its members employed several techniques that allowed them to grant thousands of slot times in the first days of the emergency, while still preserving some ability to adjust to changes in priorities. Most importantly, the team withheld 10 percent of all slot times available until they had to release them or risk having parking spaces go empty at MTPP. These "withholds" allowed the HFOCC to accommodate late requests of suitable priority, while minimizing the chance that they might restrict the flow of relief supplies. In exceptional cases, the HFOCC also canceled previously awarded slot times to let very high priority missions slip in. If it actually became necessary to divert aircraft in flight, the HFOCC usually sent US military aircraft away. Military operators, HFOCC staffers reasoned, were better able to handle the financial and operational impacts of having to go home and wait for another turn into the field. When, two weeks into the operation, the no-show rate of aircraft with assigned slot times began to approach 25 percent of total sorties scheduled, the RAMCC also began to call all no-shows and to confirm all jumbo jet arrivals 48 hours in advance. Overall, the HFOCC team later assessed that these procedures markedly reduced wasted parking slots, kept the relief airlift operating at least in rough conformity with generally agreed- upon priorities, and further increased the capacity of MTPP to 170 large aircraft arrivals per day.[29]

Not everyone respected and/or cooperated willingly with the slot system and US control of it. Determined to deliver their specific loads, and always aware of the competition for impact and visibility among civil relief agencies, many NGOs, PVOs, and other governments chafed at the need to request and accept slot times from an American military control agency. Some organizations, government

27   United States Air Force News, "Mobility Planners Move 82nd Airborne Unit to Haiti", *Air Force News*, 28 January 2010. Available at: http://www.af.mil/News/ ArticleDisplay/tabid/223/Article/117815/mobility-planners-move-82nd-airborne-unit-to-haiti.aspx [accessed 7 May 2014]; Donna Miles, "82nd Airborne Soldiers Begin Haiti Deployment", *American Forces Press Service*, 14 January 2010. Available at: http://www. defense.gov/news/newsarticle.aspx?id=57522 [accessed 29 January 2012]; Wallwork et al., *Operation Unified Response*, 51. For a summary of criticisms of the 82nd Airborne deployment see Hertz, "WikiLeaks Haiti".

28   Mendoza, M., "Haiti Flight Logs Detail Early Chaos", Associated Press Release, 18 February 2010.

29   Longmire et al., "612 Air and Space Operations Center/Air Mobility Division Haiti Flight Operations Coordination Center After Action Report", 45–46 and A–12; Wallwork et al., *Operation Unified Response*, 32.

leaders, celebrities, and politicians simply went "up channels" to find a senior government official or military commander to impose their requested slot times on the HFOCC staff. A few arrived at Toussaint and simply left their planes and crews sitting in a parking spot while they went into the city to do their business. During the first couple of days after the HFOCC began operating, a few relief organizations failed or refused to obtain slots and, if no parking spaces were available, were sent away by air traffic control in accordance with President Préval's guidance. The Haitian government also retained control over some of the parking spots on the Toussaint ramp for its own purposes, mainly to accommodate aircraft used by the President, his wife, and other favored individuals and groups. This practice, while legitimate enough, did make it difficult at times for the HFOCC to ensure that the reserved parking spots did not go to waste.[30]

Within the bounds of their status as subordinate military organizations, AFSOUTH, the HFOCC, and American troops on the ground did their best to dispel fears that they were prioritizing access to Toussaint unfairly or for imperial purposes. This took a team effort of many parts. At the center, Colonel Graff ensured that all slot-time allocations were defensible in terms of established priorities or the ad hoc needs of the relief effort. For his part, Brigadier General Robert Millmann, the AFSOUTH Director of Mobility Forces, gave candid interviews to explain the "good, bad, and ugly" of the operation. Only three days into the HFOCC's slot-time regime, 52 percent of the planes going into Toussaint were:

> from US and International civil relief organizations, 22 percent from the US military, and 18 percent from individuals and organizations approved directly by the GoH, and the rest from unidentified or unidentifiable sources.[31]

Air Force controllers also did their best to let as many smaller aircraft into Toussaint as possible, so long as they could park in grass areas, rather than on the fully occupied paved ramp controlled by the 818 CRG. In reality, the whole control system was based on managing the utilization of that precious ramp space, not on the ability of Haiti airspace to handle aircraft – a point often missed by organizations interested only in getting in their specific, "top priority" cargoes. The crowded airport is seen in Figure 5.1 opposite.

Nevertheless, the presence of a large US military force on the ground and the involvement of an Air Force headquarters staff in the allocation of slot times prompted a firestorm of complaints from governments and organizations philosophically and/or politically unfriendly to the United States. The leaders of Venezuela, Bolivia, and Nicaragua declared that the slot times and the growing numbers of US troops on the ground indicated that, in the words of Venezuelan President Hugo Chavez, "the United States was taking hold of Haiti over the bodies

---

30   Mendoza, "Haiti Flight Logs Detail Early Chaos".
31   Ibid.

**Figure 5.1   A US helicopter leaves the crowded Port-au-Prince airport
with relief supplies**

Source: UN Photo 425706, 20 January 2010.

and tears of its people".[32] Individual French and Italian officials also criticized the US "occupation" of Haiti and its alleged refusal to grant landing rights to several relief agencies.[33] In one case, the HFOCC was forced by a full ramp to refuse landing permission to a Doctors Without Borders flight arriving without a slot time. Over the next several days the organization claimed that five of its flights had been turned back and that some of its patients had died as a consequence.[34] Things were not helped by *Time Magazine*'s well-meaning declaration that:

> Haiti, for all intents and purposes, became the 51st state at 4:53 p.m. Tuesday [15 January 2010] in the wake of its deadly earthquake. If not a state, then at least a ward of the state – the United States – as Washington mobilized national resources to rush urgent aid to Haiti's stricken people.[35]

It appeared that, in the case of the HFOCC and the US intervention in general, the "Law of Unintended Consequences" was quite active.

The political heat and challenge of determining priorities led AFSOUTH and AFNORTH to take the unprecedented step of inviting the United Nations into the RAMCC to coordinate slot times. In recognition of the usually awkward relationships between American military personnel and the civilian relief agencies, American doctrine seeks to keep their interactions discreet and on an as-needed basis. Normally Joint Force Commanders establish Civil–Military Operations Centers as meeting venues for relief workers and military personnel to exchange information and coordinate planning. American military doctrine stipulates that Civil–Military Operations Centers be kept physically separate from military headquarters to spare civil personnel the necessity of appearing to be entangled with the military effort. So the notion of bringing a specific civil relief organization into an operational headquarters was something new, even if it involved an organization with the credibility and expertise of the United Nations. But circumstances were

---

32   Associated Press, "US Rejects Latin American Claim it is 'Occupying' Haiti", Press Release, 22 January 2010. Available at: http://www.bt.com.bn/news-world/2010/01/23/us-rejects-claim-it-occupying-haiti [accessed 7 May 2014].

33   *The Telegraph*, "Haiti Earthquake: France Criticises US 'Occupation'", *The Telegraph*, 18 January 2010. Available at: http://www.telegraph.co.uk/news/worldnews/centralamericaandthecaribbean/haiti/7020093/Haiti-earthquake-France-criticises-US-occupation.html [accessed 24 July 2012]; Williams, D. "U.S. Criticised Over Haiti Relief Effort as Italian Disaster Expert Brands it 'Badly Managed'", *Daily Mail* (London), 26 January 2010. Available at: http://www.dailymail.co.uk/news/worldnews/article-1245969/Haiti-earthquake-U-S-relief-effort-badly-managed-claims-Italian-disaster-expert.html [accessed 24 July 2012].

34   Doctors Without Borders, "Doctors Without Borders Plane with Lifesaving Medical Supplies Diverted Again from Landing in Haiti", Press Release, 19 January 2010, *Doctors Without Borders/Médecins Sans Frontières*. Available at: http://www.doctorswithoutborders.org/press/release.cfm?id=4176 [accessed 12 July 2011].

35   Thompson, "The US Military in Haiti".

pressing and no civil organization was more qualified to integrate its personnel into the RAMCC than the UNHAS. Its personnel were experienced with planning and conducting crisis and routine airlift support of peacekeeping and humanitarian operations. Welcoming the opportunity to improve the "synchronization" of civil and military airlift efforts, the World Food Programme sent in Philippe Martou, Deputy Chief of the UNHAS, and two assistants, Mike Whiting and Albert Rieger, to the HFOCC, which was located at Tyndall AFB in the Florida Panhandle, on 24 January 2010.

By the time the UNHAS team arrived at Tyndall on 24 January, the airlift crisis was past its peak, but there still were plenty of problems. Outside, complaints of American intentions and bias continued, while no shows and empty parking slots were becoming major concerns. The UNHAS and HFOCC personnel took over slot-time coordination and continued the practices of following up on no-shows and calling jumbo jet operators 48 hours prior to their scheduled arrivals. The UNHAS also became the HFOCC's conduit to UN relief teams in Haiti, the GoH, and the NGO/PVO community to coordinate priorities and assess the value of specific loads. Although late to come together, Philippe Martou assessed that:

> this unprecedented relationship ... significantly added to the unity of effort between civil and military aviation ... [and as] a template for future combined civil–military aviation operations would enhance rapid response capability ... and ensure synchronized processes.[36]

In the end, the overall relief airlift effort was successful. USAF and Royal Canadian Air Force units opened several additional airfields. The USAF 615th Contingency Response Element opened the Dominican Republic's San Isidro Air Base on 16 January to handle limited air cargo deliveries for onward movement into Haiti by road. The Air National Guard's 123 Contingency Response Element opened Maria Montez Airport on the 23 January 2010. Meanwhile, the Canadian Forces opened up Jacmel Airport on the south coast to CC-130 operations on 21st. In total, the Canadian Forces, in what it called Operation Hestia, airlifted 2,600 t of cargo and 5,447 passengers into and out of Haiti by strategic airlift and another 250 tons by helicopters deployed into the country or operated from offshore frigates.[37] These accomplishments and those of the overall international relief effort undoubtedly saved hundreds of thousands of Haitian lives. That this record was not marred by a single aircraft accident in the hazardous operating environments of Haiti's airspace and airports is further testimony to the effectiveness of the American

---

36   601st Air and Space Operations Center, 2.

37   Canadian Expeditionary Force Command, "CEFCOM International Operations Fact Sheet: Operation HESTIA and Joint Task Force Haiti", Canadian Expeditionary Force Command Canadian Forces, Canadian Department of National Defence and the Canadian Forces, 2010. Available at: http://www.forces.gc.ca/en/operations-abroad-past/op-hestia. page [accessed 7 May 2014].

control teams, the 818th CRG's ramp operations, and the professionalism of the pilots involved.

Implications

This present assessment earlier asked whether normalization of the USAF–UN relationship in humanitarian airlift operations would take the form of institutional, doctrinal, and/or personnel changes and improvements. It pursued this question by, first, examining the normal state of UN–USAF relations and then examining the specific circumstances and events of the Haiti relief effort. In sum, these examinations suggest that the overall organizational and human preparations for such operations worked well for both organizations. But they also revealed problems in coordination and prioritization that might best be addressed through a combination of modest reforms in all three areas.

Organizationally, neither the United Nations nor the USAF is likely to see a need to make major internal organizational adjustments in response to the 2010 Haiti experience. Despite devastating losses to its personnel in Haiti at the time of the quake, the UN's mechanisms for disaster response functioned effectively. Similarly, the US DOD, including the USAF, demonstrated alacrity and effectiveness in mobilizing units designed for combat operations and applying them as an effective team to rescue and disaster relief activities. On the ground in Haiti, both organizations set up operations quickly and seem to have interfaced effectively when and where needed.

Working relations between UN and USAF personnel also seem to have been good in the realm of coordinating air transport operations. Anecdotally, some US military personnel had less than positive interactions with the personnel of other NGO and PVO organizations over things like arrival priorities and distribution procedures. But, again, no institutional or individual complaints of bad relations between the professionals of the UNHAS have emerged in the hard or soft media to challenge Phillippe Martou's report that "it was a privilege and pleasure to work with the US military ... with whom we developed a great friendship and intend to continue our collaboration".[38]

The doctrinal question remains about what to do about air transport access apportionment and priorities in complex humanitarian relief situations. Clearly, the assumption of apportionment and slot-time control by the RAMCC was expedient and helpful. But military controllers were not prepared to make such decisions with confidence. They knew or developed procedures for receiving, filling, and coordinating slot-time requests readily enough. After all, planning and controlling an airlift flow into Haiti was no more than a variation of military airlift planning

---

38   These comments are appended to a copy of the "After Action Report" forwarded to the author by Mr Martou on 5 June 2011. See 601st Air and Space Operations Center (2010), 27.

and operations in general. But the military controllers simply would have been well out of their realms of expertise had they tried to adjudicate among competing requests from a cloud of NGOs, PVOs, and others clamoring for priority treatment. Apart from the larger organizations like the United Nations they did not know who they were dealing with or precisely how important their cargos were for the relief effort at any given moment. So to provide the necessary technical expertise and to quiet the political clamor resulting from the US military's early efforts, AFNORTH and AFSOUTH invited UNHAS into the RAMCC. Phillippe Martou and his team brought much needed expertise and political savvy to the operation, but not until the worst of the crisis was passed and the airlift was stabilizing.

United Nations and USAF after-action reports tend to focus on the UNHAS' late-to-game arrival at the RAMCC. "In the future", suggested the Haiti Flight Operations Coordination Center official history:

> it is recommended that a non-biased entity ... work closely with the RAMCC to determine aircraft priorities ... [and] be brought into the fold from the onset of any operation, as the highest demand for slots is in the first few days.[39]

Similarly, Phillippe Martou suggested in his assessment of Operation Unified Response that:

> the ability of the humanitarian community to respond quickly and effectively is limited. ... Thus, ... dependence on the trained manpower and the logistics infrastructure of the military is increasing ... [so] [t]here needs to be greater engagement to better leverage their combined capabilities.

Accordingly, Martou recommended a number of civil–military initiatives to make coordination in future contingencies smoother and quicker.[40]

Based on the experience, however, it seems that these assessments miss what perhaps is the most important lesson of the UN–USAF partnership: Neither organization is constituted to arbitrate airport access priorities in a disaster. Only the host-nation government has the legal authority to set and enforce landing rights within its borders, even in a disaster situation. The air traffic control arrangements set up for Haiti during the first weeks of the emergency respected the government's sovereignty and authority, of course. But they also moved the USAF–UNHAS team to the foreground of prioritizing slot-time requests for organizations that often saw one or both as biased, lacking legal footing, competing for impact and prestige, or even as a competitor. The relationship even created tensions between the GoH and its temporary air control agents, as the president and other officials overrode or ignored the slot-time procedures established at their behest. Thus, the people working in the RAMCC and the ramp at MTPP were, in effect, "front

---

39 Ibid, 2.
40 Ibid, 27.

guys" for the Haitian government and easy targets for every individual, blog, NGO, PVO, or government official that had a bone to pick with their management of the airlift or with the United Nations or the United States in general.

Reasonably, then, an arrangement that kept the GoH in the spotlight in the day-to-day allocation of slot times would have mitigated the tensions resulting from the ad hoc setup actually used. Had the government possessed the expertise, which in this case it did not, it was best placed to prioritize the flow of airlift cargo in reflection of changing needs for water, food, rescue teams and equipment, emergency medical teams, construction equipment, communications equipment, hospitals, and all the other useful and non-useful things put forward by their sponsors for immediate delivery. Priorities and slot-time lists published and enforced under the government's direct imprimatur would have done much to cool the complaints and discourage misbehavior. Foreign governments could not fault Haiti or cry "colonialism" in the face of its obvious control of its own relief. NGOs and PVOs concerned about access would have been less inclined to ignore in the immediate term the very government that would grant it access over the long term. Last, the GoH itself would at least have to accept the consequences of and responsibility for the "end runs" it authorized for favored individuals or groups, who sometimes contributed little or nothing to the immediate relief effort.

Of course, the GoH was not ready to manage the details involved in apportioning and allocating airport access during its emergency. In all likelihood, no government in a less-developed country facing a major disaster would be ready for such responsibilities. Senior officials likely would not understand the technical details of relief well enough to establish priorities. Just as likely, their governments would not contain the technical expertise and staff personnel needed to support their decision-making. So, any workable doctrinal arrangement for handling future HIRI situations should provide for establishment of appropriate advisory and technical support for the government of an afflicted state, without diluting the reality and appearance of its sovereign control of events.

Recommendation

The goal of ensuring effective control of the air transport stream into a disaster scene, without undermining the sovereignty of the receiving government, points to an organizational solution involving three parts.

a. *Host Nation Allocation Authority*: This individual likely would come from either the Department/Ministry of Transportation, or another part of the government charged with managing internal and/or disaster affairs. The president or prime minister of the host nation should promulgate a public announcement as soon as possible to set up and empower this authority. As the host nation's disaster relief representative, the Allocation Authority would direct, supervise, and validate the efforts of the Forward and Rear staff elements.

In the likely absence of an adequate local staff able to supervise these staff elements, afflicted host nations might well charter the United Nations Humanitarian Operation and Coordination Centre or equivalent to organize and supervise the Forward and Rear staffs and operate them as a "Logistics Management Center" in support of the Allocation Authority.

b. *Humanitarian Operation and Coordination Center – Forward Staff*: This staff would be located as close physically to the Allocation Authority and/or the disaster site as circumstances allow. Collocation will facilitate the Forward Staff's efforts to help the Allocation Authority assess requirements, prioritize access (slot-time) requests, and promulgate slot-time schedules. The Forward Staff also would coordinate with and validate the requirements and capabilities of military and civil relief individuals and organizations present at the disaster scene. To perform these functions, the Forward Staff would have at least two sub-teams:

   – *Access Team (Forward)*: UNHAS personnel to develop requirements, priorities, and finalize slot-time schedules for the Allocation Authority's approval;

   – *Planning Team (Forward)*: civil and military air transportation experts to coordinate slot-time requests coming through the Rear Staff, coordinate and validate on-scene requestor requirements and capabilities, and draft slot-time schedules for review by the Access Team. The Planning Team also would host daily meetings of local port authorities and relief organizations involved in airport operations to minimize the gap between plans and reality in managing the air transport flow.

c. *Rear Staff*: This staff could be located anywhere in the world, so long as it had adequate communications to the Forward Staff and was accessible electronically to organizations requesting access to airports in disaster areas. This staff's primary role would be to receive, process, and communicate slot requests to the Forward Staff. Once the Allocating Authority approved the slot schedules developed by the Forward Staff, the Rear Staff would communicate them to the requesters and coordinate them with appropriate air traffic control and other involved agencies. To the extent possible, the Rear Staff also would handle complaints from requesters, conduct media relations, and assess and report operational results. This staff also might have two sub-teams mirroring those in the Forward Staff:

   – *Access Team (Rear)*: Representatives from UNHAS and/or other appropriate civil or military organizations to conduct direct communications with organizations requesting access to airports in the disaster area, handle complaints, and coordinate with the Access Team

(Forward) regarding relief priorities, offers of support, airspace and air traffic management issues.
-   *Planning Team (Rear)*: A team of civil and/or military experts to receive, organize, coordinate, and communicate access requests. Also supervises operational performance of slot-time users and assesses and reports on operational results.
-   *Allocation Authority Representative*: The host nation Allocation Authority likely would assign a direct representative to the Rear Staff to serve as a liaison officer and spokesperson for the host nation's management of the relief effort.

The excellent performance of the AFNORTH RAMCC (renamed HFOCC during the recovery operations) in hosting what amounted to the notional Rear Staff (above) suggests the possibility of assigning it the role more or less permanently. Doing so would require: gaining US and international relief community agreement on the matter; coordinating appropriate doctrines and procedures; and staffing and equipping the RAMCC to deploy some personnel and equipment to the Forward Staff. If the US government committed to this mission, and since such forward deployments would impose some risk to other AFNORTH homeland missions, the USAF might choose other options. These could include strengthening the existing RAMCC, establishing a second one, or embedding the mission in one or more of the Air Mobility Command's contingency response groups/elements.

Since the Haiti relief effort, the USAF has taken several actions to enhance the capabilities of its RAMCC concepts. In February 2011, it updated Air Force Doctrine Document 3–52, "Airspace Control", to formalize RAMCC roles and organization, though its provisions are more pertinent to combat environments than to humanitarian relief operations.[41] The air components of most joint combatant commands also have plumbed the Haiti experience and taken different degrees of action to establish core RAMCC staffs or at least train key personnel in RAMCC operations. Accordingly, several air force training organizations and programs have expanded their publication and syllabus treatments of RAMCC subjects. Thus, while USAF authority over airspace access and slot times will remain problematic for many international organizations, its ability to lead or at least augment such activities has increased markedly in recent months.[42]

Regardless of how the international relief community works out the details, the experience of the Haiti earthquake relief operation points to a clear need for a well-planned, coordinated, and exercised international organization to control

---

41   Air Force Doctrine Document 3-52, "Airspace Control" *United States Air Force Doctrine Document* 3-52, 2 February 2011: Appendix 1.

42   The assessment of recent USAF developments was made by Major J.J. Grindrod, who participated in the interview and also served as the 601 RAMCC/HFOCC Chief of Plans during the Haiti Relief (Colonel Warren Hurst et al. (2011), Interview by Robert C. Owen, 10 August 2011, tape 1, side A, index 523–610).

air transport flows in HIRI circumstances. The need for such a flow control organization is greatest in the first days, literally the first hours, of a disaster. So the international community must replace existing ad hoc practices with flow control arrangements based on national sovereignty and able to be activated within a few hours of notification. To work so quickly, this organizational arrangement must be understood by the international community and have a permanent existence, at least in terms of a web presence and a small staff to develop plans and documents, conduct training and exercise activities, and maintain facilities and equipment in readiness. UN disaster relief agreements with potential host nations also should identify the local government organizations, facilities, and personnel needed to host and support the Forward and Rear staffs. For the afflicted citizens sitting amid the carnage and rubble of future disasters, the effective workings of those staffs often will be matters of life or death.

Chapter 6

Flying Humanitarians:
The UN Humanitarian Air Service

A. Walter Dorn and Ryan W. Cross

Though the United Nations is often and rightly criticised for a lack of coordination and cooperation among the disparate family of UN agencies, the United Nations Humanitarian Air Service (UNHAS) provides a strong counter-example. It serves not only the World Food Programme (WFP), the Rome-based organization which originally created it, but also a large number of UN agencies and a plethora of non-governmental organizations (NGOs). UNHAS has achieved a positive reputation, though it is not without its critics, within the global humanitarian community. However, it is little known and understood outside its immediate users. By most accounts it performs remarkably well, perhaps explaining why so little attention is paid to it in the media and the wider public literature. The academic literature suffers a lacuna on the functions and operations of UNHAS; this introductory article is meant to help fill that vacuum.[1]

Services and Mechanisms

UNHAS is the world's main transporter of humanitarian personnel and aid during natural disasters and complex emergencies (that is, those involving human conflict). It provides aviation logistics to the fringes of the world, in places and situations where normal air carriers refuse to fly. UNHAS can thus be considered the "airliner for humanitarians", often going to places unreachable, for all practical purposes, by timely ground or sea routes.

UNHAS' mandate is to provide "safe, efficient, responsive and cost-effective" air transport.[2] It provides its services to the global humanitarian community via a common pooling of aircraft, flying thousands of aid workers, relief specialists,

---

1   Aside from publications of the United Nations, the World Food Programme and their employees, the authors could identify few academic publications on UNHAS, which typically focus on humanitarian logistics.

2   World Food Programme. "United Nations Humanitarian Air Service" (WFP, 2012). Available at: http://www.wfp.org/content/united-nations-humanitarian-air-service-unhas [accessed 15 April 2013].

doctors and critical supplies into locations "where no one else goes".[3] Originally established by the WFP in the 1980s as the WFP/Air Service, to carry its food and non-food items, the first operations opened humanitarian air corridors in Ethiopia, Somalia, Angola, and Sudan. Over the years, UNHAS gradually came to serve the wider community. Waste and numerous inefficiencies became readily apparent among humanitarian and development agencies in trouble spots as they competed for limited airstrips and logistics facilities while working towards the common goal of sustaining lives and alleviating human suffering. In Somalia in 1996 the WFP was assigned to lead the first "UN Common Air Service".[4] Given this successful initiative, in 2003, the United Nations High Level Committee on Management – part of the United Nations Development Group, a group designed to oversee the family of UN agencies – mandated the WFP to "manage aviation services for all UN agencies, non-governmental organizations and implementing partners".[5] With this official directive, UNHAS was born.[6] But since peacekeeping operations continued to have such a large and long-standing air component, they remain serviced by the UN Secretariat departments in New York, instead of the Rome-based UNHAS.

Rather than owning aircraft, UNHAS contracts over 50 aircraft of many sizes and types. For example, Figure 6.1 shows a large Ilyushin Il-76 airdropping WFP food bags and Figure 6.2 shows the Mi-8P, the workhorse of helicopter humanitarian delivery.[7] Both are Russian-built. UNHAS uses an online bidding process for a shortlisted pool of contractors. It provides the bulk of the humanitarian community's air transport into the world's hot spots. For instance, at the end of 2012, UNHAS operated in Afghanistan, Chad, the Central African Republic, Côte d'Ivoire, the

---

3   Cole, A. "WFP Aviation: Meeting the Air Transport Needs of the Humanitarian Community", *ICAO Journal* 66(6) (2011), 18.

4   Ibid. Given the UNHAS relationship with the WFP, plus the fact that the WFP runs some of its own aviation separate from UNHAS, there are several variations on the name given to aviation support coordinated by WFP: "WFP–Aviation"; "WFP–UNHAS"; "UNHAS" (the term used for this chapter); or "WFP–Aviation/UNHAS". While closely synonymous they are not identical in meaning. For the purposes of this chapter, however, the distinction between them is not made.

5   Quinn, E. "Logistics for Food Assistance: Delivering Innovations in Complex Environments", in *Revolution: From Food Aid to Food Assistance – Innovations in Overcoming Hunger*, edited by Steven Were Omamo, Ugo Gentilini, and Susanna Sandström (Rome: World Food Programme, 2010), 311.

6   Cole, "WFP Aviation", 19.

7   Aircraft used by UNHAS include: jets like the Bombardier B1900D, and Dornier J-328; propeller aircraft like the Beechcraft B1900C, Cessna C208B (Caravan), de Havilland Canada Dash-8, Embraer 120, Let-410 (Kunovice), and Pacific Aerospace Pac 750XLV; and helicopters like the Bell 212 Twin Huey and Mil Mi-8T/8P/8MT. The cargo aircraft include: the Antonov An-12, An-24, An-26, and An-124; Boeing 747; Ilyushin Il-76; and McDonnell Douglas MD-11. At times, UNHAS has been called "UN Humanitarian Air Services" as can be seen from the wording on the side of the helicopter in Figure 6.2.

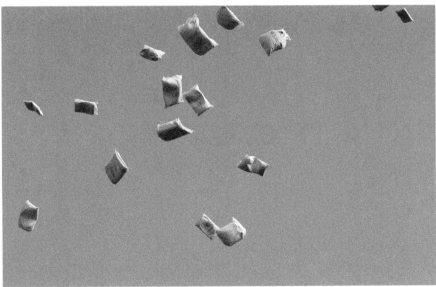

Figure 6.1   A World Food Programme/UN Humanitarian Air Service food drop in Upper Nile State, Sudan

Source: UN Photos 161581 and 161582, F. Noy, 14 November 2007.

Democratic Republic of the Congo, Ethiopia, Mauritania, Niger, Somalia, South
Sudan, Sudan, Uganda, and Yemen.[8] The aviation service covers several broad
areas: air support for WFP, including emergency airlifts and food drops; common
air services for the humanitarian community; strategic airlifts, that is, ad hoc cargo
flights for UN agencies, NGOs, donors, etc.; medical and security evacuations
on request from the UN's Department of Safety and Security; and third party
services "as able or required".[9] The service is often the *only* air carrier available for
humanitarian operations, as commercial airlines are unable to tolerate the physical
security risks of operating in violent and unstable conditions – precisely those
areas where the humanitarian need is greatest. So it can be dangerous work. For
example, in November 2010, three UNHAS crewmembers were abducted in the
Darfur region of Sudan and were held for over a month.[10] In 2010, nearly 200
of the 240 UNHAS destinations were considered no-fly zones by commercial
airlines.[11] The aircraft of UNHAS fly into and out of, for example, dangerous

**Figure 6.2   A UN Humanitarian Air Service Mi-8P helicopter in El
               Geneina, West Darfur, Sudan**
Source: Wikimedia commons, July 2007.[12]

---

8   World Food Programme. *WFP Aviation Review 2011* (Rome: World Food
Programme, 2012).

9   Maslyukov O. "WFP Aviation: The Global Leader in Humanitarian Air Support"
(PowerPoint presentation at the 4th Global Humanitarian Aviation Conference, Dead Sea, 9
October 2012), 4. Available at: http://annualghac.com/wp-content/uploads/2012/04/D1-S2-
WFP-WFP-Aviation.pdf

10   Abdelmoneim Abu Edries Ali. "Gunmen 'Kidnap Latvia Helicopter Crew'
in Darfur". Agence France-Presse, 5 November 2010; United Nations News Service.
"Three UN Air Service Crew Members Freed from Captivity in Darfur". United Nations
News Centre. 8 December 2010. Available at: http://www.un.org/apps/news/story.asp
?NewsID=36994 [accessed 7 May 2014].

11   World Food Programme. *WFP Aviation Review 2010* (Rome: World Food
Programme, 2011), 2.

12   Image from Melting Tarmac Images. Available at: http://commons.wiki
media.org/wiki/File:UNHAS_(Heli_Air_Services)_Mil_Mi-8P_MTI-3.jpg [accessed 31

airfields in Mogadishu, Somalia, and Faizabad in Afghanistan; austere airstrips such as Pweto in the Democratic Republic of the Congo; remote locations in Sudan with no runways, where airdrops are necessary; and severely damaged areas like Haiti following the 2010 earthquake.

The United Nations divides its aviation resources between those for peacekeeping (an immense undertaking) and those for other purposes. Aviation in support of ongoing UN peacekeeping operations is the responsibility of the Air Transport Section within the Department of Field Support (DFS) – see Chapter 16 in this volume. Aviation responsibility for "humanitarian and other" purposes lies mainly with UNHAS. The reason for this separation is primarily for political independence. Typically, aid agencies do not want to be associated with military operations of any kind, including UN peace operations. As one WFP aviation official put it: "keeping humanitarian and peacekeeping operations separate (along with minimizing reliance on aircraft supplied by host governments, especially in places experiencing civil conflict) is vital to maintaining credibility and independence".[13] Of course, these two sets of operations often need to be closely coordinated by the United Nations (see the last section of this chapter for a description of the difficulties).

Within the broader international humanitarian community there is a division of responsibilities into a series of "clusters", each of which is led by a specific agency, for example, the United Nations High Commissioner for Refugees (UNHCR) for emergency shelter, and UNICEF for water/hygiene, nutrition and education. Within each cluster humanitarian agencies work together (in theory, harmoniously), including the United Nations itself and its operations or agencies (such as the Food and Agriculture Organization, the UN Population Fund, UNHCR, and the World Health Organization), and major international NGOs, such as the International Committee of the Red Cross, and Médecins Sans Frontières. For example, when a major operation is required to supply refugee camps and internally displaced persons camps, an effort is made in concert with the UNHCR and the International Organization for Migration, as well as any peacekeeping operations in the area. The cluster attempts to prioritize the needs. Additionally, donor nations and their development agencies may also use UNHAS to deliver goods and services. For the entire effort, the United Nations Office for the Coordination of Humanitarian Affairs (OCHA) is tasked to provide central coordination.[14]

December 2013]. Also available at: http://www.airliners.net/photo/United-Nations-(Heli/Mil-Mi-8P/1297724/L/ [accessed 7 May 2014].

13   Cole, "WFP Aviation", 19.

14   United Nations Office for the Coordination of Humanitarian Affairs. "Cluster Coordination", OCHA, n.d. Available at: http://www.unocha.org/what-we-do/coordination-tools/cluster-coordination [accessed 7 May 2014].

Above and beyond air transport, WFP was designated the global lead organization within the international humanitarian community for logistics.[15] As air operations are a key component aspect of humanitarian logistics – moving people, critical supplies, life-saving equipment, and specialized goods in a timely manner – WFP was given wider responsibilities. UNHAS operations are usually launched upon a request to WFP from the humanitarian country team or the humanitarian coordinator to set up and manage a common air service in a specific country on behalf of the larger humanitarian community.[16]

When a peacekeeping mission also operates in the area, UNHAS aviation activities must be coordinated, often using common landing strips, air traffic controllers, and ground storage facilities. In most of the regions of the world where UNHAS operates, civil aviation authority is non-existent or lacks capacity or oversight abilities. To help ensure the safety of UN air movements, DFS, UNHAS and WFP's Aviation Safety Unit work with the International Civil Aviation Organization (ICAO). UN aviation standards were developed in conjunction with ICAO. The Montreal-based organization became a member of the WFP-led Logistics Cluster to help create a culture of safety with system-wide operational standards:

> These standards range from checking that licences, insurance and civil aviation credentials are current to whether potential operators have a good track record on safety and are not supporting illicit activities in between UN flights. This partnership has, over time, spawned additional collaborations, ranging from projects to rebuild airstrips in Sudan … to planning how to maintain operations in the event of pandemic illness.[17]

Aircraft safety can be challenging in many areas where UNHAS flies because of politically unstable environments, sub-standard airfields (which are often unprepared gravel airstrips), lack of air traffic control, and limited or no weather forecasts. Through the monitoring and auditing done by WFP's Air Safety Unit, UNHAS seeks to achieve a basic standard of "fully equipped" aircraft.[18] This includes aircraft with traffic collision avoidance systems, enhanced ground proximity warning systems, and Global Positioning System (GPS)-tracking alongside a host of cutting-edge communications technologies. The latter include automatic indicators of location and aircraft conditions in cases of crashes, and

---

15   The WFP also manages and coordinates the cluster responsible for emergency telecommunications.

16   World Food Programme Global Logistics Cluster. "Air Operations", *Global Logistics Operational Guide*, "Operational Environment", 2012. Available at: http://log.logcluster.org/operational-environment/air-operations/index.html [accessed 15 April 2014].

17   Cole, "WFP Aviation", 19.

18   Donoghue, J.A. "Safety on the Frontier", *AeroSafety World* 6(2) (March 2011), 27.

the flight data recorder or black box.[19] Likewise, in recent years coordination has increased globally on humanitarian air safety, with conferences held annually.[20] UNHAS has been able to maintain an enviable safety record in some of the most conflict-riddled countries in the world. It has been credited for raising the awareness of the aviation risks in the areas where it operates – notably in Africa where safety standards and culture as well as regulatory environments are generally weak.

Two brief cases of UNHAS operations help illustrate the service in action.

Case: Libya Operations

When the "Arab Spring" spread to Libya in mid-February 2011, it quickly evolved from civil protests to violent conflict. Within a month the UN Security Council passed Resolutions 1970 and 1973 (2011), the latter establishing a no-fly zone over Libya and authorizing enforcement of an arms embargo (see Chapter 15 in this volume). It also demanded that the Libyan authorities "ensure the rapid and unimpeded passage of humanitarian assistance".[21] As the situation escalated, UNHAS was activated in May 2011 as commercial operators halted their operations in the face of the no-fly zone and regional insecurity.[22] Though commercial service was halted, the demand from humanitarian workers remained and UNHAS began to move relief workers into Libya and neighbouring countries.[23] Over a period of eight months, UNHAS moved passengers from 150-odd UN agencies, NGOs, and donor organizations. From its operational base in Malta, it routed aircraft through Cairo, Benghazi, Tripoli and Djerba (in Tunisia). UNHAS also transported many international reporters so they could report on the humanitarian situation.[24] The editor of BBC World News later commented:

> It's no exaggeration to say that we couldn't have run our operations in Libya without the [UNHAS] assistance … Many of my colleagues were spared the ordeal of long, often dangerous, journeys thanks to the UNHAS flights.[25]

---

19 "Pierre Carrasse: Helping Humanity from the Air", *Flight International* 180(5306), (30 August 2011), 67.

20 See Global Humanitarian Aviation Conference. "Global Humanitarian Aviation Conference", n.d. Available at: http://annualghac.com

21 United Nations Security Council Resolution 1973 (2011). UN Doc. S/RES/1973 (2011), 17 March 2011, para. 3.

22 Global Logistics Cluster. *North Africa Crisis: Logistics Cluster Operations Report 2011* (April–December 2011). (Rome: World Food Programme, 2012), 3. Available at: http://reliefweb.int/sites/reliefweb.int/files/resources/Full_Report_3363.pdf

23 World Food Programme, *WFP Aviation Review 2011*, 22.

24 Cole, "WFP Aviation", 20.

25 Ibid.

On return trips, UNHAS aircraft took evacuees out of Libya, primarily labourers from nations lacking the capacity to evacuate their own citizens.[26]

As Libyan airspace was controlled by the North Atlantic Treaty Organization (NATO), which was enforcing the no-fly zone shortly after the adoption of Resolution 1973 (2011), UNHAS flew in NATO-controlled approach corridors.[27] Within NATO's Libya operation headquarters, the WFP-led logistics cluster was given observer status, which helped it ensure inter-agency coordination with an eye to the evolving military and political situation.[28] UNHAS operations ran from May to November 2011, ending at the same time as the NATO operation. In its six-month Libya operation, UNHAS moved some 4,700 passengers.[29]

Case: Haiti Post-earthquake

Immediately following the 12 January 2010 earthquake in Haiti, UNHAS launched an operation to facilitate transport of humanitarian personnel, food, medicines and other relief items to areas inaccessible by surface transport.[30] This occurred alongside intensive international efforts, led by the United States and its air force – see Chapter 5 in this volume. WFP coordinated the "logistics cluster" for the humanitarian community, and UNHAS not only transported emergency material but also made damage assessment flights to determine the areas most needing assistance.[31]

The service coordinated its work out of the Santo Domingo airport in the capital of the neighbouring Dominican Republic, specifically to reduce the burden on the collapsed Port-au-Prince airport. This helped keep non-essential logistics personnel outside the disaster zone so they would not encumber the on-site effort. This emergency operations centre in Santo Domingo became the coordination unit for humanitarian air services, as a large number of humanitarian logistics personnel were moved into the Caribbean nation.[32] Given the immense international attention focused on the disaster, at the initial stages funding was

---

26   Ibid.

27   World Food Programme, *WFP Aviation Review 2011*, 22.

28   Cole, "WFP Aviation", 20.

29   Global Logistics Cluster. *North Africa Crisis: Logistics Cluster Operations Report 2011* (April–December 2011), 3.

30   World Food Programme. *WFP Aviation Review 2010*, 16.

31   In general, for views on the humanitarian logistics response, see Whiting, M.C. "The Haiti Earthquake, January 2010", *Logistics & Transport Focus* 12(4) (April 2010), 26–9; Heraty, M. "Haiti: The Logistics Response and Challenges", *Logistics & Transport Focus* 12(5) (June 2010), 42–5; Heraty, M. "Logistics Response and Needs in Haiti – a Field Perspective", *Logistics & Transport Focus* 12(5) (May 2010), 35–8.

32   Whiting, "The Haiti Earthquake".

less of an immediate concern, allowing UNHAS to provide cost-free services to humanitarian personnel.[33]

While some in the humanitarian community and media were critical of the American "control" of the Haitian air relief operations, the flight logs indicated that this claim can hardly be supported by the available evidence.[34] Sir John Holmes, Under-Secretary-General for Humanitarian Affairs and the Emergency Relief Coordinator, publicly commended the American effort:

> The Americans taking over the Port-au-Prince airport was absolutely crucial; … clearly there were some glitches. But I don't think there was any intention to favor military flights over humanitarian flights. It was simply quite difficult to set up a system that included genuine real-time priorities.[35]

The close interaction between senior UNHAS personnel and US Air Force controllers suggests that there was mutual respect in the emergency, hardly an example of American "imperialism". One of the most senior WFP aviation logistics managers was placed at the Tyndall Air Force base (Florida), which housed the Haiti Flight Operations Coordination Center. The goal was to integrate the military response efforts and the humanitarian response efforts at a senior level.[36] Civilian aviation management personnel provided guidance to the US military. They were specially tasked to apportion airspace so that priority humanitarian needs, as identified by the humanitarian leadership in Haiti, could be met.[37] Granted, the integration of these two distinct communities (military and civilian) was a challenge, but coordination was essential to save lives. As one of the WFP deployed officials noted, "humanitarian relief is and should remain a predominantly civilian function; however foreign military assets can play a valuable role in natural

---

33 Thompson, J. "The Santo Domingo Operations Center", World Food Programme Logistics Blog, 31 January 2010. Available at: http://www.wfp.org/logistics/blog/santo-domingo-operations-center [accessed 15 April 2014]. Thompson, J. "Daily UNHAS Shuttle from Santo Domingo to Port-Au-Prince Runs Like a Well Oiled Machine", World Food Programme Logistics Blog, 28 January 2010. Available at: http://www.wfp.org/logistics/blog/daily-unhas-shuttle-santo-domingo-port-au-prince-runs-well-oiled-machine [accessed 15 April 2014].

34 Mendoza, M. "Haiti Flight Logs Detail Early Chaos", Associated Press Newswire, 18 February 2010. Available at: http://www.utsandiego.com/news/2010/feb/18/ap-impact-haiti-flight-logs-detail-early-chaos [accessed 15 April 2014].

35 Ibid.

36 McHale, M. "UN Aviation Expert Assists 601st AOC with Haiti Ops", *United States Air Force News Service*, 2 February 2010. Available at: http://www.acc.af.mil/news/story.asp?id=123188335 [accessed 15 April 2014].

37 Whiting, M.C. "Military and Humanitarian Cooperation in Haiti's Air Operations", News, Humanitarian Logistics Association, 2010. Available at: http://www.odihpn.org/humanitarian-exchange-magazine/issue-53/military-and-humanitarian-cooperation-in-air-operations-in-haiti [accessed 7 May 2014].

disaster relief".[38] The US military could bring to bear an unsurpassed capacity for mass airlift. A civil–military aviation framework, with protocols and guidelines, was developed to ensure that both civilian and military flights were "properly prioritised, synchronised, and executed".[39] WFP's Haiti Logistics Cluster reports show that approximately 10 days after the disaster, the US military was "assisting with clearance of relief cargo, the ferrying of goods for airlift and loading of organisation's trucks as required".[40] The Logistics Cluster placed a civil/military liaison officer in Miami with United States Southern Command, the lead military command for the US military response to the Haitian disaster. UNHAS' Chief Air Transport Officer also worked in Miami for five crucial days to prioritize flights and establish the "slot mechanism" with Southern Command.[41]

By the end of 2010, UNHAS had moved some 20,000 passengers, alongside 2,600 tons of cargo for 162 agencies in its Haitian response. This included aviation transport throughout the country on a scheduled basis, utilizing small airfields and helicopter landing zones and "piggy-backing" onto the logistics and infrastructure of the United Nations Stabilization Mission in Haiti, which had been established in 2004. UNHAS provided regular air-transport schedules and mission-specific services, such as airborne damage assessments and, in conjunction with the emergency humanitarian telecommunications team, it supported the establishment of radio networks for the humanitarian community.[42] Although UNHAS concluded its own operations in Haiti at the end of March 2011, its Logistics Cluster continued to explore the "options for air transport, including commercial companies, which can be offered to Logistics Cluster participants as an alternative" to UNHAS services.[43]

UN Humanitarian Air Service Accomplishments and Challenges

The scope of UNHAS operations is impressive: in 2011 alone, for example, UNHAS moved over 350,000 people between 350 destinations, as well as about 3,500 tons

---

38   Ibid.

39   Ibid.

40   World Food Programme Haiti Logistics Cluster. "Logistics-Cluster Haiti: Consolidated Situation Report", Logistics Consolidated Situation Report (Port-au-Prince and Santo Domingo, 23 January 2010), sec. 3. Available at: http://logcluster.org/document/situation-report-consolidated-23-january-2010 [accessed 7 May 2014].

41   Ibid.

42   Thompson, J. "WFP's UNHAS Transports Emergency Telecoms Team to the Remote Mountains of Ile de la Gonave, Haiti", World Food Programme Logistics Blog, 29 March 2010. Available at: http://www.wfp.org/logistics/blog/wfps-unhas-transports-emergency-telecommunications-team-remote-mountains-ile-de-la-gonave-haiti

43   United Nations Office for the Coordination of Humanitarian Affairs. "Humanitarian Bulletin for 25 March – 10 April 2011", OCHA Weekly Haiti Humanitarian Bulletin (Port-au-Prince, 25 March 2011), 4. Available at: http://reliefweb.int/sites/reliefweb.int/files/resources/report_0.pdf [accessed 7 May 2014].

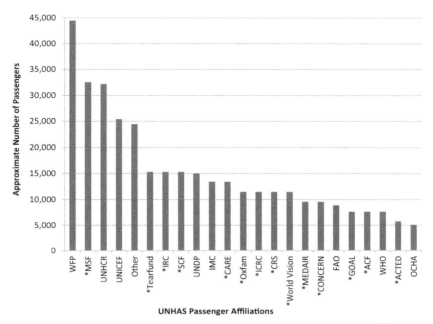

Figure 6.3   UN Humanitarian Air Service passenger numbers by UN agency and non-governmental organization (NGO), 2011

Note: NGOs marked with *
Sources: World Food Programme reports.[44]

of humanitarian cargo. A total of 870 specific agencies and organizations were served: the majority (54 percent) of the users were NGOs, 40 percent were UN agencies, and 6 percent were donor (national) missions and the media. Overall, UNHAS provides some 80 percent of global humanitarian aviation. Figure 6.3 provides the approximate number of passengers by agency and NGO moved by UNHAS. UNHAS also helped capacity-building by providing training for flight dispatchers and crew.

UNHAS spent nearly US$200 million in 2011, with an average cost per flying passenger of roughly US$450. The UNHAS system is based on a cost-recovery scheme: it does not usually provide free service to its users, though it may subsidize the costs.[45] Chartering and operating aircraft is an expensive endeavour

---

44   World Food Programme. *Annual Report of the World Food Programme for 2011 to the Economic and Social Council and the Council of the Food and Agriculture Organization of the United Nations*, UN Doc. E/2012/14, 29 February 2012, para. 36; World Food Programme, *WFP Aviation Review 2011*, ii, 4.

45   World Food Programme, *WFP Aviation Review 2011*, ii, 1–4; Channel Research, John Telford, and Robert Thomson. *Evaluation on the Provision of Air Transport in Support*

at the best of times, and often UNHAS works in poor, landlocked locations, with major changes in demand happening quickly and often, especially for its NGO "clients" with limited budgets. To survive, UNHAS must rely on voluntary donor (national) contributions, in addition to some funding from the United Nations itself. Cost recovery is increased by using flight management tools, working with partners to combine services, optimizing routes, reducing cargo load, and using the least expensive but still safe aircraft. Yet because UNHAS usually flies where commercial operations cannot or will not for security or economic reasons, so economies of scale may not apply and costs can be large.[46]

A constant challenge, also typical of NGO operations, is to obtain long-term funding.[47] An external review found that UNHAS':

> weak fundraising capacity does it no favours. Better long-term planning, a common fund for common services combined with multi-agency support and better WFP/UNHAS fundraising approaches and materials, are required in order to provide a degree of much needed stability to the service.[48]

UNHAS has been described as a "shoe-string organisation that, out of necessity, places all emphasis for safety on the shoulders of the air carriers", that is, the commercial vendors hired by UNHAS.[49] This can be dangerous because these carriers, seeking to save costs and maximize profits, often take short cuts that may not meet UN and ICAO standards. To prevent this, UNHAS seeks to vet carriers through visits to company headquarters prior to registration and to carry out UNHAS inspections during operations. But UNHAS' few staff and small operational footprint does:

---

of Humanitarian Operations, Contract Report for the European Commission (Ohain, Belgium: Channel Research, 29 March 2010). Available at: http://ec.europa.eu/echo/files/evaluation/2010/Air_Transport_Evaluation.pdf [accessed 15 April 2014].

46   Cole, "WFP Aviation", 20.

47   The following reviews of UNHAS capture some of these challenges: Channel Research, Telford, and Thomson. "Evaluation on the Provision of Air Transport in Support of Humanitarian Operations"; Channel Research, John Telford, and Robert Thomson. "Review on the Provision of Air Transport in Support of Humanitarian Operations", Contract Report for the European Commission (Ohain, Belgium: Channel Research, 29 March 2010). Available at: http://ec.europa.eu/echo/files/evaluation/2010/Air_Transport_Review.pdf [accessed 15 April 2014]. Joint Inspection Unit of the United Nations System. Review of the United Nations Humanitarian Air Service, UNJIU Report 2008-3 (Geneva: United Nations, 2008). Available at: https://www.unjiu.org/en/reports-notes/archive/JIU_NOTE_2008_3_English.pdf [accessed 15 April 2014]. See also the annual *WFP Aviation Review* reports.

48   Channel Research, Telford, and Thomson, Evaluation on the Provision of Air Transport in Support of Humanitarian Operations, 6.

49   Conversations and emails from UN aviation officials with experience at headquarters and in the field, including an email of 30 April 2013.

little to overcome the obvious risk of operating with some of the lowest cost operators (particularly freight carriers) in some of the most demanding areas with the least effective regulation.[50]

UNHAS is forced to give its contractors more freedom than do the UN peacekeeping departments (DFS/Department of Peacekeeping Operations (DPKO)). UNHAS officials have even balked at the greater regulation and quality assurance that are put in place by the United Nations in peace operations. This might mean UNHAS developing standard procedures such as the weighing of overly heavy carry-on baggage, imposing a maximum number of flying hours on crews, restricting flights when fire-service levels drop, or halting operations when no fire service is available. In another example of the more tolerant regime of UNHAS compared to DFS/DPKO, UNHAS has shown a greater willingness to use single-engine aircraft. Despite their good safety record, such aircraft are inherently more dangerous in cases of engine failure, especially in locations like Congolese jungles where safe crash-landing is almost impossible.

Since UNHAS does not have the capacity for major infrastructure projects, UNHAS culture has grown accustomed to living without it. It must fly to airports that are nothing more than rough landing strips. In emergency relief situations there is no time to resolve such things but later, once a routine service has been established, the situation should improve, though it often does not. Still, UNHAS does repair runways when necessary to fly, and does so "much faster and with much less fuss and bureaucracy than the United Nations".[51] It has gained a reputation as a "can do" service. UNHAS personnel get to the emergency zones first and "do what is needed to get the job done". The departure lounge might be a tree and the office might be a Toyota but this low-budget approach is familiar to (and often appreciated by) other aid workers. By contrast, the United Nations appears more regulated, more bureaucratic and prevented by its safety culture from taking higher risks, which is sometimes necessary during operations in combat zones with inadequate funding. Unlike UNHAS, the UN's DFS and DPKO can oblige UN member nations to pay for peacekeeping aviation and so can afford greater regulation, which improves its safety record. UNHAS, to its credit, does work with the UN's DFS and DPKO to ensure service delivery is appropriate, given the funding, political situations, and risk profiles. Sometimes UNHAS takes responsibility to fly certain routes, while DFS/DPKO flies others in a complementary fashion, though parallel services sometimes occur.[52]

Since DFS/DPKO usually do not charge aid workers, these persons typically prefer to fly on peacekeeping mission flights, even if they are not guaranteed a seat. Thus, UNHAS finds that some of its clientele is diverted to these aircraft, making passenger manifests and cost efficiencies less predictable.

---

50   Ibid.
51   Ibid.
52   Ibid.

Planning is particularly difficult for long-term needs: the sudden requirements in Haiti from the unexpected 2010 earthquake could not be forecast, for example. Financial pressures often result in service drawdowns driven by financial, rather than humanitarian needs. Likewise, once commercial air services become established UNHAS usually ends its operations, making long-term planning dependent on the precarious commercial aviation sector.[53] The service is also open to other criticisms, such as being:

> perceived as UN-centric (the "One UN" approach blurs the lines among development, humanitarian and political actors); unclear governance and policy development mechanisms; non-standardised systems throughout its various operations; a high cost base; relatively limited cargo transport capacities; problems of staff motivation and grading; and, finally, a reluctance to engage in, or provide guidance for infrastructural rehabilitation, such as of airstrips.[54]

That said, the critics recognize that the service has built "a reputation for a strong safety and security ethos, based on internationally recognised, professional modalities" while operating in complex operational environments and often facing politically "constraining official requirements".[55]

Officialdom and bureaucratic administration are enablers and, at the same time, the bane of field workers, including UNHAS personnel. The humanitarians sometimes feel excruciating psychological pain as innocent people die during conflicts and natural disasters while administrative procedures hold up lifesaving deliveries. Still, those procedures make possible aid services such as UNHAS.

Two large policy issues could be debated in the future: (1) should the UN continue with a dual-track airlift program (peacekeeping and humanitarian) or consolidate the operations to reduce some of the administrative overlap and redundancy, possibly resulting in savings and better coordination?; (2) should UNHAS continue with its "pay-as-you-go" financial posture while attempting to get internally a more stable funding base from member states and users?[56]

In any case, UNHAS' wealth of aviation experience will remain a valuable reservoir of expertise to address a critical component of humanitarian action.[57] The WFP's UNHAS is currently filling a vital interagency role, having become not only the UN's common air service but also an important service provider for NGOs and the media. Because of its committed staff, with honed experience in multisector coordination, its wide mandate, and its economies of scale, it can

---

53   World Food Programme, *WFP Aviation Review 2011*, 8.

54   Channel Research, Telford, and Thomson, "Evaluation on the Provision of Air Transport in Support of Humanitarian Operations", 6.

55   Ibid.

56   The authors are indebted to Robert Owen for raising these overarching policy issues and for his more general review of the paper reproduced in this chapter.

57   See, for example, Donoghue, "Safety on the Frontier", 25–7.

attract contributions from nations, as well as the organizations it serves.[58] It complements the airlift for peacekeeping operations that is provided directly by the UN's DFS. By making the timely and cost-effective delivery of humanitarian goods its priority, UNHAS is putting into practice the UN's goal of saving lives and alleviating human suffering.

---

58 Channel Research, Telford, and Thomson, "Evaluation on the Provision of Air Transport in Support of Humanitarian Operations", 6.

PART III
Aerial Surveillance: Eyes in the Sky

To be effective, UN missions need comprehensive situational awareness, covering the range of key actors and conditions in the field. Where are the armed combatants located? Who is firing on whom? Who is attacking civilians? Who is exploiting the natural resources, legally and illegally? Where are the refugees? And if they are on the move, what direction are they headed in? What are the conditions of the roads and bridges on which the peacekeepers and the "peacekept" must travel? The answers to these pertinent questions and many more can be greatly aided by aerial reconnaissance, as is shown in Chapter 7 by A. Walter Dorn. Aircraft can usually get to the observed targets faster than ground personnel; they can cover more territory in a flyby or can loiter on station; and they offer a different but complementary (bird's-eye) view to land-based observers. Additionally, aircraft can remain at a safe distance above most raging conflicts to avoid risk to the observer, particularly with unmanned aerial vehicles (UAVs), where pilots are far from the battlefield.

The United Nations has gained plenty of experience with aerial reconnaissance over the years, though few advanced systems have been deployed. The first UN peacekeeping force, deployed into the Sinai in 1956, used small twin-propeller Otter aircraft to reconnoiter the positions of Israeli and Egyptian troops. As described by A. Walter Dorn in Chapter 8, the 1958 UN observer mission in Lebanon used aircraft equipped with rudimentary night-vision equipment to spot arms-laden convoys covertly supplying rebel forces. In the UN's next mission, a large expansion in the Congo in 1960–1964, Swedish reconnaissance jets helped the United Nations repel ground and air attacks and determine targets for air attacks in Katanga. Returning to the Congo 40 years later, the United Nations deployed armed helicopters with fourth-generation infrared cameras to help locate and target rebel forces who, despite repeated UN demands, did not cease to attack towns in the eastern Congo. The night-vision capabilities helped halt the rebel advances in 2006 and 2008. The eastern Congo was also the location for the UN's first deployment of UAVs. After decades of modern militaries using surveillance UAVs in operational theatres (including American UAVs in Bosnia in the early 1990s and North Atlantic Treaty Organization (NATO) UAVs in Libya in 2011), the United Nations is finally contracting surveillance UAVs from a commercial vendor to augment its situational awareness. Wisely, the United Nations is not contemplating armed UAVs, which have been so controversially applied by the United States in Afghanistan, Pakistan, and Yemen. Nonetheless, the UN's UAVs can serve as potent *force enablers* and *force multipliers* to make ground troops better informed and better able to defend themselves and civilian populations. In Chapter 9 David Neil gives a view from industry involved with UAVs, enhanced by his many years of military service. In summary, aerial reconnaissance is essential for any UN operations that seek to be robust, effective, and efficient in using its military, police, and civilian capabilities.

Chapter 7

Aerial Surveillance: Eyes in the Sky

A. Walter Dorn

Military patrols by foot, jeep, and armoured personnel carriers are the norm in UN peacekeeping. Fixed observation posts and road checkpoints also contribute to missions. Such ground-level surveillance is obviously indispensable for gaining situational awareness, but there are distinct advantages to observation from the air.

While the United Nations has conducted aerial reconnaissance in some of its operations, the use of observation aircraft in peacekeeping has been ad hoc and unsystematized in both doctrine and practice. Dedicated observation aircraft were employed in the United Nations Emergency Force in the Sinai (1956–1967)[1] and the United Nations Operation in the Congo (ONUC) in 1961 after it was discovered that pilots conducting transport flights observed important activities on the ground during their journey. This prompted ONUC to begin mandatory debriefings of pilots. Later the mission deployed specialized reconnaissance aircraft, including jets (see Chapter 2). In Lebanon (1958), Yemen (1963–1964),[2] and Central America (1989–1992); and in several other locations helicopters were important tools for observation, as well as transportation. The current mission in the Democratic Republic of the Congo (DRC) is believed to have the largest and best heliborne capability in UN history, now complemented by unmanned aerial vehicles (UAVs). However, current commanders complain that UN capacity is still far from adequate for the mandated task.

There is, unfortunately, no systematic record of the UN's aerial observation experience, nor is there any listing of the aerial imaging equipment used in past UN missions.[3] Furthermore, there are no studies comparing the reconnaissance

---

1   The United Nations Emergency Force (UNEF, 1956–1967) used dedicated aircraft for surveillance: single-propeller Otter aircraft from Canada. They helped maintain a vigil along the Armistice Demarcation Line and the international frontier between Egypt and Israel.

2   The United Nations Yemen Observation Mission was mandated to observe an agreed disengagement between forces of Saudi Arabia, Egypt, and Yemen. Air patrols, carried out by a Canadian unit with a dozen or so planes and helicopters, were essential in the mountainous border region, where foot patrols could cover only very limited ground. But as in Lebanon in 1958, the United Nations came up against two limitations on UN patrols in Yemen: traffic monitoring could be done confidently only during daylight; and air-triggered ground inspections of moving caravans was difficult.

3   Air flight is one of the most regulated forms of human activity worldwide, with detailed standards and specifications for safety and flightworthiness. The United Nations

technologies in UN missions with those of other military operations, for example, the North American Treaty Organization (NATO, considered later). This chapter looks at the relative merits of ground versus aerial reconnaissance, drawing upon selected operations and experiences from the United Nations and other organizations. It also compares manned and unmanned reconnaissance flights, since the latter are increasingly used for both military and civilian applications in the wider world. The details of all such comparisons (air versus ground, manned versus unmanned) are, of course, case-specific; that is, dependent in part on objectives, terrain, weather, and so on. But the broad factors outlined here point to the relative merits and the optimum configurations for effective monitoring in a wide range of environments, while also highlighting the problems of the different approaches.

Advantages of Aerial Reconnaissance

From the earliest days of peacekeeping, UN operations recognized the advantage of observation from altitude. Static observation posts were placed on hilltops in the Middle East (Palestine, Lebanon and the Golan Heights) and in Kashmir. But they provided useful views of specific fixed areas only – hilltops, unlike aircraft, are not moveable!

The bird's-eye view that is possible from aircraft provides quicker coverage, a longer line of sight and a wider area of observation than on the ground, though usually with less resolution. Aircraft can travel with great speed and usually experience fewer obstacles blocking the way or the view for outdoor targets. Once at the site, they can adopt the observation altitude and angle for optimum viewing.

Since aircraft can move faster than ground vehicles and go directly ("as the crow flies") to their destination, airborne observers can arrive at distant areas much more quickly. In addition, more territory can be covered during the observation period. Ground vehicles (for example, four-wheel drive utility vehicles) can travel at a maximum of about 120 km/hr. Under the poor road conditions typical of many conflict areas, jeeps often move as slow as 10 km/hr with many mountainous, riverine and jungle areas being impassable by automobile. By contrast, aircraft can easily overcome such terrestrial restrictions. Jets fly at typical cruise speeds of 500 km/hr (jets), helicopters (and two-seater planes) at 200 km/hr, small tactical UAVs at 100 km/hr, and mini-UAVs at 50 km/hr. During an observation period, aircraft can slow down to linger over a particular area – circling by plane or hovering by helicopter. Additionally, cameras can be gyrostabilized to increase picture resolution by reducing the effects of aircraft vibration and wind-caused

---

generally abides by the standards set by the International Civil Aviation Organization (ICAO). UN missions also have standard operating procedures for flights and an Air Operations Manual. By contrast, the sub-activity of aerial reconnaissance is not well documented and is only briefly mentioned in the standard operating procedures.

turbulence. With appropriate software, cameras can "lock on" to their targets, that is, keep them in the centre of the picture even as the plane is moving.

Since aircraft (like ground vehicles) might be at risk of taking fire from the ground, aircraft may have to fly at higher altitudes. Fire from an AK-47 rifle, the most prevalent weapon in current conflict areas, cannot reach altitudes above 1,000 m. Even flying at much higher and safer altitudes (for instance, at 3,000 m) advanced aerial observation equipment (gyrostabilized) can still provide a resolution of 0.5 m or better, allowing tracking of groups of individuals or vehicles.

The ability to vary the altitude of an aircraft allows the pilot to control the visibility of the aircraft from the ground. Aircraft can also fly above clouds for cover or find an altitude where they are nearly impossible to spot or hear. This makes it possible to monitor some illegal and clandestine activities that would otherwise be deliberately hidden as soon as the aircraft was detected. In addition, if criminal/violent elements are aware that the United Nations can operate silently and without detection, a powerful deterrent is created, instilling fear in violators even when aircraft are not present.

If, on the other hand, a show of UN presence is desired, aircraft can fly at low altitudes. A highly visible eye-on-the-scene could deter illegal activities or make them more difficult. Aircraft could even buzz an area to create a distinct impression.[4] During Operation Artemis, which assisted the UN Mission in the Democratic Republic of the Congo (MONUC) in the Ituri province in summer 2003, a French Mirage jet on reconnaissance would deliberately break the sound barrier in the region to create a sonic boom that was clearly noticeable by all, including presumed wrongdoers. Aircraft can be painted in UN white or even with "glow" colours for greater visibility. Laser pointers/designators aboard aircraft could even notify individual perpetrators that they are being watched, by shining a laser beam on them.

Flying at higher altitudes can offer much less intrusiveness than a ground presence, when desired. At times, the United Nations must reduce the visibility of its presence to accommodate local sensitivities or because national authorities have placed limitations on the freedom of movement of UN ground vehicles, for example, with road blocks or checkpoints. While still observing national and international laws, UN aircraft can observe without being detected and move without attracting attention. Of course, take-off and landing sites are needed, but they do not need to be near the observation area and can potentially be based in neighbouring countries. Permission to enter the airspace of a country would be required, of course, unless otherwise mandated by the Security Council.

Especially at night, aerial surveillance can provide a tremendous magnifying effect. When travel by ground is difficult and vision is limited (the range of most

---

4   Even the sound of approaching aircraft can be intimidating, stimulating or warning (depending on the context). In the eastern Democratic Republic of the Congo (DRC), the mere sound of an approaching Mi-25/35 helicopter gunship caused militia forces to break up and flee.

night-vision goggles is 1 km or less), airborne forward-looking infrared (FLIR) and synthetic aperture radar (SAR) can alert the United Nations to illegal activities and movements of rebel fighters. Night flights for any purpose, however, are generally prohibited under UN rules because the United Nations rarely possesses nighttime search-and-rescue capabilities and its aircraft are usually not equipped with weather radars. In a few missions, however, UN member-state contributors are sufficiently equipped to carry out such operations. Examples can be found with Norway and others operating in the former Yugoslavia, Australia's work in East Timor, a UN-chartered company in Kosovo, and Russia's capabilities in Sierra Leone.[5] In November 2006, MONUC was able to "break the night barrier" in the DRC after gaining permission from UN Headquarters. Its Mi-35 helicopters used advanced infrared sensors to detect the movements of a renegade force advancing to attack the town of Goma. With this aerial intelligence, a combined UN-DRC force was able to halt the advance using the night-flying attack helicopters.

In the future, UAVs could be used for night surveillance, removing the applicability of the search and rescue rule. Indeed, the European Union Force (EUFOR) did fly UAVs at night in the DRC from July to November 2006 with some remarkable successes, especially in uncovering illegal shipments of arms. For instance, the FLIR cameras were able to detect imported tanks moving by rail and small arms being transferred in small boats across the Congo River. UAV video imagery was viewed at EUFOR headquarters in real time, so that commanders and analysts at headquarters could share a "common operating picture" and consider responses. Although there was no image feed to MONUC headquarters, recordings were shown to UN officials, for example, to clearly demonstrate illegal import activities clearly, thus allowing UN leaders to confront the violators.[6]

Generally reconnaissance by air is less constrained than on the ground. Host nations often insist that UN ground vehicles be escorted by the nation's troops or liaison officers, whose purpose is, more often than not, to keep an eye on the United Nations (to observe the observers) and prevent unauthorized detours, especially to areas of atrocities that the host nation does not want the United Nations to see. Air observation typically involves a lesser set of restrictions and limitations, though some may still be imposed by the host nation.

Aerial observation also has some distinct disadvantages. Aircraft may not be able to get sufficiently close for observation of individual actions. They cannot see indoors or under jungle canopy. They can sometimes be shot at and shot down and are in need of a great deal of maintenance. They also require host-nation permission for use of national airspace.

---

5 Information provided by the Air Transport Section of the Department of Peacekeeping Operations, 28 February 2007.

6 EUFOR offered to provide images extracted from its UAV video feeds to MONUC within about 1–2 hours (in near real-time).

Advantages of Integrated Systems

Aerial and ground surveillance are complementary. The combination of the two creates a more effective monitoring and response system. By air, large swathes of land can be reconnoitered separately or at the same time as by ground patrols. Advance surveillance flights can alert peacekeepers to dangers, locate them precisely through the global positioning system (GPS) and automatically update databases, accessible using laptops, with the latest imagery for immediate ground viewing. Aerial images can help peacekeepers familiarize themselves with the terrain, their objectives and the dangers. They can assist training, planning and the operations themselves, as well as post-mission evaluation. In conflict zones, where time is of the essence, ground patrols can receive advance notification of routes that are impassable or roads and bridges that are washed out, closed, or subject to militia checkpoints (or even ambush!). Lives can be saved if potential threats are identified beforehand using aerial reconnaissance. For instance, during a MONUC battle with renegade militia leader Cobra Matata in the stronghold of Tchey in May 2006, heliborne spotters warned ground troops of the militia fighters approaching stealthily. This allowed the UN forces to avoid the surprise attack and to respond with force.[7]

For UN operations to be robust, they must be situationally aware, an aspect that is much enhanced by the availability of aerial reconnaissance. Quick Reaction Forces (QRFs), for instance, need to insert themselves with great accuracy at precise locations, which requires excellent geospatial awareness. This level of information, particularly about the hideouts of rogue militias or spoilers, requires advanced surveillance, soldier briefings with detailed imagery, and cueing from aerial assets to respond to the movements and actions of hostile forces. Operating ahead of important convoys, aircraft can alert the latter to potential threats in order to avoid them through rerouting. Wide-area surveillance from aircraft can make the ground action quicker, more precise, more aware, and safer.

During robust peace operations, reconnaissance from above is especially valuable in the pre-dawn period, since attacking militia often move into position at night and wait for dawn before shooting. For instance, in the early morning of 28 May 2006, a joint Congolese–UN force walked into an ambush near Fataki soon after they began their march to search for rebel leader Peter Karim. An attack helicopter was called to suppress militia fire during their withdrawal but it arrived too late for one Nepali soldier, who lost his life in the initial shooting.[8]

---

7 Personal interview with Brigadier-General Duma Dumisani Mdutyana (Deputy General Officer Commanding MONUC's Eastern Division), Kisangani, DRC, 30 November 2006. The militia leader signed a peace agreement later that year.

8 The helicopter provided armed protection for a group of seven Nepali soldiers who became separated from the rest of the UN force, but when the helicopter went back to refuel, the soldiers found themselves surrounded by more than 300 militia and were taken hostage. After 42 days of negotiations, they were finally released unharmed.

In the eastern DRC, airborne reconnaissance has located many militiamen, deserting soldiers, and stragglers prior to their being apprehended and arrested, or becoming part of the peace process through *brassage* (that is, merging into the national army). More about the surveillance capabilities and work of the Mi-35 armed helicopters in the Congo in 2008 is found in Chapter 14 of this volume.

In summary, ground and aerial surveillance have different but complementary effects. The air provides a grand view of the terrain, whereas ground forces have the ability to interact more closely with people. A combination of air and ground surveillance permits a more persistent and precise presence over larger areas. Aerial reconnaissance acts as a force multiplier. Locations that are too distant, numerous, or dangerous for UN bases are better observed by aircraft. Various types of aircraft can be considered to optimize aerial effectiveness, including cost-effectiveness.

Enter the Unmanned Aerial Vehicle

Reconnaissance (unarmed) UAVs come in many different sizes, weights, capabilities, and configurations. The payload can include many different types of sensors. Table 7.1 categorizes and characterizes the main types of UAVs that could be used in UN peacekeeping.[9]

The smaller UAVs (especially mini-UAVs) are unstable in strong winds, making it hard to get steady video imagery, but sharp still images are possible using a fast shutter speed. Further, high-resolution devices are becoming lighter and smaller. Mini-UAVs tend to run on batteries, whereas the larger ones use gasoline or jet fuel. The petroleum-powered UAVs can attain a fuel efficiency of over 200 km/litre. Larger UAVs can support heavier payloads. SAR payloads are of the order of 50–100 kg, so they are available only for the larger, tactical UAVs.

Ever-smaller and smarter UAVs are under development. The near future might offer ultra-light UAVs (eventually, possibly, nano-UAVs) that are less than 2 cm in wingspan and less than 10 g in weight, entering the world of insect mimicry.[10] Autonomous take-off and landing UAVs are readily available, as well as self-navigating UAVs using GPS waypoints. Generally these should be used only in a well-defined territory where other aircraft are not present, though collision-avoidance systems are available for UAVs, as they are for manned aircraft.

---

9   Larger UAV systems exist, for example, the US-owned Global Hawk UAVs, but they are not appropriate for the United Nations. They are not generally commercially available, their payloads are highly classified and the cost is extremely high. For example, the price of a Global Hawk aircraft, which can fly at extremely high altitudes – over 20,000 m – is US$18–$20 million.

10   For an example of lightweight sensors for UAVs, see the Optical Alchemy, website. Available at: http://www.opticalchemy.com [accessed 10 January 2011].

Table 7.1 Unmanned aerial vehicle types and typical characteristics

	Weight (kg)	Range (km)	Max. speed (km/hr)	Time Aloft	Payload (kg)	Cost ($US)*	Example functions	Example models
Micro-UAV	0.1–2	0.1–2	2–20	30 mins	0.1–1.0	500	Aerial view of buildings below	AR2 Drone, IdeaFly
Mini-UAV	2–5	4–10	20–100	2 hrs	0.5–1.3	25,000	Perimeter surveillance of UN sites and refugee camps	Desert Hawk, Dragon Eye, Raven
Sub-tactical UAV	10–20	Up to 1,000	50–150	20 hrs	2–5.5	50,000+	Tracking humanitarian convoys; patrolling border segments	Aerosonde, Luna, Scan Eagle, Silver Fox
Tactical-UAV	120–500	1,000	100–200	30 hrs	3–200	1–10 million	Long-range patrolling, large-area surveillance, monitoring from high altitude	B Hunter, Crecerelle, CL-289, Falco, Phoenix, Shadow 200, Sperwer

Note: * Typical cost per UAV and may not include the ground station with console and launcher, if required.

Source: Survey of models on the commercial market.

The smaller UAVs have the benefit of being easier to transport (for example, carried by an individual), to launch (by hand or slingshot) and to operate (for example, with joystick controls or even from cell phones). They are cheaper to operate and to purchase (starting from under US$500 per UAV), and they usually cause less damage if they crash. They can also fly slower than large ones since the stall speed generally increases with weight. On the negative side, they have limited range, endurance, and payload capacity, as illustrated in Table 7.1.

The deployment of "mixed packages", involving different categories of UAVs, allows the different advantages of each to be exploited, including cost and capacity benefits. For instance, travelling ground reconnaissance units could control mini-UAVs flying a short distance ahead, viewing imagery, over the side of the next hill, for example, while a tactical UAV is used for more distant reconnaissance.

In early 2013, the United Nations held a bidding process for a commercial UAV to overfly the skies of the eastern Congo. Building on the lessons of previous uncompleted competitions (2007 and 2009), a successful bidder, Selex ES (a branch of Finmeccanica), was announced in August and flew the first operational flight of its Falco UAV in December 2013. The UAV system attained Full Operational Capability in April 2014, marking a significant achievement for the United Nations in the aerospace domain.

Manned versus Unmanned Aircraft

Advantages of UAVs

Unmanned flying vehicles are generally smaller, lighter, and more fuel-efficient than manned aircraft.[11] The greatest benefit of UAVs – also called "remotely piloted vehicles" or "drones" – in peace operations is that there is little danger to pilots or other crew, because there are none on board! This makes it possible to fly over high-intensity conflict zones that would otherwise be considered too dangerous for aerial surveillance.

To control UAVs, remote pilots cam remain at distances of 200 km or even further using repeater stations (which may be on the ground or in other UAVs in the air). With satellite communications, remote operators can even be on the other side of the Earth. The controllers can vary the altitude, direction and speed of the aircraft, as well as the angles and zoom of the onboard camera(s). The imaging suite can include devices to capture visible light, infrared and radar signals.

---

11   From Finmechanica's specifications, the Falco UAV is 5.25 m long, with a wingspan of 7.2 m and a height of 1.8 m. It can attain a maximum speed of 216 km/hr and fly as high as 6,500 m (21,325 ft). Its range is 250 km and endurance is 8–15 hours. It can carry a payload 70 kg for a maximum takeoff weight of 420 kg. The payload can be electro-optical sensors (visible and infrared) or synthetic aperture radar. Specifications available at http://www.selex-es.com/documents/737448/3702599/body_mm07806_Falco_LQ.pdf [accessed 7 May 2014].

Autonomous (pre-programmed) UAVs exist, but this feature is less likely to be used in peacekeeping in the near future, except for take-off and landing.

For night flying, UAVs offer tremendous advantages. As mentioned earlier, the United Nations generally does not allow its planes to fly at night for fear of crashes. UN aircraft are typically not equipped with weather radars, which help spot approaching rains, stormy winds or other hazards at night. Nor does the United Nations have nighttime or combat search and rescue capabilities to react properly and quickly at night or in heavy conflict areas. With downed UAVs, recovery operations are not as time-sensitive. Consequently, UAVs do not have the same stringent night-flying rules applied to them. Given the current lacuna in night surveillance in peacekeeping operations, UAVs offer a powerful tool to enhance effectiveness and security after dark.

UAVs are generally harder to detect and shoot down than manned aircraft, given their smaller size and decreased noise. Battery-powered UAVs make hardly any noise at all; certainly nothing detectable above the din of battle. For example, at 2,000 ft above ground level, some smaller UAVs can be neither seen nor heard.[12]

If a UAV crash does occur, in daytime or at night, the costs are much less than for a plane, most importantly in terms of human life. In terms of fiscal loss, UAVs are much less expensive to purchase or replace than manned aircraft. A mini-UAV with its control system typically costs less than US$25,000; subtactical UAVs are available for US$50,000 or less. And costs are decreasing while capability is increasing each year. Requirements for licensing, clearance, and flight planning are also decreasing as the technology proliferates.

Though UAVs still need remote pilots and a crew for launch, control, and maintenance, the number of such support personnel is less than for manned aircraft. Typically, a five- to ten- member crew is needed to form a "flight" of two or three tactical UAVs – much less for mini-UAVs. UAVs also require less training. Some mini-UAVs can be flown and operated successfully with only a few minutes of training (like model aircraft).

UAVs can also be launched from many more locations than standard planes. Short runways are sufficient for most UAVs and some take off vertically. UAVs are also easier to transport: most mini-UAVs are human-portable; that is, they can be carried in a case (or even a backpack) by a single individual. Subtactical UAVs can be transported in a minivan or on top of a utility vehicle (jeep), whereas tactical UAVs usually come with their own transport vehicle. UAVs are also easier to store, maintain and repair. All these features mean that UAVs have a "smaller operational footprint" in the field, but the area coverage can be larger than for manned aircraft.

UAVs also offer benefits to observers and analysts. In manned aircraft, onboard observers can easily become fatigued. Having more space and a greater ability to rotate personnel, ground-based observers at convenient locations can study

---

12   It should be noted that a Belgian UAV was shot down by a hunter in the Congo in 2006; however, this was considered a highly improbable hit.

monitors on large screens for longer periods of time, though not unlimited, given observer fatigue. The endurance for human observers on a plane is typically four to six hours, and most midsize planes need refuelling in even less time. UAVs can fly for longer periods because they are lighter. They can be controlled by ground personnel on rotating shifts at a safe base to support longer flights – any number of personnel can observe the video feed from the UAV, not just the crew.

Most UAVs are capable of longer loiter periods than traditional planes, not only because they have greater fuel efficiency but also because they can achieve lower stall speeds, as low as 30 km/hr (16 kt) for mini-UAVs, compared with 80 km/hr (43 kt) for small manned aircraft. Of course, rotary-wing aircraft have no stall speed. This "loiter on station" capacity is particularly useful to observe a localized activity closely for extended periods of time.

Advantages of Manned Aircraft

Unlike UAVs, the use of manned observation aircraft has historical precedence in peacekeeping. The United Nations has considerable experience in manned aerial operations, but (until this volume) little of it was described or analysed. The first (and perhaps only) reconnaissance jets were used in the Congo as part of ONUC in the early 1960s. The subsequent mission in the DRC (MONUC) in the 2000s has, remarkably, less reconnaissance capacity, though the need is as great. MONUC has four Alouette helicopters with a "glass bubble" for visual observation but no recording equipment except any still or video cameras that might be carried aboard.[13] The Mi-35 helicopters have considerably more capacity: a variable field-of-view, low-light television, and a FLIR recording system, as well as a helmet-mounted sighting and display system. But, being a prized national asset (Indian and Ukraine) whose exact resolution is kept classified, the fourth-generation FLIR video imagery is not generally shared with the rest of the mission. Only freeze-and-crop frames are provided to highlight certain observations, although a live feed would be technically possible for remote viewing. The Mi-35 FLIR cameras proved very useful during combat in spotting militia and allowing the helicopter gunship to engage them with weapons systems slaved to the reconnaissance devices. More on these systems is provided in Chapter 14.

The greatest benefit of manned aircraft over UAVs is their multipurpose capability for transportation and combat, as well as observation. Soldiers can become familiar with the terrain from the air and be dropped close to their target, particularly with helicopters. Commanders can direct ground movements from helicopters, as they have done in the Congo. This dual use of manned aircraft

---

13   Given the lack of permanent observation equipment onboard, when the Lama helicopters were deployed in Kinshasa in 2006 to observe crowd movements, the television cameras from MONUC's public TV unit and from Radio Okapi were used to produce some higher-resolution imagery. Personal interview with François Grignon (former chief of Joint Mission Analysis Centre, MONUC), Toronto, Canada, 4 February 2007.

allows cost efficiencies such as carrying out reconnaissance during or after the transportation of personnel or materiel.

Manned aircraft generally can fly at higher altitudes than most commercial UAVs. Also a typical operational range of 1,000 km is greater than most UAVs can sustain, except American UAVs such as Global Hawk, which are well beyond the current means of the United Nations. Some aircraft, such as the Cold War-era U-2 spy plane (used by the United Nations for weapons inspections in Iraq), are designed to fly and photograph at very high altitudes of over 60,000 ft.

Aircraft also travel at greater speeds and offer a more commanding presence. As has been mentioned, UAVs can provide a modest "show of presence", but a jet aircraft can streak rapidly and impressively above conflict areas; UAVs could not break the sound barrier as the Mirage jets did in Ituri.

Pilots in manned aircraft also have a better feel for their aircraft than for any UAV they may fly, since they benefit from direct flight sensations (such as vibrations and engine sounds), unlike ground-based pilots. That is one of the reasons manned aircraft have a much lower crash rate than UAVs, where the pilot's safety is not at risk.

Finally, direct observation from inside aircraft has advantages over remote viewing through computer screens of UAV imagery. Onboard personnel have three-dimensional and wide-angle (panoramic) views that cannot be achieved on computer screens. Furthermore, onboard cameras and computer systems can greatly increase the capacity of the unaided human eye for closer observation and for recording.

Like the complementary ground and aerial systems, the integrated use of unmanned and manned aircraft can offer the advantages of both types. And still other aerospace platforms are also available for synergistic use.

Aerospace Platforms for Reconnaissance

Overhead imaging can also be carried out from balloons and satellites. These offer some comparable advantages to the aerial platforms already examined. For instance, satellites can travel freely in outer space, permitting them to observe virtually any area of the Earth legally, without national consent. The relative merits of each aerospace platform are presented in Table 7.2. Each is evaluated on eight basic characteristics: six beneficial ones, and two undesirable ones.

The strengths and drawbacks are easily compared in Table 7.2. Simply put, they are: the high costs of manned aircraft; the limited payloads of unmanned aircraft; and the very limited manoeuvrability of balloons and satellites, which follow given trajectories. One advantage of satellites is that they cannot be shot down, at least not by the types of weaponry found in peacekeeping areas.

For some UN purposes, aerial manoeuvrability is not always needed. For instance, tethered balloons can be useful for observing important areas, corridors or choke points on a near-permanent basis. Cables keep the observation

Table 7.2   Comparing different types of aerospace surveillance

	Range	Endurance	Speed	Altitude	Manoeuvrability	Payload Capacity	Cost (US$)	Vulnerabilities
Fixed-wing aircraft (manned)	HIGH (up to 10,000 km)	Medium (max. 15 hr)	HIGH	HIGH (up to 20,000 m)	HIGH (but cannot fly as slowly)	HIGH (up to 250,000 kg)	HIGH (for purchase, maintenance, fuel and personnel)	Possible fatalities, needs airfields for takeoff and landing
Rotary aircraft (manned helicopter)	Medium (300 km)	Low (typically 3 hr)	Medium (up to 350 km/ hr)	Medium to HIGH (up to 10,000 m)	VERY HIGH (easy turns and stationary capacity)	Medium (up to 10,000 kg)	HIGH (for purchase, maintenance, fuel and personnel, incl. onboard pilots)	Possible fatalities
Unmanned aerial vehicles (UAV)*	Low to HIGH (from 1 km to 1,000 km)	Low to HIGH (from 15 min to 20 hr)	Medium (from 40 km/hr to 300 km/hr)	Low to Medium (from 50 m to 5,000 m)	HIGH	Low (from 1 kg to 150 kg)	MEDIUM (lower than manned aircraft, though dependent on type of UAV)	Can be shot down; weather-dependent (esp. wind conditions)
Balloons (free or tethered)	Low (up to 100 km a day)	HIGH (10 or more days)	Stationary or very low	Medium (up to 5,000 m)	Very Low (wind-dependent)	Low to Medium (up to 500 kg)	Low	Easily targeted
Satellite	VERY HIGH (but has fixed trajectory)	VERY HIGH (years, but revisit time can be days)	VERY HIGH (25,000 km/hr)	VERY HIGH (100 to 1,000 km)	Low (only certain types)	Medium (up to 5,000 kg)	HIGH (expensive to build and launch, imagery can be purchased cheaply)**	Limited availability at specific time and place

Notes: *   Subtactical UAVs are considered.

    **   A high-resolution imaging satellite can cost over US$1 billion to build and US$50 million to launch. Satellites of much lower cost, such as micro-satellites, are now coming into the market.

platforms in place and allow for the conveyance of electrical power and data signals. These large balloons can also serve as visible markers of borders or ceasefire lines, as navigation aids, as communications relays and as radio-station transmitters. Of course, these static objects might also be favourite targets for frustrated combatants. If shot at, however, they come to the ground smoothly because of their separate compartments. This allows the equipment to be saved and the platform to be repaired and reflown quickly and cheaply. Some aerostats are rapidly deployable (or redeployable) in as little as 10 minutes from the back of a pickup truck.

Radar-equipped aerostat (balloon) systems are currently employed on several international borders (for example, on the United States–Mexico border) as part of national interdiction programmes for drugs and human trafficking. Held at a typical altitude of 500 m, the view can extend for several kilometres. In Afghanistan, the 14-m long Rapid Aerostat Initial Deployment aerostats are tasked with area surveillance and force protection against small arms, mortar and rocket attacks. They can stay aloft for weeks. The Canadian Forces deployed a Persistent Threat Detection System in southern Afghanistan. When shots were fired, the acoustic sensors on the aerostat would automatically trigger camera movement toward the area of fire. This was of immense help before and during the dispatch of Quick Reaction Forces to the area.

In addition to working with ground systems, aerial systems can be multilayered and hybrid to complement each other. Although aerospace reconnaissance provides unique advantages over ground reconnaissance, the best option is an integrated system to detect threats and explore opportunities for peace and stability. Multiple layered information sources are needed to corroborate and probe sensitive and uncertain information in dangerous environments found in many peacekeeping operations.

Comparison with the North Atlantic Treaty Organization

To get a sense of the relative "poverty" of aerial UN operations, one need only make a comparison with the North Atlantic Treaty Organization (NATO). When that military alliance deployed peace operations, as in Bosnia or Kosovo in the 1990s, or counter-insurgency operations, as in Afghanistan in the 2000s, it does so with a plethora of reconnaissance aircraft. For instance, in its first peace operation, in Bosnia (1995–2004), NATO took a proactive approach in an attempt to achieve "information dominance", to show the former warring parties that the NATO mission could watch closely what they were doing. The aerial surveillance component employed an impressive array of aircraft. Apache and Kiowa helicopters provided imagery from video cameras that relayed images automatically to command posts within 90 seconds, a feature not possible with the UN's most robust platform, the Mi-35 helicopters. The NATO helicopters also had thermal infrared sensors capable of monitoring troop movements several kilometres away. Aerial surveillance was also achieved with high-altitude U2 aircraft, P-3 maritime patrol aircraft and the RC135 reconnaissance aircraft. Perhaps most significantly, the sophisticated Joint Surveillance and Target Attack Radar System aircraft provided high-resolution imagery of the ground, including synthetic aperture radar images both day and night and in virtually all weather conditions. SAR, in the Doppler mode, was especially effective at detecting moving targets.

UAVs have gathered signals intelligence and provided imagery in near real time in NATO operations. For instance, a Predator UAV was able to display the faces of people opposing US entry into the town of Han Pijesak. Ground units deployed their own shorter-range UAVs such as the US Army's Pioneer UAV. Remote Video Terminals allowed soldiers across the mission area not only to view UAV imagery but also to control the onboard camera angle and zoom in order to "zero in" on desired objects and people.

Complete awareness of the airspace was achieved with Airborne Warning and Control System (AWACS) aircraft. NATO's E-3A Sentry is the "world's only integrated, multinational flying unit, providing rapid deployability, airborne surveillance, command, control and communication for NATO operations".[14] All flying objects within a radius of over 300 km could be tracked: a single AWACS aircraft could monitor the entire Bosnian airspace.

NATO had an even greater aerospace reconnaissance capability in Afghanistan. Though not necessarily achieving success in the 15-year operation, NATO made good use of aircraft to give a much better operational picture of the situation on the ground; International Security Assistance Force commanders would not want to operate without the observation provided from the air.

While the United Nations need not take such a sophisticated and costly approach to aerial reconnaissance, its record of technology leverage is dismal, though improving. In 2013 it finally deployed UAVs to the field (DRC mission), following proposals first made in 2005. The United Nations only achieved real-time image transmission from helicopter cameras in 2010, when the Haiti mission achieved this goal. The observation helicopters used in the Congo (Lamas) did not even have gyrostabilized cameras; recordings were simply made by hand-held cameras brought on board by crew. The United Nations has not deployed AWACS aircraft to monitor vast areas of airspace. Nor has it used aerostats (tethered balloons), which could effectively monitor conditions around UN bases, refugee camps, border areas and other trouble spots. Taken together, this shows how aerially ill-equipped the United Nations has been while it tries to succeed in the enormous task of keeping the peace. There is much room for improvement in the air above the ground-based peacekeepers!

---

14   North Atlantic Treaty Organization. "NATO Airborne Early Warning and Control Force, E-3A Component". Available at: http://www.e3a.nato.int [accessed 18 January 2011].

Acknowledgments

The author thanks United Nations University Press for permission to use material from Chapter 5 of A. Walter Dorn, *Keeping Watch: Monitoring, Technology, and Innovation in UN Peace Operations* (Tokyo: United Nations University Press, 2011).

Chapter 8

UN Observer Group in Lebanon:
Aerial Surveillance During a Civil War, 1958

A. Walter Dorn

The perpetually unstable Middle East was especially chaotic and conflict-ridden in the 1950s. The presence of Gamal Abdel Nasser, president of the union of Egypt and Syria in the short-lived United Arab Republic (UAR), assured that armed force was used both overtly and covertly in the region. To deal with allegations that Nasser was fomenting rebellion in Lebanon – Syria's small western neighbour – the United Nations created a peacekeeping operation in that country.[1]

During its relatively brief six-month existence the United Nations Observer Group in Lebanon (UNOGIL) made significant efforts to deploy aerial assets. Its successes and failures in observation provide some valuable lessons, especially as the United Nations still tries to break the "night barrier" and peer into the world of illicit arms transfers conducted under cover of darkness. Since sanctions-monitoring is frequently mandated by the UN Security Council in modern multidimensional missions, the early mission's aerial monitoring experience is especially worth exploring. Reports and cables obtained from UN archives provide valuable excerpts and insights for a case study.

The trigger for the 1958 Lebanese civil war was the announcement made by Lebanese President Camille Chamoun, a Maronite Christian in Muslim-majority Lebanon, that he intended to amend Lebanon's constitution to permit himself re-election for a second term. Disturbances quickly erupted, spreading to assume the proportions of a rebellion. Chamoun accused the UAR of fomenting this rebellion by supplying large quantities of arms to subversive forces, infiltrating armed personnel from Syria into Lebanon, and conducting a violent press and radio campaign against the Lebanese government. On 22 May 1958, Chamoun's government brought the situation to the attention of the UN Security Council "as a threat to international peace and security". To some UN members, it was a case of

---

1 The term "peacekeeping" was not yet in widespread or official use by the United Nations. The previous missions had been called "observer", "observation", or "supervision operations/organizations", with the exception of the United Nations Emergency Force (UNEF), a "peace and police force" in the words of the proposer, Lester B. Pearson of Canada. UNEF had been created in response to the 1956 Suez Crisis, when Israel, France, and the UK had invaded Nasser's Egypt to seize and secure the Suez Canal. UNEF was established by the UN General Assembly with Nasser's permission to allow those nations to withdraw from Egyptian territory. The term peacekeeping became common only at the turn of the decade.

alarmism from a weak and desperate government. To others, including the United States, it reflected a genuine threat emanating from Nasser and the militarist pan-Arab republican movement.

The Early Mission

Pursuant to Lebanon's request, UNOGIL was created on 11 June 1958 by Security Council Resolution 128 "to ensure that there is no illegal infiltration of personnel or supply of arms or other *matériel* across the Lebanese borders".[2] Despite the ambitious mandate, the mission was strictly limited to an observation role (as opposed to enforcement) to determine whether the alleged infiltration was, in fact, taking place from UAR into Lebanon – and hence to deter such infiltration.

Already on 12 June, the first UN observers arrived, having transferred from another peacekeeping operation (the United Nations Truce Supervision Organization) but they found their freedom of movement was very restricted. UN Secretary-General Dag Hammarskjöld himself flew to Beirut and on 19 June he chaired a meeting exploring the methods of observation to be employed by UNOGIL. To supplement military observers in observation posts and jeeps, aerial reconnaissance was to be conducted by light planes and helicopters, the former being equipped for aerial photography.[3] The observers were to be headed by Major General Odd Bull of the Royal Norwegian Air Force. Hammarskjöld's home country, Sweden, was to play a major role in UNOGIL's aviation service.[4]

The leaders of the mission understood that there were many sensitivities and potential problems with aerial reconnaissance. UNOGIL identified one of them in a cable to New York headquarters on 22 June:

> There are, of course, psychological problems in using aerial observation. This kind of activity must be carried out in such a way as to be and appear to be concerned with infiltration at frontier and not military movement within Lebanon as such. Misunderstanding by insurgents as to real purpose of aerial reconnaissance could create additional obstacle for our penetration [of] insurgent areas.[5]

The mission's efforts to determine the extent of UAR material support to rebels in Lebanon was immediately hampered by a number of practical factors, both ground-

---

2   United Nations, Security Council Resolution 128 (1958) of 11 June 1958. Available at: http://undocs.org/S/RES/128(1958) [accessed 7 May 2014].

3   United Nations. *The Blue Helmets: A Review of United Nations Peacekeeping*, 2nd edition (New York: United Nations Department of Public Information, 1990), 177.

4   Forsgren, J. "Keeping the Peace: The United Nations in Lebanon, 1958", *The Aviation Historian* 4 (2013), 68–81.

5   Cable from UNOGIL's secretariat head David Blickenstaff (Beirut) to UN Under-Secretary-General Ralph Bunche (New York), 22 June 1958, UNOGIL 90. UN Archives.

and air-based. UNOGIL's first report to the Security Council, dated 3 July 1958, pointed to difficulties in gaining access to the eastern and northern frontiers held by opposition forces, who (at least initially) resisted a UN presence. These areas could only be patrolled by aircraft. At this point, two UNOGIL helicopters were carrying out aerial reconnaissance, four light planes had just arrived and another four were expected soon with an aerial photography capability.[6] The United Nations had asked Sweden, if possible not to send Harvards since these were also in the Lebanese Air Force, but the Swedish Air Force had no other suitable aircraft to provide.[7] The United States provided the majority of other aircraft, though they were flown and maintained by personnel from Sweden, Norway, Denmark, Italy, Burma, Canada, and a few other states. Apparently, the United States charged the United Nations only US$17.25 per flying hour for loan of the aircraft.[8]

The initial mission was for day and night flights over the border areas, observing with binoculars and taking photos with handheld Hasselblad cameras. It was soon clear that the group had too little personnel and too little equipment to carry out its intended duties.

Based on the target areas that could be monitored, the mission could provide no substantiated or conclusive evidence of major infiltration at that point. The Lebanese government immediately criticized these "inconclusive, misleading or unwarranted"[9] conclusions, particularly in view of the inability of the observers to monitor the entire frontier. The Lebanese letter complained that:

> with a view to patrolling the border areas, [aerial reconnaissance] has not yet really begun so far as this Report is concerned ... Thus, whatever information can be gathered by this device has not yet been gathered. But even if this aerial reconnaissance were fully operative, it would still have two limitations: it cannot

---

6   United Nations Security Council, UN Doc. S/4040, 3 July 1958, "First Report of the United Nations Observation Group in Lebanon", 6. The two helicopters were Bell H-13 E helicopters provided by the United States, transported on a US cargo plane to Beirut on 22 June. US technicians instructed Norwegian pilots on the operation of the aircraft. The four Harvard (Sk 16) reconnaissance planes were provided by Sweden and flown from that country. The Cessna L–19 reconnaissance planes were loaned by the United States. Later the United States provided more Cessnas to replace and augment the aircraft already in the fleet. The UNOGIL aircraft were provided with K-24 cameras by the United States. The deployment of night cameras is denied in the source. Source: Wainhouse, D.W. "The United Nations Observer Group in Lebanon (UNOGIL, 1958)", in *International Peacekeeping at the Crossroads: National Support – Experience and Prospects* (Baltimore, MD: Johns Hopkins University Press, 1973), 116–17.

7   Blickenstaff early on noted (25 June 1958, UNOGIL 110) that since Harvards were used by Lebanese Air Force in military operations "this creates psychological and perhaps security problems if we use the same type of plane. It is important that our operations have outward appearance of peaceful character corresponding to their true nature".

8   Wainhouse, "The United Nations Observer Group in Lebanon", 122.

9   United Nations, *The Blue Helmets*, 178.

spot out all infiltration during the day, and it can hardly spot out anything during the night.[10]

Despite its problems, the mission was having a salutary effect. US intelligence agreed with UNOGIL's assessment that infiltration from Syria was not as great as Lebanese President Chamoun was claiming. Furthermore, US Secretary of State John Foster Dulles told Lebanese Foreign Minister Charles Malik on 30 June that "[t]he activities of the UN and Hammarskjold have brought about a large cessation of infiltration".[11]

Still, UNOGIL sought to get a clearer picture, despite the problems of ground accessibility. On 13 July 1958, Major General Odd Bull cabled the Secretary-General that:

> efforts are hampered by persistent refusals to let Observers enter Northern districts at night in a normal manner ... even day patrols in area had to be severely curtailed due to opposition's resistance.[12]

The Group decided it was too dangerous to send out regular ground patrols at night to check possible infiltration routes. Even when an arrangement was concluded with local commanders, a lighted UN jeep "came under heavy fire and was hit several times".[13]

Aircraft were hit by bullets on many occasions, twice with non-fatal injuries to Swedish personnel aboard.[14] Already on the second day, a bullet badly damaged the engine of a Swedish Harvard. Due to this risk of rifle fire from the ground, the pilots were later ordered not to fly below 600 ft.

Despite the hazard, General Bull outlined the results of aerial reconnaissance that covered nearly all hours of darkness from 6–12 July and involved 21 UNOGIL flights for 47 hours of flying time in total. The aerial monitors examined motorized traffic along three roads, all of which led from Syria into Lebanon across the latter's northern border. General Bull provided valuable insights, but without full proof of infiltration:

---

10   United Nations Security Council S/4043, 8 July 1958, "Letter dated 8 July 1958 from the Permanent Representative of Lebanon Addressed to the Secretary-General", 6.

11   After UNOGIL started reporting, US Secretary of State John Foster Dulles told Lebanese Foreign Minister Charles Malik on 30 June that US intelligence shared UNOGIL's assessment that infiltration from Syria was not as great as Lebanese President Chamoun was claiming. Furthermore, he pointed to a "considerable reduction or termination of infiltrations across the border. ... The activities of the UN and Hammarskjold have brought about a large cessation of infiltration" (Dulles–Malik meeting, 30 June 1958, *Foreign Relations of the United States*, XI ("Lebanon and Jordan"), document 111), 185–90.

12   Incoming Code Cable to Secretary-General from Bull, Beirut, 13 July 1958, 2115 hours, Number: UNOGIL 359.

13   UN Document S/4069, 30 July 1958, 114.

14   Forsgren, "Keeping the Peace", especially 79–80.

A considerable amount of south going night traffic has been observed every night. They creep along at slow speed, as if vehicles were heavily loaded ... The first night more than 50 vehicles were observed here [on the Braghite–-Halba Road] and on subsequent nights aircraft discovered with certainty convoys of at least 20, 10, 25, and 25 vehicles respectively. All this traffic can only have come from Syria. It seems to branch off from the Homs–Lattakia road, which is located inside Syria ...

The traffic along same three roads has proved to be very much heavier at night than during daytime, and large majority of vehicles observed were moving south. Only some very few have been seen going back and north at night ...

In spite of the almost permanent aerial observation of area during hours of darkness, it cannot be assumed that all existing traffic has been seen. The reason for this is that convoys move with great care and precaution. They apparently switch off lights before entering Lebanon, and turn them on – if at all – well inside the border. Unlighted vehicles cannot be spotted by aircraft at night from heights of 1,000 to 3,000 meters at which they usually are patrolling. Furthermore convoys are now employing an alarm system with flashlights on hilltops, to warn vehicles to switch off lights when aircraft are approaching. Planes have also been under light machine gun fire in this rebel-held territory at least two times.[15]

The spotting of illicit convoys was made more difficult by the UN's own sense of duty to inform Syria and other neighbouring countries when the flights were made in proximity to their borders.[16] Notwithstanding the challenges, a subsequent cable concluded that aerial reconnaissance is "a most valuable adjunct to the group's ground observation".[17]

US Invasion

Over time, accessibility improved. UNOGIL's Interim Report of 15 July stated that the mission had obtained full freedom of access to all sections of the Lebanese

---

15   Incoming Code Cable to Secretary-General from Bull, Beirut, 13 July 1958, 2145 and 2242 hours, Number UNOGIL 358 and 359, 1–2.

16   Notification of neighbouring states about flights near their borders was made "as a matter of prudence and courtesy". Higgins, R. "Part 3: United Nations Observer Group in Lebanon (UNOGIL), 1958", in *United Nations Peacekeeping, 1946–1967: Documents and Commentary, Vol. 1: The Middle East*, 565 (Oxford: Issued under the auspices of the Royal Institute of International Affairs by Oxford University Press, 1969).

17   Code Cable, UNOGIL 401, 2, UN Archives.

frontier, a breakthrough in relations with the rebels.[18] UNOGIL proposed to expand the cadre of unarmed observers to 200, along with additional aircraft and crews.[19]

The US government was not pleased, however, that UNOGIL could not offer conclusive proof of the UARs infiltration of men and materiel, especially weapons. It complained that UNOGIL did not have sufficient night coverage. The US ambassador to the United Nations and the Central Intelligence Agency directly criticized the mission.

The geostrategic environment changed drastically in mid-July. The 14 July Revolution in Iraq overthrew that country's Hashemite monarchy. The United States saw again the hidden hand of Nasser, as well as that of Soviet communism more generally. President Chamoun called for a US intervention to save his government from a similar fate. President Dwight Eisenhower ordered 14,000 US marines into Lebanon for the "preservation of Lebanon's territorial integrity and independence". Most of the forces were concentrated in and near the capital, Beirut. In his message to the US Congress on July 15, Eisenhower stated:

> It was our belief that the efforts of the Secretary-General and of the United Nations observers were helpful in reducing further aid in terms of personnel and military equipment from across the frontiers of Lebanon. There was a basis for hope that the situation might be moving toward a peaceful solution, consonant with the continuing integrity of Lebanon, and that the aspect of indirect aggression from without was being brought under control.[20]

For the United States, the situation following the Iraqi coup now meant that the measures in Lebanon "so far taken by the United Nations Security Council are not sufficient". The landing of US marines was obviously resented by the rebel forces; however, they could not militarily challenge such a strong force.

The US invasion caused problems for UNOGIL. Rebels feared that UN airfields would be used by US "invaders". Sections of the airfield at Akkar Plain in northern Lebanon were blown up and mined to render it unusable.[21] It would take the United Nations over a week to re-establish the air station and even longer to rebuild the trust of locals.

---

18   United Nations, *The Blue Helmets*, 179.

19   Ibid.

20   Dwight D. Eisenhower, "Special Message to the Congress on the Sending of United States Forces to Lebanon", 15 July 1958. Available at: http://www.presidency.ucsb.edu/ws/index.php?pid=11132&st=&st1= [accessed 7 May 2014].

21   Curtis, G.L. "The United Nations Observation Group in Lebanon", *International Organization* 18(4) (Autumn 1964), 757.

Sustaining a System

UNOGIL's second report to the Security Council, dated 30 July 1958, shows that UNOGIL had weathered the storm. The mission stayed impartial, not associating directly with the intervening US military forces. It also could not confirm the Lebanese government allegations of infiltrations, even urgent ones said to be occurring at the time. "Air patrols were dispatched as soon as possible, but when they arrived on the scene they found nothing to observe". Suspicious night convoys seemed to take measures to avoid detection by UN aircraft.[22] The report proposed a bold plan for a constant aerial watch to cover Lebanon's eastern border with Syria. It also sought occasional air patrols along the Mediterranean to guard against possible infiltration from the sea.[23]

Due to pressure from the United States and negotiations with the United Nations, President Chamoun agreed to new elections in which he stated he would not run. Just prior to the election of the new President, General Fuad Chehab, UNOGIL reported on 14 August a noticeable reduction of tension and clashes throughout the country, including between government and opposition forces. After two months, the mission was moving into full swing and air operations were expanding: flight personnel increased from 20 to 24; a further eight L-19 Cessna ("Bird Dog") observation aircraft arrived; and six additional Bell OH-47 observation helicopters were expected soon. With the new aircraft, UNOGIL envisioned air patrols on a 24-hour basis.[24] The new report added that UN aircraft had frequently been fired upon and were hit on four occasions, fortunately without injury to the crew.[25] Additionally:

> coordination between air and ground observation has been further intensified and improved. Air patrols have been closely checking the results of ground observation and vice-versa, and direct radio contact between air patrols and stations has greatly increased the effectiveness of the combined operations.[26]

---

22   United Nations Security Council, "Second Report of the United Nations Observation Group in Lebanon", UN Doc. S/4069, 30 July 1958, 7–8. Earlier that month UNOGIL requested the "dispatch of an aircraft with night photographic capability of Douglas RB26 type. If possible, plane should be equipped for infra-red photography". Code Cable Blickenstaff to Bunch, 11 July 1958, UN Archives.

23   Ibid, 19.

24   United Nations Security Council, "Third Report of the United Nations Observation Group in Lebanon", UN Doc. S/4085, 14 August 1958, 4.

25   After the mission, the military Commander, Major General Odd Bull, wrote in his memoirs: "planes had been fired on 59 times and nine hits registered; in two of these cases the pilot was injured, though not seriously". Bull, O. "Lebanese Overture", in *War and Peace in the Middle East: The Experiences and Views of a UN Observer* (London: Leo Cooper, 1976), 19.

26   Ibid.

Finally, there was no further evidence of the flashing (signal) lights mentioned in the second report or of trucks dimming or extinguishing their lights on the approach of aircraft.[27]

In its fourth report, UNOGIL stated that its air personnel had further increased from 24 to 73, of whom 37 were maintenance personnel. According to Everstål, the strength increased to nearly 100 at the peak. At the same time, the whole setup and coordination between air and ground became much more efficient:

> The pilots were now assisted by special observers. The crews were given much better intelligence before each mission and had the opportunity to themselves give more detailed reports. The reports could be collated and edited. It became possible to keep aircraft in the air around the clock. A special operations center was always manned and was in radio contact with the airborne aircraft and with the radio equipped jeeps of the ground observers, who were now present in all parts of the country. A radio direction finder in the aircraft made it possible for the pilots to locate a particular jeep, whenever needed.[28]

The number of aircraft in use was 12 Cessnas, with six additional aircraft with night photographic equipment planned. The force's original complement of four Harvards and two helicopters was kept in reserve but these aircraft were soon phased out.[29] Apart from the political issues of resembling Lebanese aircraft, the Harvards were simply not very suited to observation flights since they were low-wing.

These resources permitted a continuous 24-hour aerial watch over the entire area with cooperation between air and ground affected by planning and radio communications. It was possible for stations and ground patrols to contact aircraft in flight in their vicinity, and thus to direct each other in their search for information.[30] Finally, the number of fixed-wing and heliborne sorties was tripling or quadrupling over this time.[31] This greater frequency of patrols on a continuous basis enabled UNOGIL to state with greater confidence that there was little traffic near the frontier which had not been reported to it by its previous observation. A Swedish analysis showed that:

> The pilots soon learned to recognize the villages and the activities in them. What were initially perhaps reported as recurring supply caravans, which might be transporting weapons across the borders, they soon learned to report as daily water collection caravans from some village to a well 10 to 20 km distant!

---

27   Ibid, 8.

28   Everstål, S.E. "Svenskt flyg i FN-tjänst", *Ett år i luften 1959–1960* (Malmö, Allhelms Förlag 1959), 163.

29   United Nations Security Council S/4100, 29 September 1958, "Fourth Report of the United Nations Observation Group in Lebanon", 4.

30   Ibid, 5.

31   Ibid, 6.

They also learned to recognize different vehicles and could therefore easily establish when some village was visited by strangers. With the aid of binoculars, it was even possible to recognize the appearance of some of the people on the ground.

Roads and caravan trails in the mountains became so familiar that it was possible to follow them from the air even in the dark, and establish if there was any traffic on them.[32]

It seems that Nasser had decided to end his campaign against the Lebanese government and abide by the Security Council resolution. The UN peacekeeping mission with its aerial observation capacity seemed to have served a deterrent after all, though the presence of US troops near Beirut had likely made a more forceful impact. Later the mission helped facilitate the withdrawal of US forces from Lebanon by providing airlift in October. It also assisted the withdrawal from Jordan of British forces, which had also intervened in that Middle East country during the July turmoil.

UNOGIL was beefed up in preparedness for possible unrest once the US troops left 25 October but things remained quiet. UNOGIL could therefore be wound down and was officially terminated 26 November. Its total cost was only US$3.7 million.

Lessons and Conclusion

There are many useful lessons, both positive and negative, from the UNOGIL aviation experience that are worth appreciating and preserving for modern peacekeeping operations.

The UNOGIL mission reinforced an important right, pioneered two years earlier with UNEF, that was to become key in future peace operations: freedom of movement for UN personnel within their area of operation "as necessary for successful completion of the task". This included the "right of over-flight over the territory of the host country".[33]

Overflights were considered a necessary part of the toolkit of the operation. UNOGIL was from the start strongly reliant on observation aircraft. Before ground observers were deployed, aircraft could reconnoitre the situation, particularly in areas hard to reach. UNOGIL acquired a fleet of twelve reconnaissance planes

---

32   Everstål, "Svenskt flyg i FN-tjänst", 160.

33   United Nations Secretary-General. "Summary Study of the Experience Derived from the Establishment and Operation of the [UN Emergency] Force: Report of the Secretary-General", in "Report of the Secretary-General to the United Nations General Assembly", UN Doc. A/3943 of 9 October 1958. Available via the United Nations Official Document System at: http://undocs.org/A/3943 [accessed 16 May 2013].

Table 8.1   Military personnel in the UN Observer Group in Lebanon

	26 June (S/4038)	3 July (S/4040)	17 July (S/4052)	14 Aug. (S/4085)	29 Sept. (S/4100)	17 Nov. (S/4113)
Ground observers	94	(no info)	113	166	214	501
Air observers	—	—	20	24	73	90
Manned outstations	6	8	150	22	34	49

Source: UN Security Council documents (numbers provided).

and six observation helicopters.[34] This complemented fixed observation posts, checkpoints and ground patrols by jeep, foot, horse, and even mule. Of the 591 military personnel in the mission at one point, 90 individuals, or 15 percent of total personnel, were part of the air section.[35] UNOGIL's personnel strength, for ground and air, grew as the mission went from initial to final operating capability, peaking in October, as shown in Table 8.1 above.

Aerial missions also increased from 160 sorties and 360 flying hours in July to 305 sorties and 767 flying hours in October. A typical flight lasted two hours. In total, the mission chalked up 2,850 operational flying hours.[36]

The air component proved to be a "valuable adjunct" to the ground mission. Air and ground observers were able to synergize though direct, real-time communication links that were established after the first month. Early on, the mission was able to raise suspicions about cross-border road traffic observed from above, but without ground units or stations to check more closely, those suspicions could not be confirmed. Even so, the early aerial reconnaissance proved useful, being quickly implemented before the ground observation posts were established, thanks to the loan of US helicopters. The aircrews flying over new territory did not need to find local accommodation, meet with local leaders, establish supply routes and arrange for logistics in the area, as ground observers would have to. Also the mountainous terrain typical of much of Lebanon meant that ground travel was difficult and that observation posts would have a limited view.

To back up the information gained by air and ground observers, including air photo imagery, it was necessary to have an interpretation/intelligence centre. The mission secretariat cabled Under-Secretary-General Ralph Bunche in New York on 23 July to say that "Intelligence, which here means collation and evaluation, is the weakest point in present military establishment".[37] At the suggestion of General

---

34   Bull, "Lebanese Overture", 17.

35   Wainhouse, D.W. "The 1958 Middle East Crisis: United Nations Observer Group in Lebanon", in *International Peace Observation: A History and Forecast* (Baltimore, MD: Johns Hopkins University Press, 1966), 377.

36   Bull, "Lebanese Overture", 19.

37   United Nations, Code Cable from Blickenstaff to Bunche, UNOGIL 460, 23 July 1958, UN Archives.

Bull, Lieutenant-Colonel Bjorn Egge of Norway was assigned to UNOGIL as an "Intelligence Officer" to set up the system (later to do the same for the mission in the Congo). The Beirut headquarters soon developed an "Evaluation Branch", so named to avoid the word "intelligence" but which was nevertheless called "G2" in regular military fashion. It was assigned the task of collection, collation, evaluation, and dissemination of information from all sources, ground and air, mission, and non-mission.

During its half-year existence, the UNOGIL mission proved to be an important and impartial observer to the Lebanese conflict, helping to sort out deadly claims and counter-claims. The mission was able to throw doubt on the extravagant allegations made by the Lebanese government of massive foreign (Nasserite) importation of men, arms and materiel, as well as to question the absolute denials made by the opposing forces. The United States also had to readjust its view on infiltration in the region after UNOGIL started reporting. Furthermore, the mission probably caused a reduction in arms/material importation as UNOGIL became more capable of detecting the illicit movements.

In the UNOGIL detection effort, air power proved essential to detecting cross-border convoys and keeping watch along the 300-km border with Syria. As the United Nations continues to be involved in multidimensional conflict and ceasefire monitoring in the twenty-first century in this region of the world, it would be wise to take note of the aerial experiences from this important use of air power in the Lebanese mission of 1958.

Acknowledgments

The author thanks Ryan W. Cross, Leif Hellström, and Robert Pauk for their inputs and feedback during the drafting of this chapter.

Chapter 9

Unmanned Aerial Vehicles Supporting UN Operations: A Commercial Service Model

David Neil

Situational awareness is fundamental to the success of any military operation. In the twenty-first century, unmanned aerial vehicles (UAVs) have proven to be extremely valuable assets in this regard. The ability of UAVs to provide commanders at all levels with persistent, day/night, high-resolution imagery has made them a staple of modern Western forces engaged in contemporary military operations.

A measure of the growing importance of UAVs to militaries around the world was developed for the European Commission by global growth consultants Frost & Sullivan. The company determined "that between 2004 and 2008, the number of UAVs deployed globally on operations has increased from around 1,000 to 5,000 systems";[1] a fivefold increase in a period of only four years.

US forces have placed greatly increased reliance on UAVs to support operations. UAV hours flown by US armed services increased exponentially between 1996 and 2006. When overlaid against US military campaigns, one sees a direct correlation to increased UAV employment and the rapid rise in the operational tempo engendered by the launch of Operation Enduring Freedom in Afghanistan in 2001 and Operation Iraqi Freedom in Iraq in 2003. In terms of flight hours, the United States broke 50,000 hrs in 2004 for the first time (not including man-portable unmanned aerial systems) and by 2008 had exceeded 350,000 hrs.

The Canadian experience with UAVs exhibits a similar pattern. The first operational use of a UAV by the Canadian Forces (CF) was during Operation Grizzly with a leased I-Gnat flown by General Atomics Aeronautical Systems, Inc., the UAV manufacturer. The I-Gnat provided part of the security umbrella established to protect international heads of state during the two-day G8 summit in Kananaskis, Alberta (a region near the Canadian Rocky Mountains), in June 2002.[2] A fixed wing UAV with a pusher propeller, the I-Gnat can be equipped

---

1 Frost & Sullivan and European Commission Enterprise and Industry Directorate General. *Study Analysing the Current Activities in the Field of UAV*, ENTR/2007/065 (2008), 5.

2 Van Bavel, G. *A Survey of Experimental UAV Squadrons in Exercise Robust Ram and Operation Grizzly*, Operational Research Division–Directorate of Operational

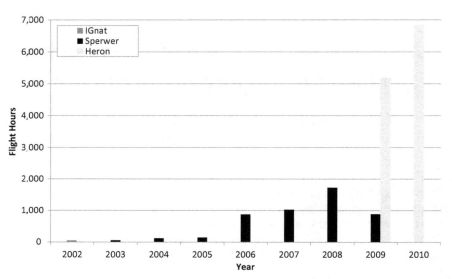

Figure 9.1   Canadian Forces unmanned aerial vehicle yearly operational flight hours

Source: Sperwer data from Canadian Department of National Defence Flying Hours Report generated by the Directorate of Flight Safety; Heron data from MacDonald, Dettwiler and Associates Ltd. flying records. Does not include small UAVs, for example the Scan Eagle that Canada flew. Heron UAV Operations ceased on 7 July 2011.

with electro-optic (EO), infrared (IR) and electronic warfare (EW) sensors and radar. It is designed to take off and land conventionally from a hard surface.[3] The I-Gnat had been leased for a Canadian Forces Experimentation Centre trial during Exercise Robust Ram at Canadian Forces Base Suffield prior to being deployed to Kananaskis during the summit.

The remainder of Canadian operational UAV employment has all been in Afghanistan. Operations began with the Sagem Sperwer and then progressed to the Israel Aerospace Industries (IAI) Heron. CF UAV operational flight hours from 2002 to 2011 are depicted in Figure 9.1. In addition to the United States and Canada, all major military powers conducting operations in Afghanistan, including Australia, France, Germany, Italy, and the United Kingdom, employed UAVs.

Research (Joint), 2003/05, Defence Research and Development Canada, 3. Available at: http://cradpdf.drdc-rddc.gc.ca/PDFS/unc10/p520035.pdf [accessed 9 September 2012].

    3   Unmanned.co.uk. *I-GNAT – Unmanned Vehicles (UAV) Specifications & Data Sheet*. Available at: http://www.unmanned.co.uk/autonomous-unmanned-vehicles/uav-data-specifications-fact-sheets/I-Gnat-specifications/ [accessed 9 September 2012].

Contribution to Military Operations

UAVs come in all shapes and sizes and have been optimized to perform numerous functions including, but not limited to: environmental monitoring, reconnaissance and surveillance, communications relay, cargo delivery, and weapon delivery. Arguably, however, it is their ability to contribute to the situational awareness of commanders at all levels and national and international leadership that represents their greatest benefit so far.

The importance of situational awareness to any military operation cannot be overstated. According to the Canadian Forces Aerospace Doctrine Manual:

> Decision superiority is the competitive advantage enabled by ongoing situational awareness … In essence, Sense is about providing a perception of the "state of the world" to a commander in order to enable him to make decisions and to optimize the other functions. Sense ultimately provides commanders the knowledge necessary to direct their forces to achieve the most appropriate effect on the operational environment.[4]

The CF Aerospace Doctrine Manual identifies "Sense" as one of the primary functions of the Royal Canadian Air Force (RCAF). It is clear that a commander's ability to effectively direct forces is highly influenced and instructed by his or her knowledge of the current state, which will seldom be static. Removing uncertainty promotes the ability to make optimal decisions. While UAVs cannot enable perfect awareness or provide enemy intent, they can significantly enhance a commander's understanding of the current state and provide him or her with a sense of how effectively their direction is shaping the battle space or the environment.

Situational awareness is not only applicable to dealing with armed adversaries. It is an essential and fundamental element in the success of all operations undertaken by military forces. Whether the mission is search and rescue, peacekeeping, disaster relief, or humanitarian assistance, success will be highly unlikely without adequate situational awareness. Like most militaries, the majority of missions undertaken by the CF do not involve combat and the adversary over which they need to achieve decision superiority may be any combination of factors, including: the environment; the terrain; a lack of infrastructure or the inability of a state or region to provide the necessary organization or coordination of relief efforts; or even conflict itself.

---

4   Chief of the Air Staff. *Canadian Forces Aerospace Doctrine*. 2nd edition (B-GA-400-000, Winnipeg/Trenton: Air Force Doctrine and Training Division / Canadian Forces Aerospace Warfare Centre, 2010), 37.

Advantages for UN Operations

UN military operations do not necessarily involve combat; however, as per the preceding paragraphs, mission success, irrespective of the nature of such operations, relies on having an adequate situational picture. As UAVs can significantly enhance situational awareness, it follows that they would be a very useful capability for supporting a wide range of UN operations. While UAVs should not be regarded as a panacea, they are a true force multiplier. In addition to complementing the traditional tools and techniques employed on UN operations, UAVs can allow commanders to employ their allocated troops more effectively. This assessment is not limited to lower rank forces but also extends to modern, professional, highly technical, well-trained, and well-equipped armies. In some cases, UAVs may even be able to reduce the resources required to conduct a mission.

UAVs can provide persistent high-resolution coverage of areas while employing a range of sensors including day/night full motion video and synthetic-aperture radar (SAR). They can move freely virtually anywhere within a commander's area of responsibility (AOR) and provide real-time information on the situation on the ground. UAVs can be dynamically retasked to loiter over areas of interest or developing hot spots and they can do so without exposing personnel to weapons fire or other potential risks. In fact, operators could be positioned hundreds or even thousands of kilometres away from the surveillance area. If needed, UAVs can also apply force.

UAVs also offer several advantages over manned surveillance aircraft. They can remain on task much longer than manned aircraft, and the deployment of UAVs can help to minimize the exposure of aircrews to risk. Aircrews and ground support personnel can operate from the relative security of a well-defended operating base far removed from any combat operations or potential armed clashes. Due to limitations in UN search and rescue capabilities, it is understood that night flights of manned aircraft are generally prohibited.[5] UAVs would not require a 24/7 Search and Rescue and air evacuation capability on standby in the event of a crash. Not only would this reduce mission costs and complexity but it would significantly extend the duty cycle into the night for the conduct of aerial surveillance and reconnaissance on UN missions.

The following paragraphs offer an overview of how UAVs could contribute to several broad types of potential UN operations. The categories chosen are meant to reflect the range of UN operations from traditional peacekeeping to more complicated peace support operations, which may include peacekeeping, combat, and humanitarian support elements. Both overland and maritime enforcement scenarios are discussed.

---

5 Dorn, A.W. *Keeping Watch: Monitoring, Technology and Innovation in UN Peace Operations* (Tokyo: United Nations University Press, 2011), 65.

Humanitarian Assistance Operations

UAVs could be very useful for missions involving the delivery of disaster relief or the rendering of humanitarian assistance. They can provide commanders with real-time information on the extent of a disaster, possible access routes and landing zones for relief convoys and humanitarian flights, potential sites for aid distribution points or the establishment of relief services. UAVs can move through the area of operations unimpeded by floods, the destruction of transportation links due to natural disasters, a lack of infrastructure in developing nations, or inhospitable or impassable terrain, though they can be limited by bad weather such as storms, or reduced ceilings or visibility at the take-off and landing point. UAVs can also survey an afflicted area without exposing operators to hazards such as fires, smoke, radiation, or toxic chemicals. They could also be used to establish communications relays to facilitate command and control links where communications infrastructure has been damaged or is non-existent. Certain UAVs such as the MMIST Snow Goose could actually be used to deliver urgently needed supplies directly to the disaster area.[6]

Peacekeeping Operations

On relatively benign peacekeeping operations conducted under Chapter 6 of the UN Charter, forces have relied heavily upon static observation posts and vehicle patrols for situational awareness. These tools impose limits on what a commander can observe within the AOR. UAVs, on the other hand, are not limited to fixed fields of view as are static observation posts. They can gain access to areas that may be remote or inaccessible by road, or to locations to which access has been denied by belligerents or by the erection or emplacement of obstacles prior to a ceasefire, for example, minefields.

UAVs can provide a commander with real-time streaming EO video of unfolding events or situations, thereby enabling more informed and timely decision-making. IR sensors can provide similar information during darkness. Both EO and IR sensors allow commanders to see events unfolding in real time and to rapidly understand what needs to be done, enabling them to act in a timely fashion. This can give commanders the initiative in dealing with escalating situations and allow them to make time-critical decisions when lives may be on the line.

Synthetic-aperture radar could also be used to collect imagery in all weather conditions during day or night. Such high-resolution imagery could alert commanders to actions such as the massing of forces, the positioning of heavy weapons, the construction of defensive positions, or the migration or concentration of refugees. Radar equipped with a ground moving target indicator mode could

---

6   MMIST. *SnowGoose UAV*. Available at: http://www.mmist.ca/pg_ProductsSnow GooseOverview.php [accessed 9 September 2012].

be used to monitor vehicle movements and traffic patterns under the same environmental conditions.

While UAVs cannot replace boots on the ground, they can provide more effective real-time situational awareness over a greater proportion of a commander's AOR. They can loiter over developing situations or trouble spots, allowing commanders to deploy and direct their limited manpower more effectively. UAVs do not expose aircrews to undue risks and they can also provide warnings to ground troops of impending dangers beyond their line of sight, thereby reducing the risk of armed confrontation and enhancing personnel safety.

Complex Overland Operations

Since the end of the Cold War, the conduct of traditional peace support operations has been largely supplanted by more complex operations:

> The transformation of the international environment has given rise to a new generation of "multidimensional" United Nations peacekeeping operations. These operations are typically deployed in the dangerous aftermath of a violent internal conflict and may employ a mix of military, police, and civilian capabilities to support the implementation of a comprehensive peace agreement.[7]

In step with the evolving nature of conflict in a less-stable post-Cold War world, the international community seems to have become more inclined to embrace an interventionist agenda. The recent deployment of North Atlantic Treaty Organization (NATO) air power to Libya under the UN's "Responsibility to Protect" doctrine is a case in point. Consequently, the fundamental principles of the traditional peacekeeping approach – "consent of the parties, impartiality, and non-use of force except in self-defence and defence of the mandate"[8] – have become inadequate for the majority of contemporary UN operations. The need to resolve conflict and establish an enduring peaceful and stable society in such situations demands a whole-of-government or comprehensive approach, incorporating defence, development, and diplomacy elements working together in a coordinated fashion. Typical complex operations can involve the delivery of humanitarian assistance, the conduct of peacekeeping, and engagement in combat operations. These three distinct missions could all be happening simultaneously within the same AOR.

The International Security Assistance Force (ISAF) formed under UN Security Council Resolution 1386 of 20 December 2001 for stability operations in Afghanistan provides an example of a complex overland peace support operation. Canada's participation in ISAF, dubbed Operation Athena, began in Kabul in

---

7 United Nations Department of Peacekeeping Operations. *United Nations Peacekeeping Operations Principles and Guidelines* (New York: United Nations, 2008), 21.
8 Ibid., 31–4.

July 2003. During this initial phase, ISAF was charged with providing security to the Afghan Interim Authority and the United Nations. Phase II, which began in August 2005, saw Canadian troops redeploy to Kandahar, where they conducted the longest-running CF combat mission, which concluded in July 2011. Coincident with the redeployment to Kandahar, Canada signed the 2006 Afghanistan compact, which outlined "a wide-ranging program of activity based on three "critical and interdependent" areas of activity: a) security; b) governance, rule of law and human rights; and c) economic development".[9]

The applications for UAVs articulated for the previous two types of UN operations would be equally applicable to the conduct of complex operations. The addition of combat operations and the inclusion of civilian aid workers and other experts involved in capacity-building activities would present additional security challenges to which UAVs could be applied.

With their capacity for persistent day/night surveillance, UAVs are ideally suited to provide over-watch, convoy escort, direct support to troops in contact, and battle-damage assessment. Real-time streaming video can be downlinked to the UAV control station and thence to an intelligence centre for analysis, or directly to troops in the field via a remote video terminal such as the Remotely Operated Video Enhanced Receiver (ROVER) system.[10]

Recent conflicts in failed and failing states such as Afghanistan have been more likely to see stabilization forces facing irregulars engaging in asymmetric warfare than conventional forces. In both Afghanistan and Iraq, the improvised explosive device (IED) has been the insurgents' weapon of choice; IEDs have inflicted the greatest number of casualties on Canadian troops in Afghanistan. Of the 158 combat and non-combat-related deaths sustained by the CF in Afghanistan from the start of operations to October 2011, 97 or 62 percent, were the result of IEDs.[11]

UAVs can make the insurgents' task of planting IEDs much more difficult. Without UAVs, casualty rates in Afghanistan both for military personnel and aid workers could have been significantly higher. Colonel Christian Drouin, Commander of the Canadian Air Wing based in Kabul, made the following statement about the CU-170 Heron UAV:

> January 2009, it started flying operationally as our eye in the sky. It sits very high and gives us the ability to see what the enemy is doing so we can manage the battlefield properly. It's a very reliable platform and it's saving a lot of lives.[12]

---

9   Canadian Expeditionary Forces Command. *Operation Athena.* Available at: http://www.forces.gc.ca/en/operations-abroad-past/op-athena.page [accessed 7 May 2014].

10   Kenyon, H. "Video Streams to the Tip of the Spear", *Signal Magazine*, April 2009.

11   CBC News. *In the Line of Duty: Canada's Casualties [in Afghanistan].* Available at: http://www.cbc.ca/news2/interactives/canada-afghanistan-casualties/ [accessed 7 May 2014].

12   MacLeod, C. "Flying Above Afghanistan", *The Maple Leaf*, 3 February 2010.

The evolution of more sophisticated counter-insurgency techniques have seen UAVs employed in more effective campaigns to protect military forces and the civilian population from IEDs. For example, the employment of pattern-of-life analysis has been a key enabler in defeating IED systems, including bomb makers and distribution networks, as opposed to individual devices.

Unmanned Aerial Vehicles for Maritime Operations

Maritime operations executed under a UN Security Council resolution typically comprise interdiction operations by military vessels and aircraft. Two examples of such operations are Operation Sharp Guard[13] and Operation Unified Protector.[14] Operation Sharp Guard was conducted by NATO and Western European Union[15] naval forces between 1993 and 1996 in the waters of the Adriatic Sea off the coast of the former Yugoslavia. Operation Unified Protector, which took place in 2011, was undertaken by NATO forces in the waters of the Mediterranean Sea off the Libyan coast. Both operations involved the monitoring and enforcement of arms embargos. In the case of the former Yugoslavia, the embargo was initially authorized under UN Security Council Resolution 713,[16] and in the case of Libya, operations were conducted under UN Security Council Resolution 1970.[17] Other examples of maritime interdiction operations sanctioned by the United Nations include anti-piracy operations such as those conducted off the coast of Somalia under UN Security Council Resolution 1851.[18]

On typical maritime interdiction operations, ships are assigned geographical boxes in which they challenge all transiting traffic. Organic helicopters[19] are

---

13   NATO. *NATO/WEU Operation Sharp Guard*. Available at: http://www.nato.int/ifor/general/shrp-grd.htm [accessed 9 September 2012].

14   NATO. *NATO Arms Embargo against Libya Operation UNIFIED PROTECTOR.* Available at: http://www.nato.int/nato_static/assets/pdf/pdf_2011_03/20110325_110325-unified-protector-factsheet.pdf [accessed 9 September 2012].

15   The Western European Union is a now defunct military grouping whose functions transferred to the European Union's Common Security and Defence Policy.

16   United Nations Security Council (1991), *Security Council Resolution 713 (1991)* ["Continued hostilities in Socialist Federal Republic of Yugoslavia"], S/RES/713 (1991), 25 September.

17   United Nations Security Council, *Security Council Resolution 1970 (2011)* ["Situation in Libya/Peace and security in Africa"], S/RES/1970 (2011), 26 February 2011.

18   United Nations Security Council "Security Council Authorizes States to Use Land-based Operations in Somalia, as Part of Fight Against Piracy Off Coast, Unanimously Adopting 1851 (2008)", UN Doc. SC/9541, 16 December 2008. Available at: http://www.un.org/News/Press/docs/2008/sc9541.doc.htm [accessed 9 September 2012].

19   "Organic" is a term defined as "assigned to and forming an essential part of a military organization" (US Department of Defense Military Dictionary) such as all helicopters assigned to a Task Force. The author's military experience includes participation

normally used to assist in this task. Any suspicious vessels are boarded and inspected by ships' boarding parties. Once inspected, vessels are either cleared to proceed or seized and escorted to a secure port. Once in port, seized vessels are handed over to national and international authorities co-operating under the UN Security Council resolution in effect.

Maritime interdiction operations normally employ a surveillance aircraft overhead to provide the fleet with situational awareness and coordination, particularly of the air assets within the naval commander's AOR. During Operation Sharp Guard this indispensable support was normally provided by NATO Airborne Warning and Control System (AWACS) aircraft.[20] When AWACS was not available, Long Range Patrol Aircraft such as CF CP-140 Aurora aircraft were employed in this capacity.[21] While information concerning Operation Unified Protector remains operationally sensitive, it is reasonable to assume that a similar approach was employed. Both NATO AWACS and CP-140 aircraft were deployed to the region.[22]

The airborne over-watch and coordination function described above could, in present times, have been performed by a UAV with sufficient endurance and fitted with the right sensor and communications package. A medium-range, long-endurance (MALE) UAV such as the IAI Heron has endurance in excess of 24 hours and can carry multiple sensors including EO/IR, SAR and EW systems, as well as a communications relay package. The relay capability would allow UAV operators to communicate directly with ships and aircraft in the operations area. The single-engine, fuel-efficient UAV could be operated by a crew of two to three personnel. On the other hand, NATO AWACS, which is based on the four-engine Boeing 707 airliner, has an endurance of "10+ hours" and carries a crew of 17.[23]

---

in Operation Sharp Guard, where a Naval Task Force deployed to the Adriatic Sea to enforce UN Security Council Resolutions. In this context, "organic helicopters" were those assigned to the Task Force; they were military helicopters from contributing nations that were embarked in the ships of the Task Force, and under the operational control of the Task Force Commander.

20   NATO. *NATO Airborne Early Warning and Control Force E-3A Component.* Available at: http://www.e3a.nato.int/eng/html/organizations/wing_ow.htm [accessed 9 September 2012].

21   Maloney, S. "Force Structure or Forced Structure", *Choices*, May 2004, 11.

22   Canadian Expeditionary Forces Command, *Operation Mobile.* Available at: http://www.forces.gc.ca/en/operations-abroad-past/op-mobile.page [accessed 7 May 2014]. Military Daily News, *NATO AWACS Eye Libyan Skies.* Available at: http://www.military.com/news/article/nato-awacs-eye-libyan-skies-.html [accessed 9 September 2012].

23   NATO. *AWACS: NATO's Eyes in the Sky.* Available at: http://www.nato.int/cps/en/natolive/topics_48904.htm [accessed 7 May 2014].

A UAV is much more cost effective to employ than a large multicrew aircraft and can operate without exposing aircrew to enemy fire. In the case of operations Sharp Guard and Unified Protector, UAVs based in Italy could have been used to effectively coordinate maritime interdiction operations. This would have allowed the more sophisticated and costly NATO AWACS and maritime patrol aircraft to be released for other missions more appropriate to their capabilities. Admittedly, the distances involved in the latter example would have required satellite control links for operations off the Libyan coast. While the nature of their missions remains classified, it is known that such beyond-line-of-sight UAV missions were undertaken during NATO operations in the Libyan AOR. Based in Sigonella, Sicily, the French Air Force launched UAV sorties using the Harfang système intérimaire de drone MALE (SIDM).[24] The Harfang SIDM is a variant of the IAI Heron with the capability for control and data downlink via satellite.[25] The US Air Force has also acknowledged that a Global Hawk UAV, also based in Sigonella, participated in the Libyan campaign.[26]

Smaller UAVs could also have fulfilled or supplemented the role that helicopters routinely play in maritime interdiction operations such as these. The same advantages suggested in the previous paragraph, for example, economy and security, would apply. In addition, UAVs could potentially operate in sea states and weather conditions that would preclude helicopter operations or would place aircrews at unacceptable levels of risk due to environmental conditions. Higher sortie rates could potentially be achieved by supplementing embarked helicopters with small UAVs.

Canadian Forces' International Missions

To date, Canada's international deployment of UAVs has been exclusively in support of operations conducted under UN Security Council Resolutions. Therefore, an examination of the Canadian UAV experience provides some practical insight into the potential employment of UAVs on UN operations, whether under UN command, or within an international coalition acting in the UN's collective

---

24 Air Recognition. *French Air Force deploys Harfang SIDM UVA over Libya.* Available at: http://www.airrecognition.com/index.php/archive-world-worldwide-news-air-force-aviation-aerospace-air-military-defence-industry/news-year-2011-aviation-aerospace-air-force-defence-military-industry/august-2011-2/108-french-air-force-deploys-harfang-sidm-uva-over-libya.html [accessed 9 September 2012].

25 Air Force Technology, *Harfang MALE Unmanned Aerial Vehicle (UAV), France.* Available at: http://www.airforce-technology.com/projects/harfang-drone/ [accessed 9 September 2012].

26 Flightglobal News. *US arms UAVs for Libya missions.* Available at: http://www.flightglobal.com/news/articles/us-arms-uavs-for-libya-missions-355895/ [accessed 9 September 2012].

interest. The following paragraphs summarize CF overseas experience with the operation of UAVs.

Canada's operation of UAVs abroad began with the deployment of a CF contingent to Afghanistan as part of ISAF. The Sagem Sperwer was acquired in 2003 and designated as the CU-161. It was a tactical (NATO Class II) UAV that was fitted with an EO payload and employed a hydraulic catapult launcher and a parachute and airbag recovery system.

Sperwer systems were initially deployed to Kabul in October 2003 but by 2006 all operational CF forces in Afghanistan, including the Sperwer contingent, were redeployed to Kandahar. In Kandahar, the CF became much more reliant on Intelligence, Surveillance and Reconnaissance (ISR) to support combat operations and to minimize casualties due to the growing use of IEDs by the insurgents. After two years of fighting and with the security situation in southern and eastern Afghanistan continuing to deteriorate, Canada's role in Afghanistan was foremost in the minds of Canadians.

To provide advice to parliament and instruct the debate on Canada's future in Afghanistan, an independent review panel was commissioned by the Prime Minister in October 2007. Chaired by former Deputy Prime Minister the Honorable John Manley, the non-partisan panel comprised former government ministers, diplomats, and senior public servants. The committee's report, which was submitted in January 2008, implied that Canada's UAV assets were inadequate for the mission and recommended that more capable systems be acquired. It stated that:

> to improve the safety and effectiveness of the Canadian Forces in Afghanistan, the Government should secure for them, no later than February 2009, new medium-lift helicopters and high-performance unmanned aerial vehicles.[27]

This recommendation was the genesis of Project Noctua. In August 2008, the Government of Canada entered into a multiyear contract with MacDonald, Dettwiler and Associates Ltd (MDA), with headquarters in Richmond, British Columbia, for a turnkey UAV service based on the IAI Heron UAV. The Sperwer System was phased out with the introduction of the Heron and the CF ceased Sperwer operations altogether in August 2009. All remaining Sperwer assets in flyable condition were subsequently sold to the French government.[28]

---

27 Her Majesty the Queen in Right of Canada, represented by the Minister of Public Works and Government Services, *Independent Panel on Canada's Future Role in Afghanistan*, FR5-20/1-2008. Available at: http://publications.gc.ca/collections/collection_2008/dfait-maeci/FR5-20-1-2008E.pdf [accessed 9 September 2012], 35.

28 Royal Canadian Air Force. *CU-161 Sperwer.* Available at: http://www.airforce.forces.gc.ca/v2/equip/hst/cu161/index-eng.asp [accessed 9 September 2012].

Figure 9.2   CU-170 Heron unmanned aerial vehicle
Source: Reproduced by permission of MacDonald, Dettwiler and Associates Ltd.

Canada's Project Noctua

In order to satisfy the recommendation of the Independent Panel on Canada's Future Role in Afghanistan regarding the introduction of high-performance UAVs, the Canadian Department of National Defence (DND) had to adopt an innovative strategy and a very aggressive implementation schedule. DND turned to industry to provide a turnkey solution that could be deployed within months, as opposed to the years normally associated with fielding a new capability of such complexity. The desired solution was to be delivered via a service arrangement whereby the selected contractor would provide the systems, maintenance, supply chain, and training, while DND would provide the operators.

A competitive tender was issued in February 2008 and on 1 August 2008 a contract was awarded to MDA and its partner, IAI, for a service based on the Heron UAV. In January 2009, only five months later, the CF were conducting operational ISR missions in Afghanistan with the Heron (Figure 9.2).

The Heron is a MALE platform (NATO Class III) with an all-up weight of 1,150 kg, a wingspan of 16.6 m and a payload capacity of 250 kg. It has a service ceiling of 30,000 ft and endurance in excess of 40 hrs. Payloads include an EO/IR turret, various EW systems and overland or maritime SAR. While both satellite relay and line of sight control systems are available, the CF variant utilized a line of sight system that supported operations out to 200 km. Extended ranges can be achieved by using another air vehicle as an airborne datalink relay station. The system is highly reliable, with redundancy built into virtually every sub-system.[29]

---

29   Israel Aerospace Industries. *Heron Family*. Available at: http://www.iai.co.il/18900-en/BusinessAreas_UnmannedAirSystems_HeronFamily.aspx [accessed 9 September 2012].

Meeting the extremely tight timelines imposed by the CF to introduce this sophisticated capability into a very dynamic operational theatre at the end of a very long supply chain was nothing short of remarkable. It was also a testament to what can be achieved by an integrated government/industry team with shared goals and a high level of motivation. In addition to the establishment of all infrastructure at the Main Operating Base in Afghanistan, individual and collective training had to be conducted in Canada and a training pipeline created to sustain the capability. Airworthiness clearances, flight permits, and frequency allocations had to be obtained, and the MDA maintenance organization had to be accredited by DND's Technical Airworthiness Authority.

A key element in minimizing fielding time was to establish requirements based on existing technologies available in the marketplace. Modifications to the Heron system were essentially limited to conversion of sensor data to standard NATO formats and the addition of a second shelter (a transportable, containerized unit, externally similar to the UAV Ground Control Station) into which sensor data was relayed to EW experts and intelligence analysts for interpretation and exploitation.

The use of experienced aircrew and former military technicians enabled individual training times to be greatly reduced. Another critical factor in minimizing the training schedule was the Heron's highly reliable Automatic Take-off and Landing (ATOL) system. The skills required to manually land a UAV take a substantial period of time to acquire and are perishable. The Heron's ATOL system obviated the need for CF aircrews to develop or maintain those skills, thereby significantly reducing training time and proficiency requirements in theatre.

Great importance was attached to collective training in Canada for the entire deploying battlegroup. It allowed the joint force to gain familiarity with the UAV capability and an appreciation for how to employ it prior to arriving in theatre. This applied to commanders from Brigade down to Section level as live streaming video from the UAV could be directly received by troops on the ground using the man-portable ROVER remote video terminal system.

Once in theatre, Heron was quickly recognized as a significant advancement from the earlier Sperwer tactical UAV. To safeguard operational security, precise details of how the Heron was used are not available in unclassified sources. However, the RCAF's unclassified website provides some insight into capabilities, general missions, and expectations for the system. It states:

> The Heron's primary functions are to gather imagery and data for use in surveillance, reconnaissance, intelligence analysis and target acquisition. It can scout out convoy routes and other ground operations areas, scan for insurgents, or observe suspicious activity, such as planting improvised explosive devices. Its capabilities will help reduce insurgent attacks, and save lives – Canadian and Afghan alike.[30]

---

30 Royal Canadian Air Force. *CU-170 Heron*. Available at: http://www.airforce. forces.gc.ca/v2/equip/cu170/index-eng.asp [accessed 9 September 2012].

Whereas previously, UAVs were routinely almost an afterthought in mission planning and an adjunct to the conduct of operations, the Heron became one of the cornerstone capabilities around which CF operations in Afghanistan were planned and executed. Commanders at all levels praised the system for its capability and its availability. Many indicated they never want to conduct operations again without this type of asset.

On a daily basis, Herons performed and contributed to a wide variety of critical tasks essential to conducting successful operations and protecting allied troops and civilian aid workers. Persistent, relatively stealthy and able to operate over the Afghanistan battle space with relative impunity, Herons were major contributors to traditional surveillance and reconnaissance missions and helped to shape new techniques such as pattern-of-life and collateral damage analysis. "The longest flight the Canadians flew in Afghanistan was 30.2 hours and the aircraft still had 4.5 hours of fuel left in the tank".[31] Herons were not armed, so strike missions were not conducted.

The Heron UAV service operated in support of the CF from January 2009 until combat operations ceased in July 2011. Flying operations were conducted every day unless prevented by weather or aircraft unserviceability. No-fly days during Canada's two-and-one-half-year Heron deployment were extremely rare: "Through 30 months of operations, the Herons logged more than 15,000 hours of flight time".[32] A procedure was also developed to operate more than one UAV simultaneously if the situation warranted this level of effort.

MDA's UAV service consistently received high praise from the Canadian military establishment for the contribution it made to operations in Afghanistan. Air Force Colonel Al Meinzinger, the last commander of the Canadian Air Wing in Afghanistan, lauded the performance of MDA's UAV service, referring to it as "an incredible capability. They really kept the commander on the high ground, operating the UAV almost 20 to 22 hours a day, providing critical information and situational awareness".[33] Meinzinger put the importance of that situational awareness into perspective when he said: "They were saving lives up to the last minute".[34]

Commercial Service Model for UN Operations

While Canada's mission in Afghanistan was clearly conducted in the context of complex overland operations, this same commercial service model is equally applicable to traditional peacekeeping, humanitarian assistance, or maritime

---

31   Ibid.

32   Marsden, W. "The Allies' Eyes in the Skies", *The Vancouver Sun*, 23 July 2011.

33   Pugliese, D. "Canada: Not Enough People to Fly New UAVs", *Defense News*, 31 October 2011.

34   Marsden, "The Allies' Eyes in the Skies".

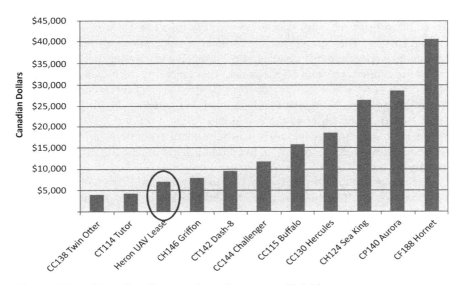

Figure 9.3 Canadian Forces aircraft cost per flight hour
Source: MDA.

operations. As has been emphasized in this chapter, the ability of the Heron to provide decision makers with enhanced situational awareness will have been as important to the effective delivery of humanitarian assistance and disaster relief as it was to the conduct of peace support and combat operations.

As reflected in Project Noctua, a commercial service can provide the UAV systems, maintenance, training, and supply chain. The customer can provide the operators or can opt for a complete turnkey service in which the contractor would also fly the UAVs, operate the sensors and provide the data to the end user. As a commercial enterprise, such services are unarmed and exist to conduct long-endurance ISR missions to enhance situational awareness.

Project Noctua demonstrated that operating Heron under a commercial leasing arrangement was a very cost effective means of obtaining persistent ISR as compared to using manned military platforms. Figure 9.3, developed by MDA, illustrates the costs per flying hour for major CF aircraft fleets and equivalent costs for the Heron UAV lease. Costs for manned aircraft operation were obtained from the DND *Cost Factors Manual*, [35] while Heron costs were derived from actual Noctua contract costs for hours flown.

While the service concept was initiated for Canada, the model is adaptable to meet the needs of other client nations. Within one year of commencing the service for Canada, MDA was under contract to the Commonwealth of

---

35 Director Strategic Finance and Costing. *Cost Factors Manual 2009–2010*, Department of National Defence, 22 June 2009.

Australia to begin training Australian military aircrews for a parallel service in Afghanistan. The service, which commenced operations in January 2010, continues to operate at full capacity and is expected to continue until Australian forces are withdrawn from Afghanistan. The capability delivered through the MDA UAV service has also garnered praise from the highest levels within the Australian military. On a visit to Afghanistan in November 2010, Air Marshall Mark Binskin, then Chief of the Royal Australian Air Force, remarked: "The Heron is an important component to the modern battlefield, providing vital situational awareness for troops on the ground. It has assisted in saving lives by identifying threats".[36]

Relying on troop-contributing nations to provide UAVs can be problematic, however. While virtually all nations can offer boots on the ground, relatively few can offer UAVs. Not even first-rank military powers such as Canada and Australia own advanced UAVs. Those that do have such assets may not be keen to offer them, as they may be committed to higher-priority missions in support of national interests elsewhere. In cases where UAVs cannot be secured from troop-contributing nations, the United Nations itself should strongly consider contracting for them using a centrally funded approach, as they do now with fixed wing and rotary wing air transport resources. This capability can be obtained today from industry and could be rapidly deployed anywhere on the globe.

The MDA service offering for Canada and Australia was based on Heron because that was the platform that best met the customers' requirements. However, virtually any system could be offered, depending on the UN mission-specific requirement. MDA has recently launched Persist-INT™, an on-demand, UAV-based ISR service for ISAF nations in Afghanistan. This complete turnkey offering was conceived to provide UAV services to nations with limited means or which do not desire or require long-term UAV support. The Persist-INT™ concept allows users to immediately access UAV services without incurring substantial capital investment or infrastructure costs. It also obviates the requirement to recruit and train specialist operators, engineers, technicians, logisticians and other services.

The Persist-INT™ approach offers a rapid and uncomplicated means of obtaining UAV services for UN missions, or in support of multinational coalitions engaged in the enforcement of UN Security Council resolutions. If a centrally funded approach were not possible, the UN or coalition leadership could appeal to nations to contribute contracted UAVs as an alternative to making a troop commitment. Depending on national or international circumstances, this could provide a more appropriate means of making a significant and highly valued contribution to a mission.

---

36   Australian Defence Force. "Chief of Air Force visits Middle East", 17 November 2010.

Challenges

There are challenges to employing UAVs; however, in MDA's experience, few if any of these are technical. While the hot, high, and dusty operating environment in Afghanistan was demanding, both equipment and contractor personnel proved fully capable of coping with all circumstances encountered. MDA was contractually incentivized to deliver system availability and reliability in excess of 90 percent. During Project Noctua, the company consistently exceeded those targets and averaged 94 percent mission availability and 94 percent mission reliability.[37] In addition, no aircraft were lost in Afghanistan. That more than 15,000 flying hours were achieved in a 30-month period is a clear testament to the technical success achieved.

Of greater concern are the regulatory challenges to employing this new technology. All aircraft, including UAVs, must be operated under some sort of an authorization for flight. Such an authorization could either be issued by the regulatory organization within the military force that operates the UAVs, or by the civil regulatory agency of the nation in which they are being operated. For example, delegated officers within the Canadian Forces Directorate of Technical Airworthiness and Engineering Support issued flight permits to MDA for UAV operations in Afghanistan and within Canadian Military Airspace. Transport Canada issued Special Flight Operations Certificates to MDA to operate in Canadian airspace when their UAVs were not on the military register.

Access to appropriate portions of the radio frequency spectrum would also need to be allocated to control the UAV and to enable the transmission of the surveillance data it collects to the ground control station. Frequency allocation is also regulated. On a military operation it would have to be built into the communications plan and allocated by the responsible military agency within the coalition. Authority may have to be sought from civil regulators within the nation in whose airspace the UAV is to be operated. Industry Canada allocated frequency spectrum to MDA for operations within Canadian airspace. In Afghanistan frequency spectrum was allocated by ISAF.

The fact that UAVs are operating in support of UN resolutions today indicates that any regulatory challenges are not insurmountable. However, they must be considered early in the planning process and addressed before flight operations commence. It is assumed that the United Nations has dealt with the issue of spectrum management for previous intervention operations and is no stranger to dealing with international air regulation issues. After all, the International Civil Aviation Organization is a UN agency. Where required, the United Nations should be able to use its good offices to obtain the necessary flight authorities from appropriate civil regulators.

---

37 Figures provided by MDA UAV Services Project Manager.

Conclusion

Situational awareness underpins the success of all military operations regardless of whether they involve combat or not. The evolution of remote sensing technology and unmanned aerial vehicles in particular has offered military users reliable, persistent situational awareness while significantly reducing risk to deployed troops. These developments have made UAVs an essential component of military forces in the developed world. Consequently, we have seen exponential growth in their use on contemporary military operations conducted during the past decade.

In addition to its obvious advantages over traditional UN surveillance tools and manned aerial surveillance platforms, the MALE UAV is a true force multiplier that can allow commanders to employ their assigned resources much more effectively. MALE UAVs would clearly constitute an extremely valuable addition to any UN mission regardless of its nature.

Today, only a handful of the world's elite military forces have MALE UAVs in their inventories. In some of those nations, UAVs may be considered too scarce or too valuable to be offered in support of non-national missions. Employment of a commercial service model could provide an effective alternative to reliance on troop-contributing nations for this critical capability. It offers a convenient, cost-effective and expeditious means of acquiring top-tier UAV mission support via a commercial arrangement and has become the template for rapidly delivering a highly technical and complex capability to meet an urgent need. A commercial service model offers the means to provide a short-term or interim capability, a long-term solution, or an approach to augment existing high demand/low density assets.

Assured access to MALE UAVs may require the United Nations to explore innovative and non-traditional solutions, just as Canada did with Project Noctua. By adopting the same centrally funded approach that has been successfully used to furnish commercial airlift services, UN operational commanders can be provided with sophisticated UAVs and the indispensable benefits they offer.

PART IV
The UN and No-fly Zones

The model for No-fly Zones (NFZs) was created by the United States after the 1991 Gulf War as a means to prevent further Iraqi aggression. First, a NFZ was established over Iraq's north to prohibit Iraqi aircraft from attacking the Kurdish ethnic minority in the region. In 1992, a NFZ was declared in the southern half of the country to help protect the Marsh Arabs and Shiite areas. The United States and its allies implemented the NFZ through operations Northern Watch and Southern Watch, which were not explicitly UN-authorized. The Americans did, however, operate in conjunction with a UN observer mission along the Iraq–Kuwait border. James MacKay elaborates in Chapter 10 on how the NFZ was enforced, including through ironic violations of the demilitarized zone that the UN mission was monitoring.

In contrast to the Iraqi NFZs, the Bosnian NFZ (1993–1995) was created by the Security Council. It sought to prevent Serb forces from using aerial superiority in the internecine war. At first the North Atlantic Treaty Organization (NATO), under Operation Sky Monitor, only reported on NFZ violations; it did not take action in response, frustrating many on the ground and in the international community. However, NATO worked closely with UN forces on the ground, including UN military observers like F. Roy Thomas, who describes his experiences "touched by air power" in Chapter 11. In 1994, under Operation Deny Flight, NATO aircraft shot down four Bosnian Serb fighter aircraft that violated the NFZ, making it the first combat engagement in NATO's history. The dual key system, under which NATO and the UN had both to "turn a key" to activate enforcement, proved confusing and inadequate for fast and effective responses. After the 1995 Srebrenica massacre, NATO was given broad authorization to carry out attacks beyond implementation of the NFZ, as described by Robert C. Owen in Chapter 13, in Part V.

The Security Council again created a NFZ in 2011, this time over Libya, to prevent the Gadhafi government from attacking its citizens and the opposition force. Furthermore, the Council authorized member states to protect Libyan civilians, leading NATO to take control of the airspace by eliminating Libyan air defences and bombing any forces targeting civilians. This resulted in the fall of Muammar Gadhafi. As this operation involved much more than a simple NFZ, the Libyan campaign is described by Christian F. Anrig in Part V, Chapter 15. But the imposition and enforcement of the NFZ remains an important application of air power.

Chapter 10

The UN Iraq–Kuwait Observer Mission and the Southern No-fly Zone, 1991–2003

James McKay

The United Nations Iraq–Kuwait Observer Mission (UNIKOM) and the Southern No-fly Zone (SNFZ) developed an unusual symbiosis. The former was conceived in April 1991 as an interpositional observation force to support a comprehensive effort to resolve the "Iraq–Kuwait" dispute. UNIKOM has not attracted as much academic attention as other UN peace support operations from the same era. This has more to do with the dramatic events in the former Yugoslavia, Somalia, and Rwanda than UNIKOM. William Durch describes it as "traditional peacekeeping in an untraditional situation".[1] Jan Bury assesses UNIKOM's entire span as being only somewhat successful in dealing with the "Iraq–Kuwait dispute".[2] In August 1992, the United States-led coalition imposed a No-fly Zone (NFZ) over southern Iraq "consistent with"[3] an earlier UN Security Council resolution in the name of preventing another crisis involving Iraq.

The mandates and nature of both operations created tensions between the UN's efforts to conduct a traditional peace support operation and the coalition's efforts at the containment of Iraq.[4] While both were intended to deal with the challenges emanating from Saddam Hussein's Iraq that could destabilize the region further, their modi operandi were at odds with one another. UNIKOM's observers dutifully

---

1   Durch, W.J. "The Iraq–Kuwait Observation Mission", in *The Evolution of UN Peacekeeping: Case Studies and Comparative Analysis*, ed. William J. Durch (New York: Henry L. Stimson Center, 1993).

2   Bury, J. "The UN Iraq–Kuwait Observation Mission", *International Peacekeeping* 10(2) (2003), 71–88.

3   The phrase "consistent with" comes from US Public Law. See: United States Congress. *Iraq and the Requirements of Security Council Resolution 687 and 688*, 5 December 1991 (Washington, DC: United States of America Congress). Available at: http://www.lawandfreedom.com/site/historical/PL102-190.pdf [accessed 9 February 2012].

4   The coalition was described by the United Nations as the "Member States cooperating with the State of Kuwait" since 1990. Over the years, its composition varied, although it was always American-led. From April 1991 to December 1998, the coalition consisted of the United States, United Kingdom, France, Saudi Arabia, Turkey and Kuwait. Its nominal purpose was to enforce both the southern and northern no-fly zones over Iraq in order to protect minority populations in Iraq. In December 1998, France withdrew from the coalition.

reported all activities along the Iraq–Kuwait border, while the post-1991 Gulf War coalition's members continued to operate in the area to contain Iraq. Within months, however, it became evident that UNIKOM needed to be more robust and act as a de facto consumer of the security provided by the presence of coalition aircraft over southern Iraq. This exacerbated the inherent contradiction between the two activities and as time progressed, the Iraqi government exploited the contradiction in an effort to dismantle the mechanisms of its containment. This chapter will explore the UN's application of the logic of the "Iraq–Kuwait dispute" through a number of measures, but UNIKOM in particular; the coalition's enforcement of the SNFZ from August 1992 to March 2003; and the effects of both on each other's operations.

The Iraq–Kuwait Dispute

The United Nations treated the post-1991 Gulf War situation as an exercise in dispute resolution between the states of Iraq and Kuwait, as opposed to the international community enforcing the decisions of the UN Security Council. By approaching the situation in this manner, it seems that the Security Council's members were trying to prevent any recurrence of the August 1990 invasion of Kuwait by Iraq. The means by which the problem would be resolved was Security Council Resolution (SCR) 687 of 3 April 1991, which maintained the regime of sanctions on Iraq until its constellation of terms was satisfied. Its preamble, however, indicated an important assumption made by its drafters:

> Affirming the commitment of all Member States to the sovereignty, territorial integrity and political independence of Kuwait and Iraq, and noting the intention expressed by the Member States cooperating with Kuwait under paragraph 2 of [UNSC] resolution 678 (1990) to bring their military presence in Iraq to an end as soon as possible consistent with paragraph 8 of resolution 686 (1991). …[5]

It assumed, and not entirely incorrectly, that the "Member States cooperating with Kuwait" would withdraw from the region once the situation was resolved.

---

5   United Nations. *The United Nations and the Iraq–Kuwait Conflict, 1990–1996* (New York: United Nations, 1996), 193 (that is, "Security Council Resolution 687"). Security Council Resolution 678 (29 November 1990) authorized the coalition to implement the plans to eject Iraq from Kuwait under Chapter VII conditions after 15 January 1991 in support of Security Council Resolution 660 (2 August 1990) (Ibid., 178). The latter condemned Iraq's invasion of Kuwait in early August 1990 and demanded its withdrawal (Ibid., 167). Security Council Resolution 686 (2 March 1991) noted the cessation of offensive operations, demanded the release of all Kuwaiti and third-party nationals detained or treated as prisoners of war, and a meeting for ceasefire talks, and maintained the previous 20 resolutions pertaining to the Iraq–Kuwait dispute (Ibid., 182–3).

It should be borne in mind that this resolution came into existence as images of Iraqi Kurds huddling in sufferance on Turkish mountainsides, a side effect of the Iraqi government's reassertion of control after its defeat, began to spur calls for action. This provided the international community with further evidence that the Iraqi government, as led by Saddam Hussein, would remain a source of strife unless contained. This, in turn, led to Operation Provide Comfort, which from April 1991 provided for the relief and repatriation of the Iraqi Kurds. In order to prevent further depredations and refugee crises, an American-led coalition of states imposed a NFZ over northern Iraq, which in US eyes, was "consistent with" SCR 688.[6]

The previous resolution, SCR 687, contained a comprehensive plan to resolve the Iraq–Kuwait dispute by removing motives for future disputes, that is, the resolution of the border dispute, a peace support operation, the creation of a demilitarized zone (DMZ), reparations, return of property and prisoners. It also sought the removal of means that could be used to threaten other states in the region – that is, nuclear, biological or chemical weapons and ballistic missiles with a range greater than 150 km. The peacekeeping operation, UNIKOM, was part of a tripartite package that cannot be understood in isolation. Section A of SCR 687 stated the Security Council's demand that Iraq and Kuwait adhere to their 1963 border agreement. The United Nations claimed that it sought to demarcate a theoretically existing boundary to convince certain governments that it was not setting a precedent by intruding into what were normally bilateral disputes. The Security Council also decided "to guarantee the inviolability of the above-mentioned international boundary".[7] This was an unusual precedent, as the United Nations does not normally guarantee the borders of any state. The UN Secretariat arranged for the creation of the United Nations Iraq–Kuwait Boundary Demarcation Commission (UNIKBDC). The commission was composed of one representative from Iraq, one from Kuwait and three independent members appointed by the UN Secretary-General.[8] The Iraqi government argued that the 1963 agreement between Iraq and Kuwait had no legal basis and that its representative would be outnumbered.[9] Section B stated that the Secretary-General would generate:

> a plan for the immediate deployment of a United Nations observer unit to monitor
> the Khor Abdullah and a demilitarized zone, which is hereby established,

---

6   Ibid., "Remarks on Assistance for Iraqi Refugees and a News Conference", 199. In resolution 688, the Security Council demanded that Iraq end the repression of the Iraqi people, including those in Iraqi Kurdistan. It insisted that Iraq allow international humanitarian organizations access to the affected areas.

7   Ibid., "Security Council Resolution 687", 195.

8   Ibid., "Report of the Secretary-General on establishing an Iraq–Kuwait Boundary Demarcation Commission, 2 May 1991 (S/22558)", 235–46.

9   Ibid., "Enclosure: Letter dated 23 April 1991 from the Minister for Foreign Affairs of Iraq addressed to the Secretary-General (S/22558)", 237–8.

extending ten kilometres into Iraq and five kilometres into Kuwait from the boundary referred to in the "Agreed Minutes Between the State of Kuwait and the Republic of Iraq Regarding the Restoration of Friendly Relations, Recognition and Related Matters" of 4 October 1963; to deter violations of the boundary through its presence in and surveillance of the demilitarized zone; to observe any hostile or potentially hostile action mounted from the territory of one State to the other; and for the Secretary-General to report regularly to the Security Council on the operations of the unit, and immediately if there are serious violations of the zone or potential threats to peace (SCR 687).

In short, the DMZ, supported by UNIKOM, would allow UNIKBDC to carry out its task of demarcating the border. After that was achieved, the DMZ, monitored by UNIKOM, would act as a cordon sanitaire between Iraq and Kuwait.

In a theoretical sense, UNIKOM represented an unusual variant of a traditional peacekeeping mission. It was both an observation and an interposition force, although the initial emphasis was on the former. It was intended to support the ceasefire through a mechanism to build confidence through transparency and by raising the potential cost of attack.[10] It was a solution to prevent a repetition of Iraq's invasion of Kuwait in August 1990; its wording, however, did not specify that its mandate be limited to only the two parties. This meant that UNIKOM would be bound to report impartially on any violations of the zone or preparations for actions mounted from the territory of either Iraq or Kuwait, leaving the decision for further action to the Security Council.

The Observer Mission Before the Southern No-fly Zone

UNIKOM's initial mandate was laid out in a report by the Secretary-General to the Security Council on 5 April 1991. Of note in this report were the "considerations relevant to the discharge of the mandate", the "concept of operation" and the "requirements". The key element in the "considerations" were the limits placed on UNIKOM's span of observation; they would observe those activities visible from the DMZ and the Khor Abdullah, the waterway separating Iraq and Kuwait, and would not take action to prevent entry of unauthorized forces.[11] This meant that UNIKOM's force of 300 observers would observe and report as opposed to enforce the DMZ. The concept of operations defined the tasks of UNIKOM in a largely land-borne sense, that is: withdrawal of any armed forces from the

---

10   For a discussion of the mechanisms at work, see: Page Fortna, V. "Interstate Peacekeeping: Causal Mechanisms and Empirical Effects", *World Politics* 56(4) (2004), 485–90.

11   United Nations, *The United Nations and the Iraq–Kuwait Conflict, 1990–1996*, "Report of the Secretary-General on the Implementation of Paragraph 5 of Security Council Resolution 687 (1991), 5 April 1991 (S/22454)", 2.

**Figure 10.1   Aircraft provided by Switzerland to the UN Iraq–Kuwait
Observer Mission for air patrol, 1 May 1991**
Source: UN Photo 72346.

zone; observation posts on the roads to monitor traffic; patrols by land or air; and investigations.[12] A few patrol aircraft were used by UNIKOM to obtain an aerial view, as seen in Figure 10.1.

There is very little to suggest that the drafters of the report considered the possibility of air traffic moving through the zone from Iraq, Kuwait, or the "Member States cooperating with Kuwait", that is, the coalition. Last, the "requirements" acknowledged that it was necessary to provide additional forces for security and disposal of explosive hazards. To this end, the report recommended that five rifle companies and a field engineer unit be loaned from other missions in the region to augment UNIKOM temporarily. The security element was justified by the transitional situation based on the presence of displaced persons, the withdrawal of the forces of "Member states cooperating with Kuwait" and the need for the Iraqi and Kuwaiti police to maintain law and order on their respective sides of the border.[13] UNIKOM headquarters would be located at the Iraqi town of Umm Qasr, within the zone, and it would maintain liaison offices in Baghdad and Kuwait

12   Ibid., 3.
13   Ibid., 4.

City.[14] The Security Council transformed the report's mandate into a formal one through its approval in SCR 689 (9 April 1991).[15]

UNIKOM had two initial tasks. First, it assumed control of the DMZ from the US 3rd Armored Division, which had established the patrol route in the zone, in early May 1991.[16] Second, and more importantly, as Iraq and Kuwait remained responsible for law and order on their respective sides of the DMZ, both states maintained a police presence in the zone. Between June and September 1991, Iraq moved a series of 14 police posts into the DMZ, including five into Kuwaiti territory. UNIKOM asked the Iraqi authorities to move them, only to be told that they:

> had been in place before 2 August 1990 and pulling them back would prejudice Iraq's position regarding the demarcation of the border. Once the demarcation had taken place, Iraq would comply with the 'reasonable distance' principle ....[17]

This was an example of Iraq's behaviour with regard to UNIKBDC and, by extension, to UNIKOM. The Iraqi government did not wish to do anything that weakened its claims of sovereignty.

By the summer of 1991, UNIKOM began to report its findings from its observations of the zone dutifully. The initial pattern also set interesting precedents. There were more Kuwaiti than Iraqi violations, although many of the land-borne violations were attributed to navigational errors or incidents pertaining to police within the DMZ. In addition, UNIKOM addressed with the Kuwaiti government a number of overflights by either American or Kuwaiti F-15 or F-16 aircraft.[18] These were perceived as relatively minor incidents and part of the process for both parties of learning to operate with the DMZ and its enforcement; this permitted the reduction of the number of UN rifle companies providing security from five to two.[19] The statistics surrounding the period from May to September 1991 show that the Kuwaitis and their allies were far more frequent violators and less respectful of the DMZ than the Iraqis.[20] The navigational challenge should not be

---

14   Ibid., 5.

15   Ibid., "Security Council Resolution 689 (1991), 9 April 1991 (S/RES/689)", 206–7.

16   Kindsvatter, P.S. "VII Corps in the Gulf War: Post-Ceasefire Operations", *Military Review* 72(6) (1992), 2–19.

17   United Nations, *The United Nations and the Iraq–Kuwait Conflict, 1990–1996*, "Report of the Secretary-General on UNIKOM, 3 September 1991 (S/23000)", 297.

18   United Nations, *The United Nations and the Iraq–Kuwait Conflict, 1990–1996*, "Report of the Secretary-General on UNIKOM, 12 June 1991 (S/22692)", 266.

19   United Nations Security Council, Letter dated 6 August 1991 from the Secretary-General to the President of the Security Council (S/22916). 9 August 1991. Available at: http://daccess-dds-ny.un.org/doc/UNDOC/GEN/N91/257/85/IMG/N9125785.pdf?Open Element [accessed 15 August 2012].

20   For the period 1 April–30 September 1992, Iraq had a total of five violations (none from the air), while Kuwait had 39 (one from the air) and the allies had 11 (all but one

underestimated and it was noted that as UNIKBDC made progress, the number of violations decreased.[21]

The Iraqi government continued to complain about UNIKBDC and by mid-1992 stopped participating altogether, claiming:

> that the Commission's work was political – that the Governments of the United States and United Kingdom in particular were seeking to deprive Iraq of its rights and justify the ongoing presence in the region and military bases.[22]

Despite Iraq's lack of co-operation, UNIKBDC demarcated the land boundary by the summer of 1992. This clarified the border and reduced the possibility of further navigational errors on both land and water.

The Southern No-fly Zone

The Iraqi government dealt ruthlessly with all armed resistance in southern Iraq after the spring of 1991. Unlike in northern Iraq, little was done inside Iraq to address this issue. The international community eventually became concerned about the human and environmental costs of Iraq's counter-insurgency. The coalition simultaneously came to believe that it needed additional forces in the region to monitor and react to the situation.

The United Nations became increasingly concerned that the Iraqi government's actions were excessively violent and showed little regard for human rights. In early 1992, the UN's Special Rapporteur of the Commission for Human Rights, Max van der Stoel, reported:

> Recent and continuing measures instituted by the Iraqi military forces against the population of the marshes (including Marsh Arabs, internally displaced persons and refugees, and army deserters), are said to include the tightening of control over food destined for the area, the confiscation of boats, and the evacuation of all areas within three kilometres of the marshlands. Further reports indicate that military attacks have been launched against the Marsh Arabs between 4 December 1991 and 18 January 1992, resulting in hundreds of deaths.[23]

---

from the air). In addition there were 20 unidentified violations. United Nations Security Council, *Report of the [SG] on [UNIKOM] (for the period 1 April–30 September 1992)*, 2 October 1992 (S/24615).

21   UN Security Council, Letter dated 6 August 1991.

22   United Nations, *The United Nations and the Iraq–Kuwait Conflict, 1990–1996*, 50.

23   United Nations, *The United Nations and the Iraq–Kuwait Conflict, 1990–1996*, "Report on the situation of human rights in Iraq prepared by the Special Rapporteur of the Commission on Human Rights on the situation of human rights in Iraq E/CN. 4/1992/31", 408.

In March 1992, the Security Council cautioned Iraq about such activities.[24] Southeastern Iraq is the site of numerous and sizeable marshes due to the confluence of the Tigris and Euphrates rivers. Despite such warnings, the Iraqi government continued its operations in the marshes through that spring and summer. From 30 June 1992, the Iraqi government blocked relief operations in southern Iraq. These operations coincided with a government-sponsored drainage of the marshes.

In spring 1992, the Iraqi government forces sought to deny Shi'a rebels a refuge and started to drain parts of the marshes as part of the "Third River Project". While the government maintained that it was creating a navigable canal, this caused fresh water to be deliberately drained, which had potentially serious environmental implications.[25] Iraqi government spokesmen, blaming the situation on Turkish and Syrian damming, claimed that it was necessary for them to drain the saltwater marshes for irrigation in support of agriculture in southern Iraq due to the reduced flow of the Euphrates River.[26] Neither of these arguments seemed credible given the visible connection between the project and Iraqi military operations. Van der Stoel stated that the situation in southern Iraq was replete with human rights violations and called Iraq's actions a threat to the UN's relief operations in the area. Even the Iraqi representative acknowledged the existence of a deliberate blockade on the marshes.[27] He later argued that such operations were necessary to get rid of saboteurs and criminals who were using the marshes as a haven.[28] Marsh drainage exposed Shi'a rebels to attack and prompted a renewed stream of refugees into Iran.[29] The international community came to believe this was a ruse for counter-insurgency operations in southern Iraq. Yet the counter-insurgency in southern Iraq was less of a concern than other issues.

---

24   United Nations, *The United Nations and the Iraq–Kuwait Conflict, 1990–1996*, "Statement by the President of the Security Council concerning general and specific obligations of Iraq under various Security Council resolutions relating to the situation between Iraq and Kuwait, 11 March 1992 (S/23699)", as well as "Statement by the President of the Security Council concerning Iraq's compliance with the relevant Council resolutions, 12 March 1992 (S/23709)", 421–5.

25   This could have led to the creation of salt marshes and the destruction of arable land. See: Murphy, C. and Boustany, N. "Iraqis Seek to Drain A Haven For Foes", *International Herald Tribune*, 3 July 1992.

26   Beschorner, N. "Water and Instability in the Middle East", *Adelphi Paper* 273(1992), 27–44.

27   United Nations Security Council, Letter Dated 3 August 1992 from the Chargé d'Affaires A.I. of the Permanent Mission of Belgium to the United Nations Addressed to the President of the Security Council, S/24386. 5 August 1992, 3–7.

28   United Nations Security Council, Provisional Verbatim Record of the [3105] Meeting, S/PV.3105. 11 August 1992, 23–9, 31–3.

29   Bruce, J. "Campaign Against Shi'ites Hardens", *Jane's Defence Weekly*, 6 June 1992; Stapleton, B. "Arabs Flee Iraq's Deadly Marshes", *The Sunday Independent*, 19 July 1992.

A series of crises in 1992 led the coalition to conclude that Iraq responded more favourably to demands when confronted with the possibility of the imminent use of force. The Security Council met with representatives of the Iraqi government in mid-March 1992 to make its concerns about weapons of mass destruction (WMD) clear to Iraq in addition to the tense situation in the marshes. Later that month, the coalition reinforced the point by issuing an ultimatum; if Iraq failed to provide the relevant information and assist in the destruction of certain WMD-related facilities the coalition would strike a week later.[30] The arrival of the USS *America* carrier group in the Gulf reinforced this warning.[31] The Iraqi government quickly provided the information and assistance.[32] It had taken almost two months for the coalition to reach this stage, at which they were politically and militarily prepared to use force. Though effective, this was considered too long to deal effectively with challenges from Iraq.

Another crisis involving the UN Special Commission on the Disarmament of Iraq (UNSCOM) occurred in early July 1992. An American-led team was denied access to the Iraqi Ministry of Agriculture in order to search for WMD-related documentation.[33] Shortly after, the coalition reached an "agreement-in-principle" about air strikes, but its members still disagreed over the issue of a fixed timetable for an ultimatum.[34] This reduced the credibility of the coalition's threat. The US government then exerted diplomatic efforts to obtain local support in preparation for the use of force. American Secretary of State James Baker visited Saudi Arabia to ensure King Fahd's support. The crews of US warships in the Mediterranean had their port leaves cancelled, and the amphibious group based on the USS *Tarawa* steamed into the Gulf.[35] This came with the implied threat of additional ground forces to bolster Kuwait, but, more importantly, it came with the possibility that the coalition could use missiles or naval-based air power to coerce the Iraqi government. The threat of force by the coalition and the UN's offer of a compromise, in which the inspection team would be made up of nationals of "neutral" European states, led to Iraqi acquiescence in late July.[36] Nonetheless, the American naval presence continued to grow with the arrival of the USS *John F. Kennedy* carrier group to add to those already on station in the

---

30   Lewis, P. "UN Gives Baghdad March 26 Deadline", *International Herald Tribune*, 19 March 1992.

31   "U.S. Sends Aircraft Carrier to Gulf in 'Signal'", *International Herald Tribune*, 14 March 1992.

32   United Nations, *The United Nations and the Iraq–Kuwait Conflict, 1990–1996*, "Third Report of the Executive Chairman of UNSCOM, 16 June 1992 (S/24108)", 442–3.

33   "Coups and Cussedness: Saddam Hussein Holds Out in Iraq", *The Economist*, 11 July 1992; Kagian, J. "New Confrontation", *Middle East International*, 10 July 1992.

34   Almond, P. "Allies List Their Targets", *The Daily Telegraph*, 23 July 1992; Walker, M. "Allies Ready for New Air War in Gulf", *The Guardian*, 24 July 1992.

35   Dettmer, J. and Bone, J. "Bush Calls Council of War over Iraqi Strike", *The Times*, 25 July 1992.

36   *Arms Control Reporter 1992*, 453.B.132.19–20.

eastern Mediterranean (the USS *Saratoga* and the USS *Independence*). A number of Patriot missile batteries also deployed to Kuwait.[37] Once the new inspection team began its activities without interference on 29 July, the *John F. Kennedy* received orders to leave the Gulf.[38] The presence of military forces was seen as the reason for the effectiveness of the threat, but the United States could not maintain such a protracted naval effort forever.

Another crisis developed in early August. The Iraqi government announced that all of its ministries were out of bounds for UNSCOM's inspection teams, but once again backed down.[39] Iraq's lack of co-operation was a major source of frustration for the chairman of UNSCOM, Rolf Ekeus, but it was not the only one. He publicly expressed his dissatisfaction with the Security Council and its lack of speed or effectiveness in dealing with the crises in late July.[40] This suggested that the US government needed to deal with Iraqi provocations in a timely manner.

Throughout the summer of 1992, Iraq was a significant irritant to the coalition. The Bush Administration's statements reflected a great deal of annoyance and frustration about Iraq's adversarial relationship with UNSCOM.[41] This called into question George H.W. Bush's ability to deal with foreign policy (his major strength as president). American voters perceived Iraq's lack of co-operation as a policy failure on the part of the president. This perception also existed in government circles.[42] Reaching for a solution, the US government considered the pursuit of another Security Council's resolution to stabilize the situation in Iraq.[43] The stability of Iraq was one of the major motives for the United States to remain involved in the region.[44] By attacking the Shi'a and others, the Iraqi government created difficult situations. Iraq's neighbours had to deal with refugee crises and it was not difficult to discern the effects on Iraq had a neighbour acted to address

---

37   "US bolsters Gulf Power to Warn Iraq: Exercises in Kuwait Planned", The Financial Times, 28 July 1992; "U.S. Sends Missiles to Kuwait To Bolster Military Strength", *The Wall Street Journal*, 28 July 1992.

38   Lewis, P. "UN Arms Inspectors Enter Disputed Site in Baghdad", *International Herald Tribune*, 29 July 1992.

39   United Nations, *The United Nations and the Iraq–Kuwait Conflict, 1990–1996*, "Fourth Report of the Executive Chairman of UNSCOM, S/24984 (17 December 1992)", 495.

40   *Arms Control Reporter 1992*, 453.B.132.20; Littlejohns, M. "UN Council Under Fire Over Iraq", *The Financial Times*, 5 August 1992.

41   Dettmer, J. "Bush Wants New UN Sanction for Military Action Over Shias", *The Times*, 30 July 1992; Graham, G. "US on Collision Course with Iraq: Congressional Leaders Give Bush Their Backing", *The Financial Times*, 29 July 1992; Walker, M. "Bush Presses UN to 'Force Iraq Issue'", *The Guardian*, 29 July 1992.

42   For an example, see Hilsman, R. *George Bush vs Saddam Hussein* (Novato: Lyford Books, 1992).

43   Murphy, C. "U.S. and Allies Act to Press Baghdad on Shiites", *International Herald Tribune*, 12 August 1992.

44   "DoD Authorization for Appropriations for FY 1994 and the Future Years Defense Program (Part 1)". Congressional Information Service Document S201-1 (1994), 317–18.

the problem at its source, that is, by intervening in Iraq as opposed to merely repatriating displaced persons. Realpolitik, presented as benign humanitarianism, did not seem to affect the international consensus on Iraq. Such concern allowed for the presence of forces sorely needed to convince the Iraqi government to co-operate.

The coalition's other members shared the American and international concern about the Iraqi government's actions with regard to the Shi'a in southern Iraq and UNSCOM. The French government, having been enthusiastic about SCR 688, wanted to do something similar for the Shi'a and Marsh Arabs to what had been done for the Kurds. Having been a major proponent and advocate of SCR 688, it is hardly surprising that the French government issued statements reflecting its desire to extend the reach of that resolution.[45] Subsequent statements revealed that it was also considering the conduct of an operation similar to Operation Provide Comfort II, launched in July 1991, which was the coalition's establishment and maintenance of a NFZ over northern Iraq to provide security for the Kurdish residents of that region.[46] The French government was, however, concerned about the legitimacy of any operation:

> They had, from the start, made it very clear what they could and could not do. The French were very even-handed in their approach and made it clear that they were there to enforce UNSC [Security Council] decisions. They were not there to punish or coerce the Iraqi government.[47]

This was a literal application of the *droit d'ingérence*, the belief in the right to interfere if a humanitarian issue is at stake, to the situation in southern Iraq.[48]

The British government was more cautious in its approach to the situation. Its statements emphasized the need to monitor Iraq and keep Iraq's government from acting inappropriately. For example, Prime Minister Major publicly stated that:

> What we have said to the Iraqi authorities is that we are now perfectly clear that they have engaged in systematic repression in the south of Iraq but that is not acceptable and that it has got to stop. What we propose to do, therefore, is to monitor the whole area from the air and whilst we are doing that to ensure the

---

45   Roland Dumas, "Interview de M. Roland Dumas, Ministre d'Etat, Ministère des Affaires Etrangères Par l'Hebdomadaire Tunisien 'Réalités'", *French Foreign Ministry Press Release*, 13 August 1992.

46   Roland Dumas, "Propos à la Presse de M. Roland Dumas, Ministre d'Etat, Ministère des Affaires Etrangères, à la suite de son audition devant la commission des Affaires Étrangères de l'Assemblée Nationale", *French Foreign Ministry Press Release*, 18 August 1992.

47   Anonymous Interview with author, 7 August 1998, 3.

48   One might argue that it is a forerunner to the Responsibility to Protect (R2P) doctrine.

security both of the Shias [sic] and of their aircraft we will instruct the Iraqis not to fly in that area.[49]

Douglas Hurd, Foreign Secretary, provided a further example by stating:

We believe in the integrity of Iraq. Iraq is one country but within that country its rulers have obligations towards their subjects, which is laid down in Security Council resolution 688.[50]

It is important to note that the coalition focused on Iraq's treatment of its citizens and not Iraq's sovereignty. The latter was a contentious issue in international forums and offered the Iraqi government a credible argument against the coalition's treatment of Iraq.

The Gulf States and other interested governments expressed concerns about any military operation in southern Iraq. Kuwait was the only state to offer unequivocal support for military operations.[51] Yet the Kuwaiti and the Saudi governments both feared the possibility of rendering the area vulnerable to Iranian fundamentalism.[52] Nonetheless, the Saudi government agreed to provide support in terms of basing and financing.[53] A number of other Arab states were opposed to a renewed Western military presence.[54] This affected the British contribution, as its government maintained close relations with the Gulf States, and they sought to delay the operation, fearing a negative reaction as a result of basing forces in the Gulf.[55]

The political constraints on a coalition force presence shaped the nature of the force. None of the Gulf States wished to see a large presence of "Western" forces in their territories due to internal security concerns. Combined with the concerns about a potential occupation of Iraqi territory and Iraq's desire to maintain its sovereignty, this factor drove the coalition to choose air power. Given that the maintenance of an aircraft carrier stationed in the Gulf on a permanent basis

---

49   "Transcript of interview given by the UK Prime Minister, Mr John Major, in London, on Tuesday, 18 August 1992", in Weller, M. (ed.), *Iraq and Kuwait: The Hostilities and Their Aftermath* (Cambridge: Cambridge University Press, 1993), 723.

50   "Douglas Hurd, Secretary of State, UK, interview with 'Today' programme, 19 August 1992", in Weller, *Iraq and Kuwait*, 723.

51   "Arab Reluctance Delays 'No Fly' Zone", *International Herald Tribune*, 25 August 1992.

52   Evans, K. "Gulf Leaders Back Allies But Fear Break-up of Iraq", *The Guardian*, 21 August 1992.

53   Anonymous interview with author, 5.

54   Binyon, M. "Arabs Hesitate Over Western Move to Close Iraq Air Space", *The Times*, 19 August 1992.

55   Philps, A. "Arab Fears Delay RAF Support for Iraq No-Fly Zone", *The Financial Times*, 25 August 1992.

required more carriers than were available in the American arsenal, any option had to be land-based.

The coalition began to put a plan in motion while the Iraqi government sought to prevent any action. President Bush implied that action was needed due to the Iraqi foreign minister's refusal to allow human rights monitors in Iraq.[56] On 20 August 1992, the Iraqi government announced that it would allow the coalition to inspect the marshes region.[57] This was a partial concession, as the Iraqi government had refused to permit the re-entry of UN personnel from Bahrain.[58] This was the same gambit they employed in vain against Operation Provide Comfort in Kurdish regions in northern Iraq.

The coalition, Joint Task Force – Southwest Asia (JTF-SWA), composed of the United States, the United Kingdom and France and hosted by Saudi Arabia, launched Operation Southern Watch on 26 August 1992. The purpose of the operation was stated clearly: "the coalition has concluded that it must itself monitor Iraqi compliance with UNSCR 688 in the south".[59] President Bush claimed that its purpose was to support SCR 688 by creating the SNFZ, thus denying the Iraqis the use of the airspace below the 32nd parallel:

> [T]he United States and its coalition partners have today informed the Iraqi government that 24 hours from now coalition aircraft, including those of the United States, will begin flying surveillance missions in southern Iraq, south of the 32 degrees north latitude, to monitor the situation there. This will provide coverage of the areas where a majority of the most significant recent violations of [UNSC] Resolution 688 have taken place ... It will remain in effect until the coalition determines that it is no longer required.[60]

The JTF-SWA flew a mix of planes to carry out this mission. There were aircraft designed for air superiority (F-14, F-15C, F-16, F-18, Mirage F-1, Mirage 2000), air reconnaissance (Tornado GR-1) and electronic warfare (F-4G, E-3, EC-135, EF111A), bombers (F-117A, F-15E) and ground attack aircraft (A-10).[61] The

---

56   George Bush, "Letter to Congressional Leaders Reporting on Iraq's Compliance With United Nations Security Council Resolutions", 16 September 1992, *The American Presidency Project*, University of California Santa Barbara. Available at: http://www.presidency.ucsb.edu/ws/?pid=21462 [accessed 16 August 2012], 1.

57   "UN Envoy to Leave Baghdad as Relief Talks Break Down", *International Herald Tribune*, 22–3 August 1992.

58   Doyle, L. "Iraq Rejects UN Olive Branch", *The Independent*, 22 August 1992.

59   "Statement Issued by the Members of the Coalition at New York, 26 August 1992, 10 a.m. (New York Time)", in Weller, *Iraq and Kuwait*, 725.

60   George Bush, "Remarks on Hurricane Andrew and the Situation in Iraq and an Exchange With Reporters", 26 August 1992, *The American Presidency Project*, University of California Santa Barbara.

61   *JTF-SWA Briefing Package*, 35. It should be noted that the ground attack aircraft became more prevalent as a result of Operation VIGILANT WARRIOR.

coalition could monitor operations by being capable of detecting Iraqi operations while maintaining air supremacy.

From its inception in the summer of 1996, there were two main activities for the JTF-SWA: it sought to demonstrate its presence and to monitor events in southern Iraq. Its patrols were organized to fulfil these roles. The "standard Operation Southern Watch profiles" consisted of four fighters that would fly from Dhahran Airbase on the east coast of Saudi Arabia and head for the Iraqi–Saudi border. South of the border, they would undergo aerial refuelling before entering the SNFZ. They would fly around the zone for 30 to 45 minutes before returning to Dhahran. Occasionally, patrols were directed to fly over specific areas to observe events, but the main purpose of the patrols was to create radar signatures to demonstrate their presence.[62] Such actions established that the coalition was present and watching what occurred in Iraq. It was inevitable that the Iraqi government, due to the nature of Integrated Air Defence Systems, would detect the presence of coalition aircraft.

A NFZ creates particular requirements for air planners. Reconnaissance and air superiority aircraft are required for the monitoring of the airspace and territory under the zone. The key to a successful NFZ is the maintenance of a perpetual presence within the zone. This translates into a series of infrastructure requirements. First of all, airfields with facilities that allow for the maintenance of modern jet aircraft are required. Second, to enforce a NFZ, the force requires a "Command, Control, Communications and Intelligence" system that can provide planning direction for its units, control them while they are in the NFZ and ensure that the airspace between the NFZ and the airfields is free of conflict. Tanker aircraft are also required, as the airfields are frequently far from the NFZ – as was the case with the SNFZ. In addition, a perpetual presence requires aircraft to be in the NFZ for protracted periods. The decision to maintain tanker tracks over northern Saudi Arabia leads one to conclude that the coalition wanted to make the best use of every sortie by increasing loiter times. Given the maintenance requirements for aircraft, it is counterproductive to send aircraft for short periods of time, as they require the same number of maintenance hours regardless of whether the sorties last one or six hours.

The coalition was very concerned about force protection and the potential for casualties associated with operations over Iraq. The decision to select the 32nd parallel as the border of the SNFZ was not arbitrary. One commander recalled that:

> It was a political decision based on my recommendation and view of the No-Fly Zone. While it was a political decision, we did not want to see aircraft shot down or airmen paraded through Baghdad. The 32nd parallel was arbitrarily chosen because it meant our aircraft could tank over northern Saudi Arabia in safety, enforce up to the 32nd, and fly further if required. To push the No-Fly Zone further north meant that the refuelling would have to take place over southern

---

62   White, P. "Crises after the Storm", Washington Institute for Near East Policy Military Research Paper no. 2, 16.

Iraq, and this was dangerous. The tankers would be vulnerable to Iraqi fighters and [Surface to Air Missiles], and so would the aircraft being refuelled. It had the potential for huge numbers of casualties. Further north would have of course meant that more airfields and other installations were subject to the zone, but it would be very dangerous. "Flying in the Box", as it has become to be known, meant that the decisions and planning took into account the need to ensure that no aircraft were lost.[63]

These comments illustrated a particular problem for the JTF-SWA: it needed to be staged from a location that allowed it to maintain sufficient coverage of the NFZ without unnecessary effort, where the JTF-SWA could react quickly in the event of an Iraqi provocation, but outside the range of Iraqi forces. An attack by Iraqi ground or air forces was considered highly improbable, but the possibility of previously a well-hidden Scud missile (or even a rocket with a range of less than 150 km) was a less a remote possibility.

So what could the coalition see from the skies over southern Iraq? As coalition aircraft flew over southern Iraq, they could also gather information in the course of monitoring. The coalition could engage in the process of target acquisition, and the coalition's reconnaissance aircraft were very helpful in this regard. Coalition forces received very realistic training as a result of such provocations and the ability to reconnoitre potential targets. One United States Air Force (USAF) officer noted that:

> Flying over southern Iraq affords us the opportunity to scout out the targets we will be tasked to hit in wartime, practice attacking them, and evaluate and refine our tactics and thereby our chances for success.[64]

Such information was necessary to make assessments of the nature of particular target sites by gauging the relative weight and type of air defence coverage, the best routes and altitudes for attack, and the suitability of targets (in terms of the possibility of collateral damage or the target's proximity to other installations such as hospitals or other facilities). The coalition could also analyse the target sets and their relationships to one another, leading to a near real-time intelligence picture of Iraq as a system of target sets. The SNFZ did not contain a significant amount of individual targets. It contained the majority of the Southern Air Defence Sector and some key transportation links, but only a small number of WMD-related sites clustered around Baghdad. Given that UNSCOM was also present and exchanged information with the coalition, whether this was intentional or not, the coalition's knowledge of Iraq increased significantly.

At first glance, the establishment of the SNFZ was effective in reducing the air threat from southern Iraq. Evidently remembering the air campaign in the

---

63   Anonymous interview with author, 7 August 1998, 2.
64   Gration, S. "Combat Smart, Inherently Safe", *Combat Edge* 4 (May 1996), 10.

Gulf War, the Iraqi Air Force promptly moved its aircraft out of the SNFZ on 26 August 1992.[65] The original commander of JTF-SWA, USAF Lieutenant General Michael Nelson, noted in early September of 1992: "We've been at this almost two weeks and he [Iraqi President Saddam Hussein] has clearly decided not to challenge the 'no-fly' zone".[66] President Bush publicly concluded that the mission had succeeded by mid-September 1992 in protecting the people of southern Iraq from attacks by the Iraqi Air Force.[67] In November 1992 there were indications of some "small-scale" activities by the Iraqi Army in the area.[68] The coalition had to remain to monitor the situation.

In legal terms Operation Southern Watch was based on the precedent set by the Northern NFZ. The logic of SCR 688 – there was a perceived need to protect the persecuted elements of Iraqi society but no action was authorized specifically by the United Nations – was applied to southern Iraq. It would seem reasonable that a state's action be considered to be justified as long as the following conditions are satisfied. There should be:

- A suitable reason to act forcefully to modify a state's behavior.
- An agreement within the international community on the ends being pursued.
- An agreement within the international community that the ends being pursued warrant the use or threat of force.
- A credibility of the belief that the ends being pursued are representative of international desires as opposed to national objectives.

The coalition's governments claimed that SCR 688 provided sufficient justification for the operation.[69] Given the Iraqi government's actions, they were not wrong. So how did this apply to the aforementioned conditions? The first condition appeared to be instantly satisfied by the general frustration with the Iraqi government's human rights record. The other conditions proved to be more contentious.

The desired ends of Operation Southern Watch were unclear and this lack of clarity had particular implications. Like Operation Provide Comfort II, it represented what could be done given a series of political limitations. It represented the proverbial "lowest common denominator" by allowing the coalition to provide for the security of the Shi'a without intruding too deeply into Iraqi affairs. In this

---

65   Dettmer, J. and Walker, C. "Iraq Moves Combat Aircraft Away From Shia Marshlands", *The Times*, 26 August 1992.

66   Bird, J. "Southern Watch", *AFT*, 28 September 1992.

67   "Bush Says Iraq Halts Raids on Shiites", *The International Herald Tribune*, 18 September 1992. Note, the Iraqi government, on a number of occasions, employed helicopters in support of its counter-insurgency operations in southern Iraq.

68   "Written Answers", 18 November 1992, *Hansard [of the United Kingdom]*, 6th Series, Vol. 214, Column 250 written.

69   Doyle, L. "UN was Bypassed over 'No Fly Zone'", *The Independent*, 19 August 1992.

case, the desired end state was the absence of counter-insurgency operations, or operations so weak that a refugee problem would not be created. Yet it offered the coalition a potential tool for supporting the containment of Iraq and this contributed to the international doubt about American motives in the second half of the decade.

What represented the will of the international community? On the one hand, Security Council resolutions assign a legal quality to what are essentially political decisions and are useful in this regard. The existence of a philosophy of intervention (that is, the *droit d'ingérence*, now called the "responsibility to protect") within international discourse could also be considered representative without requiring recourse to a political or legal authority. This was a curious situation. Both the British and French governments favoured the argument of the responsibility to protect, consistent with a philosophical outlook, but the American government consistently argued that SCR 688, a political decision with legal qualities, provided sufficient justification.[70] This argument assumed that some form of approval (even if not direct or considered as binding) was required from the international community to avoid difficulties, as the responsibility to protect was not considered to be universal or even a right by most states. To argue that SCR 688 was insufficient would have weakened the American position with regard to Operation Provide Comfort II. Eventually, the British government changed its position to match the American.[71] This argument reinforced the idea that international law is fundamentally driven by consensus as opposed to controlled by rules set by a central authority, despite the cynical use of such rules by various governments. Such arguments were therefore only as valid as the international community decided and few governments shared this interpretation of the situation. The coalition's concerns for the Shi'a, much like its concerns for the Kurds, were sufficient to create a consensus within the international community that Iraqi sovereignty could be violated if it kept the situation in southern Iraq relatively calm and saved some lives. Ironically, it breathed some life into the *droit d'ingérence*, the concept of the responsibility to protect.

The rules of engagement (ROE) further complicated the legal situation that surrounded Operation Southern Watch. A coalition spokesperson described the ROE by stating:

> No threat to coalition operations over southern Iraq will be tolerated. The Iraqi
> Government should know that coalition aircraft will use appropriate force in

---

70   Bone, J. "Security Council Members Query the Legal Basis of 'No-Fly' Zone", *The Times*, 28 December 1992.

71   Doyle, L. and Richards, C. "'No-Fly Zone" Imposed on Iraq', *The Independent*, 27 August 1992. Under the droit d'ingérence: "Limited armed action on behalf of a population in danger of being exterminated is legally justified, even in the absence of positive authorisation from the Security Council". See Weller, M. "Intervention Plans Lack Specific UN Sanction", *The Times*, 20 August 1992.

response to any indication of hostile intent as defined in previous diplomatic demarches. Inter alia, illumination and/or tracking of aircraft with fire control radars and any other actions deemed threatening to coalition aircraft, such as the intrusion of Iraqi aircraft in the NFZ, would be an indication of hostile intent.[72]

This stems from the state's right to self-defence enshrined in Article 51 of the UN Charter. As the forces conducting Operation Southern Watch were monitoring compliance with SCR 688, they needed some justification for the use of force in SNFZ enforcement. However, in the absence of de jure authorization for their presence over Iraq, this position was dubious. The Operation Southern Watch ROEs, promulgated by the commander of JTF-SWA in accordance with the agreements between the coalition members, allowed force in self-defence. Due to the nature of Iraqi air-defence weapons, the target needs to be "illuminated" by radar prior to launching the missile.[73] This led to the "illumination" of targets being perceived as a threat and, therefore, sufficient justification to attack air-defence radars and weapons systems. Larger uses of force, such as deliberate air strikes, came to require more elaborate justifications. However, coalition forces were already present over the skies of Iraq due to a de facto authorization and their ROEs permitted them to use force prior to the development of crises if threatened by Iraqi forces.

One last point needs to be considered in light of the SNFZ. UNIKOM also employed helicopters to supplement their surveillance of the DMZ by ground patrols and observation posts as well as other utility tasks such as liaison and casualty evacuation, but these were suspended on the Iraqi side of the border from December 1998.[74] Given that UNIKOM's observers reported that there were violations of the airspace, the lack of any evidence of problems in airspace coordination suggests that some form of airspace control was exercised by the forces conducting Operation Southern Watch (that is, no low-level transits of the DMZ by coalition aircraft) or between those forces and UNIKOM.

Progress Made in Terms of Security Council Resolution 687?

It was not clear in late 1992 whether the SNFZ represented progress or evidence of a lack thereof. On the one hand, UNIKOM had a significant degree of activity to

---

72   "Statement issued by the Members", in Weller, *Iraq and Kuwait*, 725. Illumination of coalition aircraft was treated as a hostile act.

73   This included the SA-2, SA-3, and Roland systems. The remainder of Iraqi air defence weapons are passive infrared guided. For technical details see Cullen T. and Foss, C. *Jane's Land-Based Air Defence 1996–1997* (Coulsdon: Jane's, 1996), 8–10, 98, 100, 102, 113, 115, 140, 247 and 250.

74   United Nations Security Council, *Report of the Secretary-General on the United Nations Iraq–Kuwait Observer Mission*, 30 March 1999 (S/1999/330).

track within the DMZ. On the other, UNIKBDC's work bore fruit, though it is not possible to draw a causal link to the SNFZ's coming into existence.

The influx of coalition aircraft over the DMZ associated with the enforcement of the SNFZ meant that UNIKOM's military observers had much more to report. Indeed, even the nature of the reporting changed to reflect which violations were Iraqi, Kuwaiti, "Allies" (read "coalition"), and unidentified.

The effect of the SNFZ imposition is discernible from September 1992. The number of allied and unidentified air violations began to increase in that period. UNIKOM made its concerns known to the relevant parties in all cases of violations. The UN Secretary-General noted that since the SNFZ came into existence, UNIKOM noted an increase in the number of flights over the DMZ; however, these tended to be too high to allow identification. He also asked those governments that declared the SNFZ to avoid the DMZ.[75] The problem, from the coalition's perspective, is that this reduced the flexibility of ingress/egress routes for its aircraft to the Saudi–Iraqi border. While this border was far larger, it would increase the degree of logistical effort required to maintain the same effect in the SNFZ.

The effect of the SNFZ on UNIKOM's reporting became more apparent as time progressed. In the fall of 1992 and early 1993 the number of violations increased, although the rate decreased over time. Some of the unidentified violations can be explained, however, as due to increased flight activity over the SNFZ in reaction to heightened tensions between the coalition and Iraq. These came to a head in January 1993.

Realization: Security Needed

In the fall of 1992, despite the Iraqi government's misgivings and complaints, UNIKBDC was able to complete its study of the 1963 border between Iraq and Kuwait. The study was submitted shortly thereafter to the Security Council for approval. The Security Council's members wished to bring this issue to a close quickly and approved UNIKBDC's finding in late 1992.[76] This, however, brought up an old point of friction. Iraq still maintained some police forts in the DMZ and the Security Council ordered their removal no later than 15 January 1993.[77]

75 Ibid.

76 United Nations, *The United Nations and the Iraq–Kuwait Conflict, 1990–1996*, "Security Council resolution concerning the work of the Iraq–Kuwait Boundary Demarcation Commission", S/RES/773 (1992), 26 August 1992, 473. "Statement by the President of the Security Council concerning general and specific obligations of Iraq under various Security Council resolutions relating to the situation between Iraq and Kuwait", S/24836, 23 November 1992, 486–90.

77 United Nations, *The United Nations and the Iraq–Kuwait Conflict, 1990–1996*, "Letter dated 8 January 1993 from the President of the Security Council addressed to the Secretary-General", Annex I to "Special report by the Secretary-General on UNIKOM,

This combination of the UN's requests and the Iraqi government's intransigence contributed to increased tensions between Iraq and the coalition over the skies of the DMZ and southern Iraq.

The crisis of January 1993 developed as a result of the of Iraqi government's testing of the international community's will to uphold SCR 687 and its supporting resolutions. It denied overflight rights to aircraft supporting the disarmament effort; stepped up its resistance in the NFZs; and tolerated, if not abetted, "riots" that crossed from the Iraqi to the Kuwaiti side of the DMZ and forcefully retrieved materiel and munitions that previously belonged to Iraq. This, in turn, led to increased activity in the SNFZ and missile strikes against targets in Baghdad on 17–18 January 1993. It was only after this point that the Iraqi government pledged to cooperate.

This crisis had two effects. One, the United Nations realized that it needed to take greater measures for its own security. As a result of the "riot" in January 1993, the Council passed SCR 806, which decreed that UNIKOM was to be augmented by three battalions of mechanized infantry.[78] No nation was willing to deploy these forces to UNIKOM, as the international community became overburdened with a series of peacekeeping missions in 1992–1993.[79] In October 1993, only the Bangladesh government came forward to offer a single, unequipped, infantry battalion, which the Kuwaiti government promptly equipped.[80] The reinforcement of UNIKOM became nothing more than a symbolic and slightly partial gesture. The battalion's companies were dispersed so that each of the three sectors could rely on security forces.

Two, the Iraqi government, having drawn attention to the matter, argued that UNIKOM was far from impartial. In April 1993, it complained about the disparity in violations, since from 1 April 1992 to 31 March 1993 there were only 29 Iraqi violations as opposed to 313 Kuwaiti and/or allied violations.[81] The number of complaints began to drop, however, in 1994. There are a number of potential explanations. First, the evidence available is thin; while UNIKOM's records are clear from 1991 to 1996 and from 2001 to 2003; there is less information available

---

S/25085, 10 January 1993, 514; Ibid., "Statement by the President of the Security Council concerning general and specific obligations of Iraq under various Security Council resolutions relating to the situation between Iraq and Kuwait", S/24386, 23 November 1992, 487.

78   Ibid., "Security Council resolution concerning UNIKOM", S/RES/806 (5 February 1993), 525–6.

79   International public opinion shifted to focus on the famine in Somalia and the impending war within the Yugoslav federation. The major commitments were the United Nations Somalia Operation (UNOSOM I), its successor the United Shield Task Force (UNITAF) and the United Nations Protection Force (UNPROFOR) in the former Yugoslavia.

80   "Letter dated 15 October 1993 from the Secretary-General to the President of the Security Council concerning the composition of UNIKOM", S/26621 (24 October 1993), UN, 596.

81   United Nations Security Council, Letter dated 26 April 1993 from the Representative of Iraq to the Secretary-General, 27 April 1993 (S/25677).

about the intervening period. Second, UNIKBDC's work was complete by 1993 and the Kuwaiti government took additional measures to address the matter of the border by constructing a series of obstacles colloquially known as the "Kuwait–Iraq border fence". This meant that UNIKOM's importance began to wane. Third, the period from February 1993 to October 1994 was one of relative calm that saw genuine progress on all of the programs mandated by SCR 687. Fourth, the reports to UN Headquarters in New York from UNIKOM were extremely brief and lacked detail.[82] Last, the nature of subsequent provocations (the June 1993 assassination attempt on George H.W. Bush, Iraq's feint or rehearsal for an invasion of Kuwait in October 1994, the September 1996 Kurdish crisis and the December 1998 inspections crisis) meant that the coalition took steps to address the "Iraq–Kuwait dispute" as it saw fit.[83]

While UNIKOM would continue to report diligently, it was becoming increasingly irrelevant and working from the relative safety afforded by coalition aircraft and, on occasion, American brigades deployed to the Kuwaiti desert. Yet the number of crises eroded the international community's will to enforce earlier resolutions and by the end of the decade, the situation was not necessarily the "Iraq–Kuwait dispute" but the "United States–Iraq dispute". Tensions continued to mount.

End of Consensus

The period from 1998 to early 2003 is best described as the heading above suggests. Prior to Operation Desert Fox in December 1998, it appeared that Iraq could be disarmed with some "encouragement". After that operation and an increase in Iraqi resistance, the United Nations opted for less intrusive and more engaging approaches towards the Iraqi government.

---

82   United Nations, *The United Nations and the Iraq–Kuwait Conflict, 1990–1996*, "Report of the Secretary-General on the UNIKOM for the period of 1 April–29 September 1994), S/1994/1111 (29 September 1994)", 664; Ibid., "Report of the Secretary-General on the UNIKOM for the period of 7 October 1994–31 March 1995, S/1995/251 (31 March 1995)", 726; Ibid., "Report of the Secretary-General on the UNIKOM for the period of 1 April–30 September 1995, S/1995/836 (2 October 1995)", 765. These refer to "limited numbers of violations involving mostly over flights". See also United Nations Security Council, *Report of the Secretary-General on the United Nations Iraq–Kuwait Observation Mission*, 24 September 1998 (S/1998/889); United Nations Security Council, *Report of the Secretary-General on the United Nations Iraq–Kuwait Observation Mission*, 30 March 1999 (S/1999/330).

83   For example, United Nations Security Council, *Report of the Secretary-General on the United Nations Iraq–Kuwait Observation Mission (1 April 1996–23 September 1996)*, 27 September 1996 (S/1995/801), 1. In the case of September 1996, UNIKOM reported seeing eight cruise missiles cross the DMZ.

As the completion of the removal of its WMD capability progressed, the Iraqi government became increasingly intransigent. The UN inspection teams found themselves increasingly unsuccessful and access to facilities hindered or denied in late 1997 and early 1998. This led the international community to gear up for a series of strikes dubbed Operation Desert Thunder that February, but a negotiated settlement prevented the operation from occurring. The preparations, however, led to UNIKOM reporting an increase in the number of air violations of the DMZ by coalition forces.

The deal struck in the early spring of 1998 held over the summer, but by November the Coalition was prepared to strike Iraq again. At issue were Iraq's lack of disclosure of WMD-related information and attempts at hindrance of inspections. The matter came to a head in December 1998, and the coalition struck before the Security Council could discuss the matter. The French government withdrew its forces from the NFZs and the Anglo-American coalition remained over the skies of Iraq, now contested by Iraq's ground-based air-defence forces. With DESERT FOX came a marked increase in Iraq's diplomatic and military resistance. While Iraq's air defenders attempted to hassle coalition aircraft, its foreign ministry delivered protest letters to the United Nations about Kuwait, the coalition, and UNIKOM's reporting of incidents.

The Iraqi government began a campaign of monthly letters to the Secretary-General in 1991, complaining about Kuwaiti collusion with the coalition's efforts and/or the number of coalition air violations of the DMZ. The letters that could be found at the time of writing dated from fall 2000 and appear on a monthly basis (if not more frequently) thereafter. The aforementioned letters were similar in tone and nature, although the details varied from letter to letter. In the letter transmitted in December 2000, the Iraqi foreign minister wrote to argue that UNIKOM was complicit in permitting the coalition to operate with impunity:

> On this occasion I wish to draw your attention once more to the fact that United States and British military aircraft continue to violate Iraqi airspace on a daily basis and to carry out acts of military aggression against Iraq, taking off from their bases in Saudi Arabia and Kuwait and from aircraft carriers belonging to their two States in the Arabian Gulf. A not inconsiderable number of those hostile military aircraft overfly the demilitarized zone in the course of the flights into Iraq which they make on a daily basis for the purpose of perpetrating acts of aggression against that country. They overfly the zone again when returning after carrying out those acts to their bases in Saudi Arabia and Kuwait. This constitutes a blatant violation of the relevant Security Council resolutions. The Mission is responsible for closely observing such violations and, in view of their seriousness, submitting immediate reports thereupon. However, close examination of the reports submitted by UNIKOM make it clear that their contents do not comply with the specifications of its mandate, namely, to observe any hostile action and determine the identity and nationality of the aircraft that overfly the demilitarized zone with a view to

mounting hostile actions against Iraq ... The pretext persistently put forward by the United Nations Observer Mission in the demilitarized zone in order to justify its inability to establish the nationality of the aircraft that violate the aforementioned zone is that those aircraft fly at extremely high altitudes, making it impossible to identify them or include that information in the Mission's periodic reports.[84]

In short, he presented the argument that the SNFZ is a violation of Security Council Resolutions and UNIKOM has the capacity to bring this to light. The Iraqi government, tracking aircraft on with their air-defence assets, drew different conclusions than UNIKOM's observers. With the monthly letters, they continued to argue their point of view. This did not exactly receive a warm reception in the United Nations. In one letter from the Secretary-General, the frustration was palpable:

[I]t is for the Security Council to interpret its own resolutions. Consequently, only the Council itself is competent to determine whether or not its resolutions are of such a nature and effect as to provide a lawful basis for the "No-Fly Zones" and for the actions that have been taken for their enforcement. Therefore, it is for the Council to address the lawfulness or otherwise of the actions to which you refer in your letter.

...

From 1999 to date, UNIKOM has recorded over 200 aerial violations of the demilitarized zone. In the majority of cases, however, it has not been possible for UNIKOM to identify the aircraft involved or to determine their nationality.

I should emphasize that the inability of UNIKOM to identify the States that are responsible for conducting such flights is in no way to be understood to constitute condemnation of them.

I would note in this regard that, in view of the fact that the United States and the United Kingdom have been conducting military air operations in the region, the United Nations has intervened with representatives of those States urging them to respect the demilitarized zone established by Security Council resolution 687 (1991) of 3 April 1991.[85]

---

84   United Nations Security Council, *Letter dated 25 December 2000 from the Permanent Representative of Iraq to the United Nations addressed to the Secretary-General*, 21 February 2001 (S/2000/1242).

85   United Nations Security Council, *Letter dated 21 February 2001 from the Secretary-General to the President of the Security Council*, 21 February 2001 (S/2001/160).

UNIKOM reported the violations as it understood them throughout the period it was active. See Table 10.1 for a summary. While the Iraqi government complained, it received little sympathy for its arguments against the SNFZ and about the

Table 10.1   Violations reported by the UN Iraq–Kuwait Observer Mission, 1999–2003

Period	Number of violations	Remarks
24 September 1999–30 March 2001 (S/2000/269)	77 violations total: 8 ground 20 weapons 1 maritime 48 air	One F-15 sighted over southern sector HQ. Iraqi SLO complained of incursion by air near Umm Qasr.
31 March–21 September 2000 (S/2000/914)	42 violations total: 15 ground 10 weapons 11 maritime 6 air	F-16 sighted over southern sector HQ. UH-60 helicopter sighted over DMZ. Gazelle helicopter sighted over DMZ.
22 September 2000–27 March 2001 (S/2001/287)	267 violations total: 101 ground 12 weapons 11 maritime 143 air	Gazelle helicopter identified three times, British Lynx identified once, F-18 twice and a pair of F-14s once.
28 March–24 September 2001 (S/2001/913)	255 violations total: 10 ground 8 weapons 74 maritime 163 air	
25 September 2001–20 March 2002 (S/2002/323)	437 violations total: 6 ground 1 weapons 9 maritime 421 air	
21 March–15 September 2002 (S/2002/1039)	278 violations total: 20 ground 4 weapons 21 maritime 233 air	
16 September 2002–21 March 2003 (S/2003/393)	714 violations total: 24 ground 14 weapons 16 maritime 660 air	18 involved helicopters and three involved unmanned aircraft.

Source: UN Security Council documents as listed.

partiality of UNIKOM. For most allied overflights, the altitude was reported as "too great for identification".

In 2002 and 2003, two trends coalesced to suggest that UNIKOM's tenure would soon end. First, the coalition's remaining members sought to address Iraq's lack of full compliance with the terms of SCR 687 once and for all. While engaging in the diplomatic preparations, coalition force levels in the region began to increase in order to use force if necessary. The increased coalition force presence led to a greater level of activity and the concomitant increase in the number of coalition violations of the DMZ's airspace. Events in the last six months led to the second point: three of the air violations in the last six months were not by manned aircraft.[86] The coalition's increase in the use of unmanned aerial vehicles (UAVs), likely intended to engage in reconnaissance over southern Iraq meant that both UNIKOM helicopters and UAVs used the same airspace over time. This suggests that there was either a degree of corroboration between UNIKOM and the coalition or that the latter acted unilaterally. Either way, UNIKOM's observers reported the violations.

"Lessons"

In hindsight, it is possible to suggest that there are potential "lessons" for others to learn about the relationships between forces operating directly on behalf of the United Nations and those operating indirectly for similar but distinct aims. The first is the potential effect the unusual symbiosis created by overlapping mandates. UNIKOM was a product of SCR 687 (1991) and the SNFZ was "consistent with" SCR 688 (1992) according to the coalition that launched it. Both missions were intended to provide security in the area, although as they came into existence for different reasons at different times, their ends, ways, and means differed significantly. UNIKOM was there to create security through its observation and reporting of incidents; this would produce transparency and stabilize the situation sufficiently to fix the border in accordance with the 1963 agreement. Operation Southern Watch, borne of the need to prevent a more widespread humanitarian crisis, was to prevent the Iraqi government from using its southern airspace as a vector for attacks on elements of its population. This, in turn, afforded the coalition the ability to monitor the situation in southern Iraq, which also meant it was present in the region and could react to crises rapidly. The presence, however, meant that coalition aircraft had to operate in and around southern Iraq; this made air violations likely if not inevitable. Based on the principle of impartiality, UNIKOM's observers dutifully recorded what they believed were violations. After 1993, however, they came to benefit from the presence of the aircraft over

---

86   United Nations Security Council, *Report of the Secretary-General on the UN Iraq–Kuwait Observation Mission (for the period from 16 September 2002 to 21 March 2003)*, 31 March 2003, S/2003/393.

southern Iraq, although the coalition's air assets never acted directly in support of UNIKOM. There was a distinct relationship between the number of coalition air violations and any deterioration in the Iraq–Kuwait situation. As UNIKOM observers came to need greater security, they received it as coalition aircraft flew overhead, but they had to report it. While dependent on the implied and actual threat of air strikes for their security, UNIKOM observers were compelled by the mandate to continue reporting in an impartial manner regardless of the cost.

Chapter 11

Observing Air Power at Work in Sector Sarajevo, 1993–1994: A Personal Account

F. Roy Thomas

No one watched the weather more closely in the fall of 1993 than the unarmed United Nations Military Observers (UNMOs) who manned the observation posts surrounding besieged Sarajevo, capital of the new state of Bosnia-Herzegovina. Every UNMO who was hoping to take a well-deserved time off from monitoring the shelling and counting the dead had to leave via the international community's air bridge to Ancona, Italy, or the UN charter flights to Zagreb, Croatia.[1] Every UNMO about to be posted out to a more benign sector had to leave through local airports. Truly *The Road to Sarajevo* that Major-General Lewis Mackenzie followed could not be taken by the 30 UNMOs who rotated into or out of Sector Sarajevo each month.[2] The life of every military observer in Sector Sarajevo was shaped to some extent by at least one of the tools of air power: they all arrived by air! Indeed, this chapter is literally made possible by air power, as it outlines how aerospace tools shaped what UNMOs in Sarajevo did from 15 October 1993 through 17 July 1994.

Fifteen minutes before landing on my first flight into the besieged city, the day that I arrived to assume command of the UNMOs of Sector Sarajevo, I put on my flak jacket and helmet. My UNMOs in Sector Sarajevo, serving as part of the much larger United Nations Protection Force (UNPROFOR), were dispersed among teams in observation posts: some inside the Bosnian city, some surrounding the Bosnian capital; a team in the so-called safe-haven in the city of Goražde, the only UNPROFOR presence there; and a team in the town of Žepa, also home to a Ukrainian UNPROFOR mechanized infantry company. The two safe-haven teams

1 Thomas, R. Testimony at Galić Trial, International Criminal Tribunal for the former Yugoslavia, IT-98-29 (2002); Thomas, R. Testimony at Milošević Trial, International Criminal Tribunal for the former Yugoslavia, IT-02-54 (2003); Thomas, R. Testimony at Karadžić Trial, International Criminal Tribunal for the former Yugoslavia, IT-95-5/18 (2010).Thomas R. Testimony at Mladić Trial, International Criminal Tribunal for the Former Yugoslavia, IT-09-92 (2012).

2 Mackenzie, L. *Peacekeeper: The Road to Sarajevo* (Vancouver: Douglas and McIntyre, 1993). An interesting historical note is that on the creation of UNPROFOR in early 1992, the United Nations decided to place their headquarters in Sarajevo, as the organization was "looking for a nice tranquil, neutral location from which to control our operations in Croatia" (Mackenzie, *Peacekeeper*, 119).

in Goražde and Žepa, in eastern Bosnia on the banks of the Drina, communicated with me in Sarajevo using capsat, a form of texting using satellite communications. "UNMO Sarajevo", as it was designated, included from 120 to 200 officers from any of 39 countries from all continents as well as up to 50 locals hired as UN interpreters.[3] "Welcome to Hell" said the graffiti smeared on the wall of a building. "Welcome", indeed, to what detail can be shared about how air power shaped UN observer teams during my nine months as the Senior UN Military Observer (SMO) for Sector Sarajevo.[4]

Situational Awareness October 1993

> There are no good guys, only villains and victims.
>
> Richard Round, 1993[5]

Attempting to label the conflicts ongoing in the former Yugoslavia was, and remains, a challenge. Canadian Colonel George Oehring suggests these labels:[6]

- in Slovenia, June to July 1991: "the War of Slovenian Independence";
- in what is now Croatia, from July 1991 to August 1995: "the War of Croatian Partition";
- in current Bosnia-Herzegovina, from April 1992 to December 1995: "the War of Bosnian Serb Secession";
- related to the above, alongside in 1993–1994: "the Bosnian Croat–Bosnian Muslim War";
- April 1993 to August 1995: "the War of the Bihać Pocket", which occurred within the above-mentioned "War of Bosnian Serb Secession".

In the case of the last three "wars" two of the three belligerent parties, the Bosnian Serbs and the Bosnian Croats, had no international standing as political entities. In

---

3   Thomas, R. "UN Military Observer Interpreting in a Community Setting", in Silvana Carr, Roda Roberts, Aideen Dufour and Dini Steyn (eds) *The Critical Link: Interpreters in the Community* (Amsterdam: John Benjamins, 1997).

4   See also Thomas, R. "Commanding UN Military Observers in Sector Sarajevo 1993–1994", in *In Harm's Way: The Buck Stops Here* (Kingston: Canadian Defence Academy Press, 2007), 1–25.

5   The best pre-deployment advice the author received was this quote from Richard Round, a Canadian officer serving with the European Community Monitoring Mission.

6   Maloney cites George Oehring, former Commander Sector South 1993–1994. The wars in Kosovo and Macedonia were yet to come in 1997. See Maloney, S. "Operation BOLSTER: Canada and the European Community Mission in the Balkans, 1991–1994", MacNaugton Paper 10(2) (1997). Pocket refers to an ethnic group surrounded by another in these cases. Some pockets were designated safe havens by the UN.

UNPROFOR's Bosnia-Herzegovina Command (BHC), the legitimate government (of Bosnia-Herzegovina) came to be associated with "Muslims" but, in fact, included both Bosnian Serbs and Bosnian Croats. Certainly 1993–1994 Sarajevo could be considered a war zone in all but name, at least in terms of the volume of fire and subsequent casualties.[7] For military personnel serving with UNPROFOR, including UNMOs, there was no identifiable enemy to be fitted in the templates used in Cold War planning: there were only enemies of the peace!

In Sarajevo itself the complexity of the political and military situation was further illustrated by the fact that a Croatian brigade had responsibility for holding part of the defensive perimeter against Bosnian Serbs in UNPROFOR. Yet only 30 km away in Kiseljak, Croatians were allied with Bosnian Serbs against Bosnian government forces and were even suspected of "lobbing" the odd shell into Sarajevo, on occasion even hitting their own forces. On many pre-war ethnic maps Sarajevo was shown as "white", that is, with no ethnic colour assigned. In 1993, in and around Sarajevo, as elsewhere, all sides were trying to paint the map with their own particular ethnic colour.

The response of the international community to the conflict was equally complex. For example, Canadian military personnel served not only in several UN military deployments to the region but also under the auspices of the European Community, the Organization for Security and Cooperation in Europe and the Western European Union. Canadians also flew North Atlantic Treaty Organization (NATO) assets deployed in support of the United Nations, giving rise to disputes about whether those Canadians were entitled, under the regulations, to receive the Canadian peacekeeping medal.

UNMO Sarajevo was one of the organizations used by the international community to tackle the Bosnian conflict. The four battalions of UNPROFOR in the Sarajevo environs, under command of French Brigadier General André Soubirou, constituted another Sector Sarajevo military component who reported, as did I, to the Lieutenant General commanding BHC. The Office of the United Nations High Commissioner for Refugees (UNHCR) was the lead non-military agency, responsible ultimately to UN Headquarters, while a host of other non-governmental organizations were also present, including the powerful Doctors without Borders and the International Committee of the Red Cross. UNMOs in Sector Sarajevo provided the only permanent uniformed UN presence, albeit unarmed, on Bosnian Serb territory around Sarajevo. UNMO reports were sent up a separate channel direct to the observer mission headquarters in Zagreb and often were copied, for example in daily Situation Reports, direct to UN Headquarters in New York. The role of the more than one thousand UNMOs in UNPROFOR, like the naval and air resources supporting the many UN Security Council resolutions, is often overlooked.

---

7   Thomas, R. Testimony at Galić Trial; Thomas, R. Testimony at Milošević Trial; Thomas, R. Testimony at Karadžić Trial.

The principal media focus of the international community in October 1993 seemed to be on Sarajevo and its siege. Therefore it is appropriate to narrow the discussion of the use of air power in so-called peace support operations during this period to this specific case, which was so well publicized by media, pundits and practitioners at the time, perhaps because pre-war multiethnic Sarajevo, host of the 1984 Winter Olympics, was a model many in the international community hoped would survive as an example for the rest of the fledgling state.

The air-power tools to be discussed in relation to UNMO Sarajevo October 1993 to July 1994 cover the spectrum ranging from aerospace surveillance, combat aircraft, military airlift and charter airlift to contingent aviation in the casualty evacuation role. The impact of these air-power tools on UNMOs will be considered in relation to the ongoing phases of the siege of Sarajevo during this period. On my assuming command in October through to December 1993 there was a continued use of "terror tactics" by all belligerents. Then in January 1994 there appeared to be an attempt to obtain international intervention, culminating with the Market massacre, which killed 68 outright and wounded hundreds. The Market massacre resulted in the February 1994 Sarajevo ceasefire and the creation of a heavy weapons exclusion zone around Sarajevo. Still within Sector Sarajevo's responsibility but in eastern Bosnia, the situation around safe-haven Goražde deteriorated to the point of an outright Bosnian Serb assault on this large enclave, which eventually ended with another ceasefire and a Goražde exclusion zone. In the meantime the situation around Sarajevo escalated as belligerents increasingly resorted to sniper fire to terrorize and counter opponents' trenching efforts. Air power played a major part in shaping what happened in most of these phases of the Sarajevo saga, as well as directly impacting on the tasks that the unarmed UNMOs carried out.

Terror Tactics, Sarajevo UN Military Observers and the Tools of Air Power

Failure to Inform

From the time the first shot was fired in the first of these wars in the former Yugoslavia there was ongoing aerospace surveillance to collect data that could have served as valuable intelligence to those on the ground, including the unarmed military observers. Such sharing did not occur, however. Indeed, not only did faulty intelligence in February 1992 lead to the ill-advised attempt to locate the headquarters of UNPROFOR in Sarajevo but also aerospace surveillance data was apparently not made available to support the epic march of Canada's Royal 22nd Regiment Battlegroup to Sarajevo in June/July 1992; and this failure to share information available from aerospace assets continued during my first few months of command in Fall 1993.[8]

---

8   Mackenzie, *Peacekeeper*, 119, 257–87.

In December 1993 during the use of terror tactics, as commander of UNMO Sarajevo I personally participated in an investigation of an alleged attack by a Bosnian "fighting" patrol on a Serb village near the Bosnian Serb Army Headquarters at Hans Pijesak, in eastern Bosnia but still within Sector Sarajevo's area of responsibility. I went myself, as SMO, because the Bosnian Serbs rarely asked the United Nations to investigate. Unarmed UNMOs and UNPROFOR troops escorting humanitarian convoys were the only uniformed UN presence in the Bosnian Serb-held areas and were subjected to extensive limitation-of-movement restrictions. I found that there had definitely been a massacre and, in my judgement, it was highly unlikely that the Bosnian Serbs had fabricated this incident. Rather, it seemed to me and my colleague that a guerrilla-type force had infiltrated through the lines and inflicted this atrocity to pass the message that the Bosnian Serbs, being short of manpower, could not protect isolated hamlets, not even one as close as this particular village was to their main military Bosnian headquarters. Exploiting the "old boy network", I did discover that indeed NATO aerospace assets indicated a pattern of as many as 20 possible destroyed villages.[9] Personally, I believe that information on the map of Serb gun positions that I saw for the first time, years later, at The Hague during the International Criminal Tribunal for the former Yugoslavia, should have been made available to me at my first briefing on assuming command in Sarajevo in October 1993, as it was available from aerial assets controlled by the same NATO nations contributing to UNPROFOR!

UNMOs had to patrol on the Bosnian Serb side to obtain information, at risk from mines, booby traps and belligerent fire often deliberately aimed at them. Yet the information in the target lists provided as a result of aerospace surveillance, which formed the basis of so many UNMO patrols after the February 1994 Sarajevo ceasefire, could have formed the basis of patrol plans during the period of terror tactics (October 1993 to December 1993). On the other hand, intelligence needed to monitor and enforce the No-fly zone (NFZ) through this period was available to NATO air-power assets, compelling compliance in Bosnian airspace, but not to UNMOs on the ground in Bosnia below.

Use of Combat Aircraft

During the terror tactics period, the major impact of combat aircraft, all controlled by NATO under the auspices of Operation Deny Flight, was the enforcement of

---

9   Richard Round, European Community Monitor Mission op cit; deployment communications with Roy Thomas; SMO Brief, "UN Senior Military Observer Brief for Force Commander UNPROFOR", SMO 029 19030A Dec 1993. Note that the second citation details how a Bosnian government patrol went hundreds of kilometres behind Bosnian Serb lines to destroy a farming hamlet. Serbs did not permit investigation of further damage inflicted but reported 19 such raids in the October–December 1993 time frame. NATO aerial coverage of this period can be interpreted to indicate damage to a string of villages. Further information available from author.

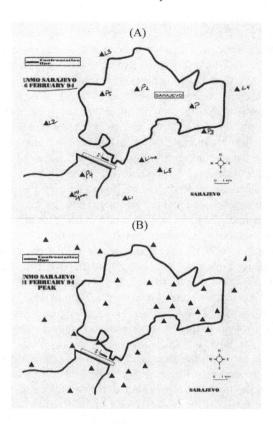

**Figure 11.1   Layout of UN Military Observer Sarajevo observation posts
before ceasefire (A, on 6 February 1994) and after (B, at peak
on 24 February 1994)**

Source: The author and the International Criminal Tribunal for the former Yugoslavia.

the NFZ over Bosnian airspace. During my nine months in Sarajevo I only heard
a suspected An-2 aircraft flying late at night, but never saw a belligerent aircraft
of any type.[10]

NATO air forces began monitoring the ban on military flights in Bosnian
airspace, including Sarajevo, in October 1992 with NATO Airborne Early Warning
and Control Force (NAEW&CF) assets under UN Security Council Resolution
(SCR) 781. This was expanded by SCR 816 in March 1993 to include all flights
not authorized by UNPROFOR with the addition of provisions for use of NATO

---

10   The An-2 is a common single-engine, Soviet-era Antonov utility biplane –
NATO reporting name "Colt" – similar to a bush plane, able to operating out of austere or
improvised conditions.

combat aircraft to enforce compliance. The time-consuming "Blue Sword" approval process that was later used for combat aircraft in close air support and air strikes was not a factor in enforcement of the NFZ. ("Blue Sword" is the name of the dual-key – UN and NATO – release system for air strikes in Bosnia. Both UN and NATO officials had to authorize strikes. It was a process which caused significant tension and problems.) Moreover, the NAEW&CF assets were in-place to detect fixed-wing violations and, to a lesser extent, helicopter violations.[11] The violators could be shot down immediately. Probably as a result of belligerent capabilities, the NFZ, by 1993, was directed primarily at one belligerent, the Serbs, in the case of Sector Sarajevo.[12]

During this period, October 1993 through January 1994, belligerent terror tactics meant that observation posts around Sarajevo were placed to detect the firing of artillery, mortars, and tanks, or to evaluate the results of such shelling (see Figure 11.1A). The UNMOs also had an established procedure not only to observe the fall of shot but also to investigate the subsequent casualties. The NFZ meant that UNMOs were not watching for terror tactics involving the use of aircraft. The NFZ also meant that the air bridge could operate without the threat of hostile air action while supplying Sarajevo and the Sector Sarajevo safe havens of Žepa and Goražde (airdrops needed in the winter), the rotation of UNPROFOR personnel, including UNMOs, and equipment, and casualty evacuations. However, ground fire remained a concern.

The UN Protection Force, Air Transport, and UN Military Observers

All UNMOs arrived in Sector Sarajevo via UN chartered aircraft from Zagreb to Sarajevo. Normally up to one-quarter of the military observers rotated into or from Sarajevo each month. En route 15 minutes before arrival at Sarajevo airport, passengers including UNMOs would be told to put on their flak jackets and helmets to protect against small arms fire. On arrival all passengers were required to carry all their own baggage.[13] No airport ground personnel, whether civilian or UNPROFOR, would risk carrying someone else's belongings. Indeed, during my tenure of command, a French officer marshalling passengers for the rush to the terminal was shot by a sniper while several new UNMOs joining my team were disembarking.

---

11   UN Security Council Resolutions 781 and 816 mandated Combat Air Support and air strikes. It should be noted that UNMOs (not from Sarajevo) were deployed to monitor Serbian airfields as directed in Resolution 786. See respectively: United Nations Security Council, United Nations Security Council Resolution 781, 9 October 1992; United Nations Security Council, United Nations Security Council Resolution 786, 10 November 1992; United Nations Security Council, United Nations Security Council Resolution 816, 31 March 1993.

12   The author can find no indication that Croatian aircraft violated Bosnian airspace.

13   Author personally rotated in and out of Sarajevo on seven such flights.

Figure 11.2   Hercules aircraft on the tarmac at the airfield in Ancona, Italy
Source: Photograph by Roy Thomas on one of his four flights from Sarajevo to Ancona on leave.

The International Air Bridge and UN Military Officers

The Sarajevo Airlift commenced 2 July 1992 and ended on 9 January 1996 after delivering 160,536 tons of supplies in 12,895 sorties – see Figure 11.2 for a photograph taken by the author. Ad hoc arrangements brought together military airlift from 21 participating nations.[14] Without this airlift it would have been unlikely that the population of Sarajevo and their political leadership would have withstood the siege. If Sarajevo had fallen there would be no reason for UNMOs being there!

UNHCR controlled the passenger list, impacting on the UNMOs. For example, a Bosnian Serb interpreter working for the UNMOs required a medical operation which he did not trust to have done in Sarajevo or Zagreb and which was not possible in Belgrade due to the embargo. While the individual travelled via the air bridge to Ancona for his initial treatment at an Italian hospital, he was denied access for the necessary medical follow-ups. UN authorities would not change their position, so UNMOs eventually flew this interpreter, who had been awarded an UNPROFOR force commanders' commendation for bravery in assisting observers, at their own expense, to Ancona commercially.[15]

---

14   United States Air Force, "USAF Humanitarian Efforts in Bosnia-Herzegovina", *Fact Sheet, National Museum of the US Air Force*. Available at: http://www.nationalmuseum. af.mil/factsheets/factsheet.asp?id=1402 [accessed 1 August 2012].

15   The author personally made four trips to Ancona, Italy, from Sarajevo and back on this air bridge.

Airdrops

In addition to participation in the Sarajevo air bridge, the United States Air Force flew 2,200 airdrop sorties to augment delivery of humanitarian aid across Bosnia.[16] In Sector Sarajevo the two safe havens of Žepa and Goražde both received supplies through airdrops, primarily in the winter. In both enclaves the UNMO teams deployed UNMOs to monitor the collection process on the Drop Zone.[17] The chaos on the drop zone in Žepa led to the collaboration between my UNMOs and local leaders, as opposed to a Bosnian military element, in an attempt to form a police force to bring law and order not only to the Žepa drop zone but also to the enclave community itself, which had not accepted the presence of more than 8,000 refugees very willingly. In Goražde, the drop zone was under the tight control of the Bosnian forces that were permitted in Goražde but not "officially" in Žepa, so the UNMO task was to ensure that only humanitarian aid was being received. A separate UNMO team, not from Sarajevo, verified contents when airdrop pallets were loaded in Frankfurt, Germany.

Medical Flights

During this period of terror tactics the threat from ground fire remained. Although fortunately no UNMO required evacuation from Sector Sarajevo during my command, an attempt to have a seriously sick child flown by a French UNPROFOR helicopter from Goražde to Sarajevo failed to take place due to lack of assurance that the air defence assets of belligerents would be not be activated. (Later in April 1994 these Goražde air defence assets did shoot down a NATO fighter.) In this particular case, UNMO military observers from the team in Goražde ended up driving the sick child and mother to Sarajevo.

Without the NFZ these airdrops and medical evacuations would have been risky because such sorties were very vulnerable to hostile air action. A ground fire threat to aircraft of the air bridge and the UNPROFOR charters remained, as landings and take-offs at Sarajevo airport involved a vulnerable flight profile over disputed urban terrain.

January 1994 Brings Additional Tasks

A series of tragic events in January 1994 leading up to the Market Massacre of 5 February suggests that outside intervention was being sought by many in the Sarajevo region. On 3 January 1994, shelling killed 15. On 22 January, shells killed six children who were sledding. On 4 February, ten were killed by shells while waiting in a bread line. At the same time, the shooting at the aircraft on the air bridges connecting the city to the outside world seemed to be almost ignored by the international media. All of these actions, culminating with the 5 February

---

16   United States Air Force, "USAF Humanitarian Efforts in Bosnia-Herzegovina".
17   Author joined UNMOs in both žepa and Goražde to observe airdrops.

killing of 68 and wounding of over 200 when a single 120 mm mortar round hit Markdale market, did result in outside intervention and a temporary Sarajevo ceasefire. There is an argument based on these events that a party or parties (from one or both sides) sought foreign action to halt the war, at least in the Sarajevo area. During this time of increased military activity, the UNMOs were extremely occupied with observing not only the conduct of the belligerents but also the small arms threat to the ongoing airlift; and the UN charter flights forced further tasks on the observer organization.

One of the observation posts/team sites (L2) provided an overview of the normal approach to Sarajevo airport – see Figure 11.1A. Observation Post L2 is the furthest left triangle. The first step was to monitor transport aircraft in their final approach. National flying regulations dictated different profiles. Additional eyes were added to this team just to observe flights. Based on evidence on where aircraft were hit, the major threat appeared to be small arms fire. This analysis led to other additional tasks for Sarajevo's military observers. An UNMO Listening and Observation Post was established 500 m from the end of the primary runway, where it was estimated that the transport aircraft were most vulnerable to ground fire. Further, two UNMOs were located in the control tower during the flights (all of which occurred during daylight hours, wind conditions permitting). It was hoped that the new observation post and the UNMOs in the tower could quickly direct UNMO patrols to the suspected location of any small arms fire directed at an incoming or outgoing aircraft. Liaisons between UNMOs and the Bosnian Serb Ilidza brigade located in the primary approach path were also instituted twice a day, specifically to discuss this small arms threat. This particular Bosnian Serb formation provided extensive assistance in deterring small arms fire. Aircraft flying over the disputed Stup suburb of Sarajevo near the airport, a much fought-over area of destroyed and damaged houses, faced a threat from small arms fire from all belligerents. During January there was constant shelling and small arms fire in the Stup area. Determining which side fired from different piles of rubble was difficult. Often aircrews were unaware that they had been hit until they landed.[18] When in January 1994 there was a possibility that the airflow would be interrupted by further small arms fire, the alternative, land-based routes and the need for vehicles created significant challenges for the UNMOs in Sarajevo at a time when some belligerent parties were increasing the casualty count in the hope of foreign intervention.[19] However, even during the height of the aircraft-shooting terror tactics campaigns, only a handful of UNMOs were deployed to Sarajevo by any means other than the UN charter aircraft.

The additional UNMO task of monitoring the flight path to Sarajevo airport further illustrates that the intelligence garnered by international (UN/NATO)

---

18   SMO Brief, "UN Senior Military Observer Brief for Commander on Actions to Watch [Sarajevo] Airport", 22 January 1994.

19   For example, SMO messages through early January dealt with the issue of alternative movement of UNMOs in case of flight cancellations.

aerospace tools was not being shared with UNMOs or, indeed, passed to UNPROFOR contingents on the ground. Yet ironically, the author was asked to help a national intelligence agency collect information on air defence assets.[20] This lack of access to NATO intelligence would change with the implementation of the Sarajevo ceasefire set in place by General Sir Michael Rose, who levered the outrage of the international community into strong pressure on the belligerents to cease firing and then to implement his Sarajevo peace plan.

Air Power in the Implementation of the February 1994 Sarajevo Ceasefire

Two additional air power tools changed completely what UNMOs did around the besieged city with the implementation of General Rose's Sarajevo peace plan. First, combat air power was used to coerce compliance with a Total Exclusion Zone (TEZ), prohibiting heavy weapons within 20 km of Sarajevo. Secondly, and much more significantly, NATO aerospace surveillance assets were used to provide information as to what heavy weapons had not been moved to the UN-secured heavy weapons collection points.

Now UNMOs had access to a flood of NATO intelligence from aerospace surveillance on a daily basis. In conjunction with French and British contingent reconnaissance assets, a new demanding task for UNMOs was to investigate on the ground why a particular belligerent heavy weapon that was in violation of the TEZ, as identified by NATO aerospace assets, should not be bombed.

Combat Power in a Coercive Role

The mandate to use combat power in close air support of UNPROFOR troops was provided by UN Security Council Resolution 836, 4 June 1993, which provided the coercive threat to influence the belligerents in moving their heavy weapons to the designated UN-controlled weapons collection points. The NATO North Atlantic Council meeting of 9 February 1994 authorized the Commander-in-Chief of NATO's Southern European forces (CINCSOUTH) to launch air strikes in reply to artillery or mortar attacks on Sarajevo, or against heavy weapons still in the TEZ that had not been placed under the control of the United Nations.[21] The threat of the first potential mission, an air strike to punish shelling, added weight to a ceasefire that the UN Commander was fashioning on that same date between the two belligerents in Sarajevo. The threat of the second potential mission, an

---

20   The intelligence derived from such efforts did not seem to reach the Operations Room of UNMO Sarajevo, although others were quick to ask for UNMO briefs. There is reason to believe that perhaps aircrews were briefed on this gathered intelligence. See Thomas, R. "Special Forces in the Service of Peace: Sarajevo and Haiti", *Canadian Defence Academy*, unpublished.

21   NATO, Decision Sheet 4, 9 February 1994.

air strike to destroy heavy weapons not under UN control, required belligerents to move their heavy weapons to weapons collection points or permit access to UN observers to validate why such heavy weapons could not be moved or were not functioning.

The belligerents in the Sarajevo TEZ were given 10 days, with one extra day of grace, until 21 February, to place all tanks, artillery, mortars, multiple rocket launchers, anti-aircraft missiles, and anti-aircraft guns within 20 km of the centre of Sarajevo, under UN control.[22]

The process of identifying weapons and either monitoring their collection or confirming that the weapons system was inoperable was only one of four parts to the February 1994 ceasefire and peace plan for Sarajevo. Monitoring and maintaining the ceasefire was perhaps more important because no party would permit weapons to be collected if hostilities were imminent and UNMOs were a major element investigating violations. The United Nations also had to quickly position troops between the belligerents, where possible. UNMOs surveyed the confrontation line as part of this process. Finally efforts were being made to create a joint commission to address the issues arising from the Sarajevo ceasefire.[23] The NATO assistance in finding heavy weapons violations was essential, as even if full freedom of movement was given by belligerents, ground reconnaissance resources to do this were limited.

Up until 9 February 1994, only UNMOs and UN troops involved in escort of humanitarian assistance had had any freedom of movement on the Bosnian Serb side. This meant initially that military observers had to work at the front lines to maintain a UN presence, investigate violations on the Serb side, and also start the process of searching for heavy weapons. UNMOs even had to monitor at least one weapons collection point because of its isolation.

By 19 February, the process had developed to the point where NATO was preparing a target list of possible TEZ violations based on aerial surveillance. This NATO list was then passed to UNPROFOR and UNMO Sarajevo, which tasked military observers to proceed to these sites within 24 hours to verify the status of the reported violation.[24] Additional military observers were deployed from other UN sectors to Sarajevo to assist in this and other military observer tasks. General Rose, as part of his Sarajevo peace plan, also deployed special teams of what came to be called Joint Commission Officers (JCOs) to help in this role. The JCO teams reported directly to General Rose, and their

---

22   NATO, Decision Sheet 4.

23   UNMOs played a specific role in implementing General Rose's peace plan. See Thomas, R. "Sarajevo UNMOs", *Esprit de Corps* 4(1) (1994), 9.

24   UNPROFOR laid out principles for use of air strikes as well as detailed mission procedures. See respectively Annex C and D in: United Nations Protection Force Bosnia-Herzegovina Command, *Headquarters United Nations Bosnia-Herzegovina Command* OPO 2/94, 19 February 1994.

communications and prior training permitted the use of these teams to call in air strikes if required.[25]

The difficulties of measuring success in peace support operations become apparent when considering the 21 February deadline set by the TEZ conditions. Shelling had stopped, but was only to be replaced by an increase in sniper fire. Some 237 so-called heavy weapons had by that time been collected in 11 sites on the Bosnia Serb side, and 10 sites ostensibly under control of UNPROFOR troops, with 1 site only monitored by the unarmed military observers. On the Bosnian government side, 47 heavy weapons had been collected at Tito Barracks in Sarajevo itself, also the home of the Ukrainian UN battalion. The NATO/UN threat to bomb translated into partial compliance on the ground.[26] The discovery of 15 armoured personnel carriers and several tanks hidden in a Sarajevo tunnel under Bosnian government control was yet to come, as were many other surprises following the deadline. As late as May, there were still 41 identified heavy weapons that were not under UN control.[27] Thus the process of identifying heavy weapons violations, monitoring them if they were not moved, and controlling weapons that had been collected did not end on 21 February, but continued.[28] There were legitimate explanations for many of the heavy weapons that remained uncollected in the TEZ after the NATO deadline: many could not be moved, either for technical reasons such as no engines in tanks, or because, in the case of some towed guns, the snow or mud prevented grouping until late spring.

Procedural difficulties were also a factor. For example, the exact centre for determining the 20-km radius for the TEZ was not at first specified. It became important, as the Bosnian Serbs had guns near Visoko, close to the edge of the TEZ, but facing away from Sarajevo into Central Bosnia where hostilities continued as the Croats there allied themselves with the Bosnian government. Another procedural issue was the definition of what constituted a heavy weapon. These technical issues demonstrate that simply having the air power to coerce is not enough on its own. UNMOs, because they lived in the communities with interpreters as part of the team, were in a position to undertake not only investigation but also the liaison necessary to bring to light these procedural and technical issues

---

25   The Joint Commission Officers (JCOs) were actually Special Air Service (SAS) teams with the communications that permitted them to guide offensive air support strikes if required. With the establishment of weapons collection points on the Bosnian Serb side, the Serbs had to agree to have armed UN troops positioned on their territory. See Rose, Sir M. *Fighting for Peace* (London: Harvill Press, 1998), 57–8.

26   UN Military Information Officer, "UN Sector Sarajevo Military Information Officer summary as of 1700 hrs. 21 February 1994" (Note that "Military Information Officer" was the UN cover title for "Sector Intelligence Officer").

27   UN Military Information Officer, "UN Sector Sarajevo Military Information Officer summary of Sarajevo Total Exclusion Zone violations dated 2 May 1994".

28   SMO Brief, "UN Senior Military Observer [SMO] Brief for Sector Commander on TEZ violations", 5 June 1994.

at the local level and seek resolution. Some difficulties, were political, however, and could not be resolved by BHC or even NATO Allied Forces Southern Europe.

In view of the well-known exemptions to NATO enforcement it is difficult to assess what role the NATO air threat actually played in compelling the belligerents to place their heavy weapons in collection points. What is clear is that the use of combat air power to force belligerents to put heavy weapons in designated collection points *required UNMOs* to validate targets identified by NATO, information which had never been shared before General Rose's Sarajevo ceasefire.

It should be noted that only on rare occasions was this UNMO task assisted by the limited use of helicopters for UNMO missions, when the snow blocked access to some target sites. In contrast, in Macedonia, where the author served for three months prior to becoming SMO in Sarajevo, his position was allocated a helicopter for monitoring tasks once a week.

The NATO combat aircraft used to enforce compliance with the NFZ ensured that the air assets of the Bosnian Serbs were not utilized to replace ground assets in the Sarajevo siege after the creation of the Sarajevo TEZ. While no belligerent air assets violated the NFZ over Sarajevo, NFZ violations were attempted elsewhere in Bosnia – near Banja Luka for example – in an incident in which four Bosnian Serb aircraft were shot down by NATO.[29]

Air Bridge/Airdrops and UN Air Transport

Restrictions on UNMO land movements by all belligerents continued into the Sarajevo ceasefire. Close air support on checkpoints hindering or limiting freedom of movement were not apparently considered.[30] Therefore the air bridge remained vital to Sarajevo's continued resupply. Airdrops also continued as weather and Bosnian Serb restrictions on movement continued to hamper surface resupply of Žepa and Goražde. The TEZ and some easing of limits on movement were only taking place in the Sarajevo area.

The United Nations continued to utilize air transport for rotations as well. The Sarajevo ceasefire was used to establish another observation post in hitherto restricted territory to better monitor the Sarajevo runway approaches. During the ceasefire the author was on a Yak-40[31] that was hit by at least 11 small arms bullets on take-off from Sarajevo airport when it was forced to fly over the disputed Stup

---

29   Four fixed-wing violators of the NFZ were shot down near Banja Luka, outside the Sector Sarajevo area of responsibility. See: NATO, "Operation Deny Flight", *NATO Allied Forces Southern Europe Fact Sheet*, Naples/Brussels: NATO, 2.

30   It is difficult to imagine a close air support mission being called in to attack a belligerent checkpoint manned by a few militiamen or irregulars, especially since the only elements with ground-to-air radio capability in Sector Sarajevo were the JCOs (i.e., Special Air Services (SAS) – see Note 37) and unknown French elements.

31   The Yak-40 is a three-engine regional jet aircraft produced from the late 1960s to the early 1980s in the former Soviet Union – NATO reporting name: "Codling".

suburb because of one faulty engine. This aircraft had not been left at Sarajevo for repairs because of a fear of further damage due to mortar fire during the hours of darkness.

Medical Flights

The TEZ applied to a 20-km circle around the Sarajevo area. This restricted the use of French contingent helicopters for medical airlift from the Sector Sarajevo safe havens on the Drina River in Žepa and Goražde. One helicopter medical evacuation of civilians from the safe haven of Žepa in March 1994 illustrates the involvement of not just UNMOs but even high-ranking UNPROFOR officers. The Bosnian Serbs would not give assurances that the French helicopters would not be fired upon in approach to that safe haven. General Soubirou, the UN Sarajevo Sector Commander, told the Serbs that he would be in the first helicopter, clearances or not! He was! However, problems did not end with the arrival of French helicopters in Žepa. This is when the UNMOs become involved. The UNHCR representative and an "outside" doctor were NOT present. When it appeared that the Žepa Pocket's only dentist – the wife of the local doctor, who was now deciding who was so seriously ill as to merit evacuation – was among those to be flown out, the UNMO team leader in Žepa had to order that she be taken off the passenger list and a valid medical evacuee be substituted.[32]

The Goražde Assault

Goražde fell within the purview of UN Sector Sarajevo headquarters, although only the SMO had military personnel in the form of a team in this large enclave.

The rationale that prompted the selection of Goražde for a Bosnian Serb attack may never be known. What was clear was that a major attack on Goražde was taking place and by 10 April it appeared that the Bosnian Serbs had secured the ground necessary to dominate the city of Goražde itself. The assessment of the Bosnian government situation in Goražde on that date by one military observer was that it was "untenable".[33]

To stop a total victory by the Bosnian Serbs in their Goražde assault a warning was given by the BHC Deputy Commander, in writing, to the Bosnian Serb political leader, Radovan Karadžić and military commander Ratko Mladić in the afternoon of 10 April 1994, threatening air strikes if the Bosnian Serb attacks continued. Attacks continued. A telephone warning was then made. When these two warnings had no apparent impact, approval was sought and received for NATO aircraft to attack Bosnian Serb tanks and artillery. Two NATO air strikes

---

32   Thomas, "UN Military Observer Interpreting in a Community Setting".
33   UNMO Goražde, "UNMO Capsat message 10:48:13 hrs. 11 April 1994".

were made.[34] Serb shelling ceased on 10 April, then resumed on 11 April. NATO aircraft made several passes, with pauses to permit UN warnings to be relayed and subsequent reflection on the part of the Bosnian Serbs to take place, before a Bosnian Serb tank was attacked. In the meantime, on 11 April, the Bosnian Serbs detained the bulk of the UN military personnel on their side of the front line as hostages. This included all the unarmed UNMOs in Sector Sarajevo deployed on the Bosnian Serb side of the confrontation line, including one UNMO en route from the Žepa team to Sarajevo to take compassionate leave.[35] Ominously, in some instances, UN military observers were moved from their accommodations to various Bosnian Serb headquarters, a forecast of the human shield technique that would be exposed to the world in 1995 when NATO launched another series of air strikes. Hostages were only released when a settlement was reached in Goražde.

In Sector Sarajevo this hostage-taking immediately impacted on UNMO operations. Several important negotiations were stopped, for example one attempting to place UNMOs permanently on a Bosnian Serb position in the area of the "sharp stone" feature, which was a favourite sniper firing position for shooting into Sarajevo.[36]

In Goražde proper, it appeared that the Bosnian Serbs were progressing in accordance with their own timetable, unaffected by any threat of NATO air strikes or UN negotiations. On 16 April, resumption of the air strikes resulted in a British Harrier being shot down. This was the last air strike near Goražde and it, too, had not stopped the Bosnian Serb advances.

On 19 April 1994, a temporary arrangement for the town of Goražde was agreed to by Bosnian Serbs and UNPROFOR.[37] That same day, UNMOs on the Bosnian Serb side near Sarajevo were given freedom of movement. On 21 April 1994 a 3-km TEZ was created around Goražde.[38] The following day, somewhat after the fact, NATO authorized CINCSOUTH to conduct air strikes

---

34 UNPROFOR Sector Sarajevo, "Sequence of Events of CAS [Combat Air Support] Incident at Goražde", 18hrs. 11 April 1994; UN Civil Affairs, "Weekly Bosnia-Herzegovina Political Assessment".

35 A report on the detention of UNMOs outlines what happened to the UN military observers held hostage between 11 and 19 April 1994 (see: UNMO Sarajevo, "Report on Detention of UNMOs by BSA in Sarajevo from 11 April 1994 to 19 April 1994 and the Lessons Learnt", 24 April 1994.). In addition to the 58 UNMOs, three UNPROFOR platoons at weapons collection points, an UNPROFOR light armoured squadron at the UNPROFOR base at Rajlovac, the 49 UNPROFOR personnel and 18 vehicles already held hostage near Hadžići, and two UNPROFOR checkpoints with 14 personnel and 2 vehicles were also being held hostage (see: UNPROFOR Bosnia-Herzegovina Command Forward, "Reporting Change in Attitude Toward UNPROFOR Locations", 131610 B April, 1994).

36 Thomas, Testimony at Galić Trial; Thomas, Testimony at Milošević Trial; Thomas, Testimony at Karadžić Trial.

37 "Memorandum of Understanding on the Temporary Arrangement for the Town of Goražde signed by a Bosnian Serb and UNPROFOR representative", 19 April 1994.

38 UNPROFOR Force Commander, "Agreement on Goražde", Fax No 5805/3510, 21 April 1994.

against Bosnian Serb heavy weapons and other military targets within a 20-km radius of the centre of Goražde.[39]

The ceasefire and creation of a TEZ around Goražde necessitated an augmentation force of UNMOs collected from the observers presently employed around Sarajevo proper and then deployed to that safe haven. The augmented UNMO Goražde team carried out validation of target tasks in the Goražde TEZ similar to those undertaken around the Sarajevo TEZ. The Goražde ceasefire also resulted in the deployment of two armed UNPROFOR battalions in that safe haven.

Medical Evacuation from Goražde

Bad weather prevented use of air power for several days after the 11 April NATO air strikes but did not stop evacuation of a seriously wounded JCO by helicopter on that day. Finally, on 18 April, the JCOs, who had acted as the Forward Air Controller in Goražde, were evacuated by helicopter at the same time as a medical evacuation of the most critically injured civilians took place.

Air Power and UNMOs During the Increase in Sniping

Although UNMO Sarajevo tasks continued to be strongly shaped by the need to validate ongoing NATO targeting and the monitoring of TEZ violations, the observers increasingly became involved in attempting to deter sniping as hostilities along the confrontation line around the city increased.[40] The presence of observers deterred sniper activity if the first shooter could be identified. UNMO activities again swung back to observe arms fire, with small UNMO teams in the belligerent trenches, especially at night.[41] Coercion by air power was not an effective option in dealing with this belligerent sniping in urban terrain. UNMOs also continued to be tasked to monitor the flight paths used by airlift aircraft.

Conclusions

The tools of air power were seen by me to have truly shaped the tasks that were undertaken by the unarmed UNMOs in Sector Sarajevo. The use of combat aircraft

---

39   NATO, "Decisions Taken at the Meeting of the North Atlantic Council April 22nd 1994", NATO Press Release 94/31 (22 April 1994). Available at: http://www.nato.int/docu/pr/1994/p94-031.htm [accessed 1 August 2012]. Note that the previous NATO air strikes were authorized under provision of the threat posed to UN personnel.

40   Thomas, "Implementing the February 1994 Peace Plan for Sarajevo", 23–4.

41   For more details on carrying out this specific task in an area near Sarajevo International Airport, see: Thomas, R. "Passion for Football led to "94 World Cup Peace Pause" (*EMC Barrhaven–Nepean*, 24 June 2010), 33.

to create compliance with the NFZ in Bosnian airspace meant that the air forces of the neighbouring states of Croatia and Serbia, as well as the fixed-wing assets of the Bosnian Serbs, were never observed as a factor in the siege of Sarajevo or operations against the two safe havens in Sector Sarajevo during my nine months as SMO.

From October to December 1993, the NFZ ensured that UNMOs only observed and investigated tank, artillery, mortar, and small-arms fire, never air attacks. The need to counter small-arms threats to transport aircraft shaped the tasks of some UNMOs. UNMO resources had to be dedicated to monitoring the Sarajevo airport flight paths, a task that continued even after the February Sarajevo ceasefire. However, that same ceasefire did bring about the further shaping of UNMO Sarajevo tasks by the tools of air power.

The threat of NATO air strikes against first use of heavy weapons fire made possible the ceasefire and hence a survey of the confrontation line by UNMO teams. More pronounced "shaping" of UNMO tasks was evident in the enforcement of the TEZ. Aerospace assets provided Headquarters Sector Sarajevo with NATO target lists of violations of the TEZ to be struck unless ground validation provided a reason for not doing so. UNMO resources were heavily committed to this task of validating each violation that was targeted by NATO. Moreover, some major exemptions such as the Bosnian Serb tank rebuild facility at Hadžići and the Bosnian government heavy weapons on Mount Igman had to be monitored through patrolling and repeated attempts for access. One heavy weapons collection point was actually monitored by UNMOs. Tanks transiting the TEZ had to be followed. The UNMO tasks related to the TEZ continued until the summer of 1995; a role indeed shaped by the tools of air power, both aerospace surveillance and combat aircraft.[42]

In Žepa and Goražde, UNMOs were more involved with the delivery of humanitarian assistance. Monitoring the drop zones was a task for UNMOs in both safe havens. This led in Žepa to UNMO involvement in the creation of a local force to police the drop zone. UNMOs also became involved in the evacuation of seriously sick or injured civilians. If the French aviation assets could not be deployed from Sarajevo, then UNMOs provided road transport. In the case of Žepa, in the absence of other UN staff UNMOs had to step in to stop a "healthy" individual being evacuated under medical pretences.

As had been the case in Sarajevo in February 1994, with the creation of a TEZ around Goražde UNMOs' tasks in that Pocket became "shaped" by investigation and monitoring of violations of the TEZ identified by aerial assets and tentatively put on a target list for NATO air strikes. UNMOs never had the communications to contact NATO Combat Air Support or other air-strike resources, so could not perform the function that the JCOs had done. Instead, some nations (among them

---

42 For an overview of the events after the author's departure in July 1994 see Ripley, T. *Operation Deliberate Force* (Lancaster: CDISS Lancaster University). It should be noted that French and American unmanned aerial vehicles were used for surveillance in 1995.

Canada) contributed teams of Forward Air Controllers/Forward Observation Officers to UNPROFOR. More immediate was the impact that the air strike of 10 April had on the tasks of all UNMOs on Bosnian Serb territory in Sector Sarajevo. All UNMO activities on the Serb side were suspended as all UNMOs were held hostage for over a week.

My conclusions drawn in an earlier paper on the influence of air power in the creation of the TEZs around Sarajevo and Goražde in the spring of 1994 still seem valid.[43] The Bosnian Serbs may well have derived military benefit from the Sarajevo ceasefire. The heavy weapons collection undertaken by the United Nations scarcely impacted on the Bosnian government's main asset: infantry.

In the case of Goražde, air power did not stop the assault, let alone the Bosnian Serbs, from apparently achieving their tactical or, indeed, operational objectives. It is hard to see air power as anything but an aerial demonstration of NATO's political condemnation of the Bosnian Serb aggression against the Goražde safe haven. In addition, the Goražde air strikes exposed UNPROFOR vulnerability: the use of UN personnel as human shields, which was to be exploited in 1995 by the Bosnian Serbs. For instance, they handcuffed a Canadian officer to a post outside a Serb munitions storage site near the Bosnian Serb capital of Pale.

If the influence of air strike threats on the belligerents is debatable, there can be no denying that the TEZ process – combat air power in tandem with aerial surveillance – shaped what UNMOs did. The unfortunate aspect of that is that the data about heavy weapons passed to the UNMOs to validate, first around Sarajevo starting in February 1994, then around Goražde in late April 1994, could have been on my desk when I became commander of UNMO Sarajevo in October 1993. If this NATO intelligence had been shared with my UNMOs from the beginning, then truly military observer tasks in Sector Sarajevo would have been shaped much more proactively by air power over the period October 1993 to July 1994 when so many civilians of all ethnic backgrounds died.

---

43   Thomas, R. "Bombing in the Service of Peace: Sarajevo and Goražde, Spring 1994", *Chronicles Online Journal*, 10 February 2000. Available at: http://www.airpower.maxwell. af.mil/airchronicles/cc/thomasrev.html [accessed 1 August 2012].

PART V
Combat: Enforcing the Peace

The United Nations is often criticized for not using enough force to enforce international law and maintain the peace. There are certainly examples of the UN's lack of robust and forceful responses to aggression and genocide, especially in the difficult period 1993–94 in Bosnia and Rwanda. But there are also cases where the United Nations and allied forces may have used excessive force, for example in Somalia in 1993. That important case is examined in intriguing detail by in Chapter 12 William T. Dean III, who introduces the reader to many US Air Force terms and concepts while providing insights into the missions and the famous "Black Hawk Down" incident, named after the US helicopters that were shot down by Somali militiamen. The US operation resulted in the deaths of 18 American soldiers and US withdrawal from the joint US/UN mission, which in turn led to the end of the UN mission and the continued suffering of the Somali people. By contrast, the response to the conflict in Bosnia, though weak at first, eventually proved successful. In Operation Deliberate Force, described in detail by Robert C. Owen in Chapter 13, the North Atlantic Treaty Organization (NATO) flew over 3,500 sorties (flights) against over 300 individual targets. This helped bring the Serbian side to the negotiating table and the Serbs' quick agreement to the 1995 Dayton Peace Accords, which finally brought longed-for stability to Bosnia. Although NATO worked closely with the United Nations in 1995, Operation Deliberate Force was a NATO-run enforcement operation using substantial air power.

It was the United Nations itself that applied armed force and combat in the eastern Congo in 2006 and 2008. After repeated warnings, the world organization was able to "engage" rebel forces with its Mi-35 helicopters, armed with Gatling guns and rocket launchers, without losing any UN personnel. Though fired upon, the armour of the aircraft was able to withstand penetration and so prevent crashes. Overall, as described by A. Walter Dorn in Chapter 14, the UN mission's use of robust aerial force against rebel groups seems to meet the just war criteria, including just cause, last resort, proportional means and right conduct. Airborne force was needed not only for self-protection but also to protect the mission mandate, the civilian population, and the tenuous peace. After the rebels accepted negotiations, the tentative peace deal allowed the fighters to reintegrate into the Congolese government forces, providing a welcome respite from fighting.

Libya 2011 was another success, if not triumph, for UN-authorized force, in this case conducted mostly by highly capable and well-equipped NATO air forces, of greater capacity than any UN peacekeeping operation had ever incorporated. While it is often stated that air power cannot achieve sustainable territorial results without ground troops, the rag-tag Free Libyan Army made effective use of NATO strikes to seize ("liberate") territory and eventually bring an end to the Gadhafi regime. Air power over Libya is examined by the Swiss expert on air power doctrine, Christian F. Anrig, who in Chapter 15 covers the NFZ as well as combat operations.

Part V of this volume shows how sometimes *force is needed to control force.* Air combat power, judiciously used, can help the United Nations maintain international peace and security; but the proper application of force needs thorough intellectual exploration.

Chapter 12

Air Operations in Somalia: "Black Hawk Down" Revisited

William T. Dean III

In response to an immense humanitarian crisis in the Horn of Africa, the United States joined with the United Nations to help secure humanitarian food distribution to the region's starving people in 1992. The mission changed in 1993 when the United States/United Nations attempted to capture the Somali warlord Farah Aideed. Throughout this US/UN operation air power played an important role, involving air mobility, close air support, aerial interdiction, medical evacuation, and psychological operations. Despite the variety of air power applications, most of the focus in this chapter will be on the kinetic use of air power. Numerous problems of command and control (C2) developed in this operation and there were serious issues of coalition cooperation, especially in air–ground operations. There was a significant disconnect between political objectives and military operations as the campaign continued. This impacted on the use of air power. The limitations of air power in US/UN humanitarian operations were starkly demonstrated; furthermore, the misuse of air power helped cause the operation overall to fail in 1993, especially after the infamous "Black Hawk Down" episode. Civilian policymakers and military leaders forgot or never understood that the use of air power must take place in a political context and that force by air, as on the ground, is a very blunt instrument of policy. Further, they did not understand the historical context that they were operating in, especially the legacy of European imperialism.

Geography and Social Setting

Today the nation of Somalia is 637,660 sq km in size, or slightly smaller than the State of Texas, USA. It has a strategic location on the Indian Ocean and the Gulf of Aden. The major cities of this beleaguered country are Mogadishu, Kismayo, Baidoa, and Berbera. Its population of almost 10 million people is mostly Sunni Muslim. Although relatively ethnically homogeneous (especially for Africa), the clan is the most important social unit. There are few rivers in this hot and arid country and the richest area for cultivation can be found south of Mogadishu. There is also some well-watered pasturage in the northwestern part of the country.

Traditionally, most Somalis are pastoral nomads.[1] It was a combination of ecological disaster and political chaos threw this society into turmoil in 1992. The terrain, climate, and lack of roads, railroads, and other lines of communication influenced military operations and particularly the use of air power.

The Collapse of Somalia

In 1991 the United States won a spectacular victory over Iraq in Operation Desert Storm and the Ethiopian dictator Mengistu was overthrown. By this point the Soviet Union had collapsed and Soviet influence and communism were no longer part of the international calculus. Islamicism had replaced socialism in the Horn of Africa. In January 1991, as the United States was starting its air campaign in Kuwait and Iraq, Siad Barre, president of what was then the Somali Democratic Republic (1969–1991), was overthrown and the central government collapsed. Soon local government was severely degraded all over the country.

In the midst of the political chaos of early January 1991, the US ambassador to Somalia, James Bishop, called for the extraction of US embassy personnel from Mogadishu, the Somali captial. Operation Eastern Exit was launched by US Marines and Navy Sea Air and Land Teams (SEALs). These teams had trained and prepared for Operation Desert Storm, but had to be diverted to Somalia. Thus, the first US operation in Somalia was a non-combatant evacuation operation and helicopters played the dominant role. Because of US basing in the Persian Gulf and the fall of the Soviet Union, Somalia was no longer a strategic priority. Added to US evacuation, numerous European nationals fled the country, as well as the UN High Commissioner for Refugees. Somalia was alone; free to destroy itself.[2]

A civil war based on clans raged throughout the country, with Farah Aideed on one side and Ali Mahdi on the other side. Even Mogadishu was divided between these two warlords. Much of the fighting took place in the southern part of Somalia, the region that was the country's breadbasket. The battles between the clans in this agricultural region would be one of the principal causes of the famine. Aideed fought with Siad, who was trying to hold onto power, doing further damage to Somalia's agriculture. To make matters worse, southern Somalis felt betrayed because the United Nations focused its humanitarian aid on the northern part of the country.[3]

Attempts by non-governmental organizations (NGOs) to alleviate the suffering were greeted by attacks in the fall of 1991. By early 1992 over 300,000 Somalis had died from famine and tens of thousands more were killed or wounded in the fighting.

---

1   See: Lewis, I.M. *A Modern History of Somalia* (Athens, OH: Ohio University Press, 2002), Chapter 1.

2   Rutherford, K. *Humanitarianism Under Fire: The US and UN Intervention in Somalia* (Kumarian Press, 2008), 9–11.

3   Ibid, 16–17.

An international Red Cross aircraft was hit by a missile on 17 September 1992 and a month later 45 Red Cross vehicles were looted and Red Cross workers were robbed. Numerous other incidents continued on into December.

In January 1992 Boutros Boutros-Ghali became UN Secretary-General and he wanted to take a more aggressive stand in peacekeeping in general and in Somalia in particular. On 24 April 1992, the United Nations Operation in Somalia (UNOSOM; later known as UNOSOM I) was authorized by the Security Council.[4] This was the first move in Boutros-Ghali's new aggressive policy. At this time, the United States was suffering from peacekeeping fatigue and also trying to enjoy the peace dividend with the major drawdown of its military forces after the end of the Cold War.[5] As food aid poured into Somalia, fighting intensified between the non-affiliated clans and at least 20 percent of all aid was stolen. By the summer of 1992 Boutros-Ghali became more insistent in intervening and in July he inserted a military observer team led by Pakistani General Imtiaz Shaheen, after the Security Council had established UNOSOM I.

A 50-man team landed in a country consumed by chaos and civil war; and this was soon followed by 500 more UN peacekeepers who would be flown in by United States Air Force (USAF) transport. The first major mission of air power in this escalating humanitarian crisis was air mobility. At this point the Joint Chiefs of Staff in Washington did not want US involvement in air mobility for humanitarian relief. But they were encouraged to do airdrops by Richard Clarke of the National Security Council in the White House. There was also growing pressure from the Congressional Black Caucus saying the United States was staying out of this crisis because of the Bush administration's racial prejudice.[6]

Operation Provide Relief, 15 August to 9 December 1992

Operation Provide Relief was under the control of US Central Command and it was supposed to be an attainable mission with measurable outcomes. It supplemented the meagre efforts of UNOSOM I to distribute aid in crisis-torn Somalia. George H.W. Bush (later dubbed "Bush the Elder") authorized this mission at the height of the 1992 presidential campaign and at a time when he was feeling great pressure to intervene in the escalating crisis in the Balkans. In accordance with Security Council Resolution 767, he authorized the immediate airlift of supplies

---

4 *Editor's note*: UNOSOM I was established by Security Council resolution 751 (1992) of 24 April 1992, to monitor the ceasefire in Mogadishu, to protect UN personnel and supplies, and to escort certain humanitarian deliveries. By resolution 775 of 3 December 1992 its mandate was expanded to protect humanitarian convoys and distribution centres throughout Somalia. United Nations, "Somalia – UNOSOM I, Mandate". Available at: http://www.un.org/en/peacekeeping/missions/past/unosom1mandate.html

5 Ibid, 21.

6 Ibid, 40–47.

to southern Somalia. The USAF would stage out of Mombasa, Kenya, but was slow to let the Kenyan government of Daniel Arap-Moi know about the operation. This took some diplomatic finesse to make the operation viable. The operation was commanded by US Marine Brigadier General Frank Libutti, which is ironic since most of the initial mission was carried out by USAF personnel.

All planes were flown by pilots from USAF, given the difficulties of handling the poor conditions of the runways in Somalia.[7] In the late summer and fall USAF would fly over 2,500 sorties with C-130s and C-141s that would deliver 28,000 tons of food aid to the starving people of this region. However, USAF had promised 28 C-141s but was only able to employ 12. This is significant because the C-141 is much larger than the C-130 and could provide strategic lift, whereas the C-130, with its much smaller payload, could do tactical airlift only. US Special Forces personnel were placed inside these cargo aircraft to provide security in case they were attacked.[8]

It soon became apparent that most of the airlifted supplies were not making their way to the starving Somalis and were taken by the warlords' forces. In fact, the warlords' political and military power grew with the introduction of more aid. As Boutros-Ghali became more aggressive, the Somali warlords became more hostile to the UN/US aircraft and ground forces. Soon USAF personnel were flying in more peacekeepers from Pakistan and Belgium. Besides a lack of success in getting the food to the famine victims, it was a very expensive campaign to maintain. To make matters worse, on 18 September 1992 one of the aircraft was shot at and soon all air operations were suspended. By this point one-third of all Somalis were at or near starvation. Nonetheless, because of the presidential campaign, the Bush administration did not want to intervene more aggressively. There was a great deal of finger pointing and disorganization at the UN peacekeeping operation (UNOSOM I). At the same time, UN force levels grew to 4,200 men, but still these soldiers could not protect the distribution of aid.[9]

Operation Restore Hope

After Bush lost the presidential election to Bill Clinton in November 1992, and with the immense difficulties that UN peacekeepers had in distributing food to the Somalis, President Bush decided to launch in December (a month before leaving office) a larger and more intrusive operation codenamed Operation Restore Hope. A great deal of international media pressure had been a factor in launching this operation. The Unified Task Force (UNITAF) was under US command and not UN

---

7 Ibid, 49.

8 Buer, Major E. "United Task Force Somalia (UNITAF) and the United Nations Operations Somalia (UNOSOM II). A Comparative Analysis of Offensive Air Support". Master's Thesis (United States Marine Corps Command and Staff College, 2001), 8.

9 Ibid, 10.

command. It would involve 28,000 US military personnel and it was expected by the United States that 10,000 more troops from other nations would also participate. The reason the force was so large was that General Colin Powell, the Chairman of the Joint Chiefs of Staff, wanted to attack the problem with overwhelming force, which was part of the previously developed Weinberger–Powell Doctrine. What Bush did not fully realize was that the United Nations had more ambitious goals of nation-building and disarming the militias. This was indicated by the fact that the United Nations was operating under Chapter VII of its Charter. The United Nations did not have any experience in running a Chapter VII operation with large numbers of troops (except perhaps decades earlier in the Congo).

Just before the operation, the US military had failed to carry out effective "intelligence preparation of the battlefield".[10] They could have gathered at least basic intelligence on current ground conditions from the NGOs.[11] Basic reconnaissance was done by US Navy SEAL teams three days before US forces landed in Mogadishu. Further, F-14As did intelligence, surveillance and reconnaissance (ISR) from the carrier battle group to supplement the SEALs' work. Throughout the US involvement in 1992, the American military was quite willing to use US Special Forces in Somalia. In fact, the Assistant Secretary of Defense for Low-Intensity Operations had been heavily consulted before this operation.[12] These clandestine maneuverings were quickly replaced by US Marines landing amidst the camera lights of the international media.

Nine hours after landing, a primitive Air Operations Center was established at Mogadishu Airport. The purpose of this center was to set up C2 of all air assets in the theater. Further, a Joint Forces Air Component Commander (JFACC) was designated to run the Air Operations Center. He was responsible for all air assets in Somalia. Most of the military air assets were US Navy or Marine fixed-wing aircraft, or US Army helicopters. The Navy SEAL team provided security for the airport.[13] From the US side, Marine Lieutenant General Robert Johnston commanded military operations and Ambassador Robert Oakley handled the political aspects of the operation.

There were many practical problems that the US military faced in trying to run air operations, even in a relatively permissive environment. The increased air traffic heavily taxed Mogadishu air traffic control. This operation was also complicated because it would be a joint, coalition, and inter-agency operation. Greater air traffic created new problems of refueling aircraft on the tarmac.[14] Logistics in general was hampered by the poor and small harbor at Mogadishu. This meant a

---

10   See US Department of the Army. *Field Manual*, 34–130. Available at: http:// www. enlisted.info [accessed 31 March 2014].

11   Bauman, R. and Yates, L. "My Clan Against the World, US Coalition Forces in Somalia 1992–1994" (Leavenworth County, KS: Fort Leavenworth, 2004), 34.

12   Rutherford, *Humanitarianism Under Fire*, 67.

13   Bauman and Yates "My Clan Against the World", 3, 8.

14   Ibid, 44.

greater reliance on airlift for supplies and personnel, which aggravated the above problems. Very quickly the limits of airlift were discovered.

Air power was also used in a variety of other ways. US aircraft engaged in psychological operations.[15] This could be done in three different ways: loud hailers, leaflet drops, or presence missions. Presence missions meant airplanes flying low and slow over potentially hostile areas to intimidate potential opponents. Radio broadcasts could also be used for information operations to assure the Somali people that the United States had no hostile intentions. On the ground, to better coordinate with NGOs and governmental agencies, a Civil Military Operations Center was created.[16]

Soon the United States pushed out from Mogadishu to expand the reach of air assets. Nine airfields were rebuilt in the southern half of the country and the airfields at Baledogle and Baidoa were taken by helicopter assault. It became quickly apparent that airlift was not a substitute for ground convoys, but the problem was that the roads and lines of communication were in very poor shape. There were also some mines on some of the roads. Despite these risks, by early 1993 US and UN forces were moving throughout the countryside.

Somalis were afraid of offensive air assets like F-18s or attack helicopters. Whenever there were meetings between US personnel and Aideed or Ali Mahdi, AH-1s flew presence missions.[17] Oakley and Johnston proved to be a good team in managing the operation.[18] They were believers of the Weinberger–Powell Doctrine and they did not see arms reduction as an objective of this operation. It soon became clear that the United States and the United Nations had different and conflicting objectives. Very quickly US efforts began to overshadow UN involvement in Somalia.

Not long after establishing air assets, the Americans saw that it would be difficult to employ air power, especially in an urban environment like Mogadishu. The United States had poor maps of Mogadishu and the weather conditions and need for water caused problems for their helicopters.[19] In general, since the first use of air power in an urban environment by the French in Damascus, Syria, in 1925, the urban environment has proved to be the most difficult for air operations. There were strict rules of engagement for the use of close air support and aerial interdiction in the Somalia campaign because the "peacekeeping" nature of this campaign also limited the capability of available aviation. US Air Force AC-130 gunships that operated out of Kenya were available, but the question arose: was this level of lethality appropriate for a peacekeeping mission? Oakley wanted to limit kinetic operations because of

---

15   For more detail on psychological operations in Somalia, see: Ibid, 47–8.

16   See: Ibid, 53.

17   Rutherford, *Humanitarianism Under Fire*, 91.

18   Buer, "United Task Force Somalia", 15.

19   Ibid, 22.

Somali public opinion.[20] He saw that kinetic air power was a blunt instrument that had to operate in a political context.

The warlords and their militias were a limited threat to US/UN air operations. These armed groups had received smuggled weapons from Kenya and Ethiopia and they knew the urban and rural environments. The key weapon of choice for the militias were the so-called "technicals", which were trucks armed with heavy machine guns. These proved to be a very mobile and elusive target. Added to the machine gun anti-aircraft or AAA guns, there were some SA-7 surface-to-air missiles from man-portable air defence systems (MANPADS).

Unknown to the United States/United Nations was the fact that they faced a more deadly enemy that was meeting in Sudan. In February 1992, in the capital, Khartoum, al-Qaeda founder Osama bin Laden, Sudanese President Omar al-Bashir and the Sudanese religious and political leader Hassan al-Turabi met members of the Iranian intelligence, to plan strategy. Osama bin Laden had spent the previous two years in Sudan turning al-Qaeda into an effective terrorist organization. Al-Bashir and al-Turabi were the political and spiritual leaders of Sudan who aided al-Qaeda. They would be useful to the Iranians, who now saw Somalia as part of their strategy. In spring 1992 bin Laden made clandestine visits to Somalia and in late spring al-Qaeda and Iranian Special Forces, known as "Quds forces", were inserted into Somalia. Volunteers from Iraq and Pakistan landed on the remote shores of Somalia. To make matters worse, the Iranians gave the Somali militia stinger MANPADS. The Iranians would help promote conflict in Somalia from June to October 1993.[21]

Throughout the late winter and early spring of 1993, there were numerous problems of coordinating US and UN C2 and air power. The UN military staff had little impact on planning air operations. There was a slow process of managing UN requests for air support. The JFACC only controlled US Navy and Marine air operations, not US Army attack helicopters, so there was a lack of "jointness" in air operations.[22] There was a slow and cumbersome cycle of Air Tasking Orders, which did everything from giving the vector of the target to deciding what type of ordnance would be used. In the case of this phase of the Somalia Operation, US communications and Air Tasking Orders had to be sent from headquarters in Mogadishu to an Aegis class cruiser to a carrier.[23] C2 was further hampered by the fact that only US personnel ran the Air Operations Center.[24] To make matters worse, there was a great deal of air activity to control.

Offensive Air Support (OAS) was focused on helping humanitarian relief. For the NGOs there were numerous examples of hijacked convoys and they made

---

20 See Ibid, 23.

21 On the involvement of Iran and al-Qaeda in Somalia, see Shay, S. *Somalia between Jihad and Restoration* (Piscataway, NJ: Transaction Publishers, 2008), 59–66.

22 On "jointness" and C2, see Buer, "United Task Force Somalia", 27.

23 Ibid, 29.

24 Ibid, 28.

this situation worse by hiring some members of the militias for security guards. OAS was flown to deter the militias and these air operations did bring more road security. UN ground operations also centered on protecting convoys. The Air Tasking Order processed at least 2,500 sorties a day.[25] At Kismayo Airport in the south, US attack helicopters went against the militias of Mohammed Said Hersi Morgan, Siad Barre's son-in-law, in conjunction with Belgian ground forces. This was an example of a successful joint and coalition operation.[26]

The Security Council requested a smooth, phased transition from UNITAF to UNOSOM II in Resolution 814 of 26 March 1993. The expanded mandate for UNOSOM II was to *continue* the work of UNITAF and *expand* on it. UNOSOM II was to sponsor disarmament and reconciliation in order to re-establish law and order. The United States maintained a separate command in Somalia, which included the US Quick Reaction Force (QRF) that became involved in the "Black Hawk Down" episode. To complicate the C2 aspect, the Deputy Commander of UNOSOM II, Major General Thomas Montgomery, also exercised "tactical control when committed" over the QRF. But General Montgomery was only informed of the 3 October operation 40 minutes before its launch.[27]

UNOSOM II

During the transition period from the Bush to the Clinton administration, Anthony Lake was named the National Security Advisor. He erroneously believed that most US military personnel would be out of Somalia by inauguration day in late January. Richard Clarke, who was part of his team, told him that they would be there several more months.[28] In Somalia, Oakley and Johnston had met with 15 warlords in the late winter and convinced them to park their "technicals". Oakley did not believe that democracy in Somalia was possible and that the best bet was to back the strongman Aideed. He thought "if you treat him like a statesman, he will act like a statesman". This political progress largely ended when Johnston was replaced by Admiral Jonathan Howe, who refused to listen to old Somalia hands.[29] He hated Aideed and was determined to get rid of him. Furthermore, Oakley and Howe disliked each other and leadership in the theater became dysfunctional. It

---

25 Ibid, 29.

26 For details on this, see: Ibid, 30–31. *Editor's note*: UNITA's mandate in Security Council Resolution 794 of 3 December 1992 "entrust[ed] to certain Member States, on a temporary basis, the responsibility for creating a secure environment for the unimpeded delivery of humanitarian assistance" in Somalia.

27 Hillen, J. *Blue Helmets: The Strategy of UN Military Operations* (Washington, DC: Brassey's, 2nd edition, 2000).

28 Sale, R. *Clinton's Secret Wars. The Evolution of a Commander in Chief* (New York, 2009), 78.

29 On the problems of Howe's leadership, see: Ibid, 80–81.

was under Howe that the Somalis quickly came to see the United States as an enemy, just like their colonial oppressors of the past. The Central Intelligence Agency (CIA) in Washington said that the Somalis would not welcome a foreign presence because of the legacy of colonialism.[30]

Under Howe (March 1993) there was a shift from humanitarian aid to nation-building and disarmament. This was far more aggressive than anything imagined by the Clinton administration, whose members wanted to forget Somalia. Meanwhile the US forces planned to draw down to a 4,000 man logistic force and a 1,300 QRF under the command of Army General Montgomery. UN Security Council Resolution 814 was passed on 26 March 1993 to authorize UNOSOM II. This mission received the enthusiastic support of US representative to the United Nations, Madeleine Albright.[31] On 4 May UNOSOM II started and there was a poor transition from the previous UNITAF regime. Now most of the military forces would be UN peacekeepers under the command of Turkish general Çevik Bir. With UNOSOM II most US OAS forces went away. The only kinetic air power available were the AC-130s in Kenya and the Army attack helicopters. The United States had demanded that the United Nations take the leading role, but there was poor coordination between Washington and Somalia and ineffective coordination between the United States and the United Nations in the field. This would only get worse during the summer.

After the US military drawdown there were very limited US/UN military air assets. All OAS shifted from proactive air operations to reactive operations.[32] To make matters worse, UNOSOM II had no control over any US air assets. This of course militated against the aggressive UN goal of disarming the Somali militias and nation-building. The UN's means did not match its strategy, but it tried to carry this policy regardless. For air mobility the United Nations had to rely on civilian aircrews. With fewer military personnel in the theater, there was a significant reduction of ground patrols, which meant the militias gained greater control of the cities and countryside. This weakened the political bargaining power of the United Nations because air power had helped bring the Somalis to the bargaining table. After July 1993, the US Air Force and Army would provide what OAS was available. Added to the problem of limited air capability was that the UN military officers were indifferent to the reality and potential of air operations. This meant that the UN forces did not have any air ISR capability and could not generate presence missions. The United Nations did not conduct psychological operations to counter Aideed's anti-US/UN radio messages.[33] This further increased the vulnerability of UN ground forces.

From the beginning UNOSOM II had real problems with C2. Further, there was divided C2 between UN forces and the remaining US military assets. American

30   Ibid, 80.
31   Rutherford, *Humanitarianism Under Fire*, 107–8.
32   Buer, "United Task Force Somalia", 33.
33   Rutherford, *Humanitarianism Under Fire*, 129.

military personnel answered only to Montgomery and not to General Bir.[34] The C2 problem would only worsen as the campaign wore on.

Aideed hated Boutros-Ghali, as well as Howe, because he believed the Secretary-General was a Christian Coptic Egyptian meddler who threatened his power and was a threat to Islam. On 5 June 1993 al-Qaeda, acting with Iranian advisers, attacked Pakistani peacekeepers and Aideed was blamed.[35] Twenty-four Pakistanis were killed and 55 men were wounded. There was no air cover available for these men and there were no Pakistani forward air controllers who could speak English.[36] This is another example of UNOSOM II's failure to understand the importance of air operations. This was a turning point in the operation. The attack on the Pakistanis showed that the policy of disarmament would be quite difficult and it appeared that Aideed was at war with UNOSOM II. In reality, the threat was far greater than the United States/United Nations realized. Local UN units tried to broker deals with local clans, but it was obvious that UN forces had no clear air or ground strategy.

Howe wanted to remove Aideed, but the rest of the US military was opposed to this. Both US and UN forces had poor human intelligence (HUMINT) – that is, intelligence gathered by interacting with people – regarding the activities of Aideed or other leaders. Howe demanded in June that the Pentagon send in US Special Forces, but Washington initially refused. On 12 June, as a measure to appease Howe and to show limited resolve to Aideed, AC-130 gunships were ordered to attack Aideed's radio station and weapons caches.[37] Part of the Ranger regiment along with Delta Force were ordered to start training for Somalia.

To further up the ante, on 12 July 1993 Howe ordered an attack on Aideed's headquarters at Abdi House in Mogadishu by helicopters and TOW missiles. This attack, codenamed Operation Michigan, was implemented by the US QRF.[38] To limit collateral damage Howe first sent helicopters to warn the civilian population of the impending attack. Nonetheless, some innocent Somalis were killed in the raid, including Somali leaders who were sympathetic to the United States/United Nations. After this the Somalis stopped talking to members of UNOSOM II. Howe failed to hit Aideed and so the admiral placed a US$25,000 reward on the warlord. It was an all-out war between Aideed and the United States.

Immediately after the attack on his headquarters Aideed went to Khartoum, Sudan, to meet with al-Turabi. Clearly, the Somali warlord wanted another increase of military aid and personnel. The Sudanese thought they were the next target, so they were willing to help as part of a defensive strategy. Added to this, al-Qaeda and the Iranians wanted to make Mogadishu a second Beirut, where

---

34   Ibid, 122.
35   Shay, *Somalia between Jihad and Restoration*, 66.
36   Buer, "United Task Force Somalia", 39.
37   Bauman and Yates, "My Clan against the World", 111.
38   On this, see Rutherford, *Humanitarianism Under Fire*, 146.

several hundred US forces died, or Kabul, where Soviet forces were destroyed.[39] In any case, the attack on Aideed's house increased outside support, united the militias against the United States and set the stage for Task Force Ranger. The US Congress and the US military supported a change in mission from helping the United Nations with humanitarian operations to capturing or killing Aideed. Admiral Howe had driven the policy in this direction.

Task Force Ranger and the Battle for Mogadishu

By late July the militias were shooting at US helicopters and in early August US Army Military Police were killed by mines on the road.[40] This was the final impetus needed for the White House to send in Special Forces. A portion of the Ranger Regiment, Delta Teams, Special Operations Aviation Regiment (SOAR) and a few SEALs were sent in. SOAR would provide the helicopter support for the operation. There were also CIA assets on the ground. Task Force Ranger was under the command of Major General William Garrison of Joint Special Operations Command or JSOC. The Rangers and Delta Force, who were sent in late August, were not the units that had trained that summer for Somalia.[41] Also omitted from this operation were the AC-130 gunships which the stateside Rangers and Delta had trained with. This was a conscious decision of Secretary of Defense Les Aspin, who wanted to avoid collateral damage. He wanted a surgical raid to capture Aideed.

The addition of Task Force Ranger to the theater made a complex system of C2 even more dysfunctional. Garrison did not report to Howe but to JSOC back in the United States and through them to General Hoar, the commander of US Central Command. Hoar had been opposed to the sending of Special Forces to Somalia and did not like Howe's aggressive policies. Further, Garrison did not report to Montgomery, who was in charge of the Army QRF. To make things worse, none of the US forces reported to UN forces. Each was its own little kingdom answering to its own master. It was a complete breakdown of coalition operations.

Early operations did not go well for Task Force Ranger. On 30 August 1993 Delta Force launched a raid to put Aideed out of business by capturing his aids and destroying his military infrastructure. The US Special Forces team repelled down ropes and mistakenly attacked a villa that housed UN development staff.[42] They thought a UN staffer was Aideed! This showed what poor intelligence Task Force Ranger had. Clearly they were not communicating with CIA personnel on

---

39 See Sale, *Clinton's Secret Wars*, 83.

40 These mines were detonated by remote control by Aideed's men. See Shay, *Somalia between Jihad and Restoration*, 68.

41 Rutherford, *Humanitarianism Under Fire*, 152.

42 Ibid, 154.

the ground. Aspin said, "We looked like the gang who couldn't shoot straight".[43] Powell was shocked by the amateurish nature of the operation. The United States had underestimated the importance of the legacy of European colonialism and Somali resistance to foreigners. In this raid innocent Somalis were killed by US helicopters. Attacks by US helicopters continued through September and these helicopters were also used to protect Pakistani UN ground forces, who were still fruitlessly trying to disarm the militias.

In September, the United States declared Aideed an enemy and Boutros-Ghali thought that UNOSOM II was becoming too militarily focused. Clearly the peacekeeping mission had failed. In Washington there was starting to be a shift among policymakers to a political solution.[44] This could be seen when on 14 September Montgomery of the QRF was refused artillery and Bradley Fighting Vehicles. This change of policy was not transmitted to US forces fighting in Somalia. The day before, US Cobra attack helicopters hit a hospital while attacking militia members.[45] In September and early October the United States launched six new missions in the theater; these resulted in the loss of two Black Hawk helicopters that were to provide cover for Delta Force and the Rangers, who were capturing some of Aideed's lieutenants. Clearly the militias, with the help of foreign fighters, had learned how to shoot down US helicopters. US air assets were proving less effective than had been hoped. There was no question at this point of using fixed-wing assets like AC-130 gunships.

In terms of C2, Garrison left Howe and Bir out of the loop. Naturally, there was very poor coordination with the UN forces. Further, there was no coordination with Montgomery's QRF since Task Force Ranger did not think they would ever need any outside help because they were such an elite force. The downfall of Special Forces was their hubris. This demonstrated another problem of using Special Forces in a peacekeeping operation.

By the beginning of October 1993, certain Somali clans and their foreign allies were ready to stand up to the United States and had figured out how to shoot down US helicopters. The foreign fighters led by Quds force, al-Qaeda, and Iraqis were the dominant element. This coalition of terrorists used Aideed as cover. A CIA team told the US military that between 150 and 200 fighters arrived per day and that an attack was being planned. CIA leader Ernie Shanklin told the military that they should do a snatch and grab mission against two tier-one Somali leaders, Omar Salad and Abdi Awale. They had been spotted 400 yds from the Delta Force compound in a tea house. Shanklin wanted this to be a small mission with just ten JSOC and CIA members.[46]

On 3 October, Garrison decided to launch a large raid with Delta and the Ranger battalion. P-3 Orions were sent up for ISR and C2 and this was the only

43   Sale, *Clinton's Secret Wars*, 84.
44   Rutherford, *Humanitarianism Under Fire*, 156–7.
45   Shay, *Somalia between Jihad and Restoration*, 68.
46   On the role of Shanklin and the CIA, see Sale, *Clinton's Secret Wars*, 85–87.

fixed-wing aircraft in the operations.[47] This platform was not really effective and JSOC's Somali-based HUMINT was only partially accurate. They did not fully leverage CIA HUMINT. Almost all of the air assets were helicopters, from Kiowas and Little Birds to Black Hawks to Cobra attack helicopters. These rotary platforms were engaging in ISR, C2, insertion of forces, air mobility and resupply, along with combat search and rescue (CSAR).

The story "Black Hawk Down" is well known in books and movies; what is important for this chapter is the role of air power. After a fairly successful snatch and grab at the Olympia Hotel, it soon became apparent that the foreign fighters and the Somalis could shoot down helicopters. The battle shifted from seizing Somalis to rescuing two Black Hawk crews; it went from decapitation to CSAR. All ground and air assets focused on rescuing downed helicopter crews. JSOC's loss of two helicopters and severe damage to another helicopter limited the robust use of rotary assets in this battle, which limited their key advantage over the insurgents. Airborne C2 proved ineffective and the ground columns got lost in the streets.[48]

One of the key accomplishments was that helicopters were providing close air support all night long, as the Rangers were low on ammunition and were under constant attack.[49] The Cobra crews became adept at night operations and were greatly aided by their Forward-looking Infrared (FLIR) radar and night-vision goggles. Paramedics from the USAF were inserted to help with the wounded. Helicopters were employed successfully for resupply during the attack of the Rangers. The QRF, which had not been part of the planning, was waved off and no AC-130s were employed.

UN forces consisting of Malays and Pakistanis in UN armored personnel carriers working with US forces came to the rescue the second day. US troops rode inside the armored personnel carriers and US attack helicopters provided cover to the column.[50] Helicopters were also used for ISR and C2, which was more effective on the second day. Elements of the 10th Mountain division from the QRF were employed. There were still problems of coalition warfare between US and UN forces, especially regarding moving through roadblocks. Aspin blamed the slow response on the UN troops.[51] Throughout the operation there was poor coordination between the CIA and all the military forces. Further, there was an intense rivalry between Delta Force and the Rangers.[52] At the political level, President Bill Clinton and Admiral Howe were left out of the loop. In the end, 18

---

47   See Bowden, M. *Black Hawk Down: A Story of a Modern War* (New York, 2000), 112–13.

48   Ibid, 170–71.

49   Ibid, 230–31.

50   Ibid, 271.

51   Rutherford, *Humanitarianism Under Fire*, 164.

52   Bowden, *Black Hawk Down*, 173.

US servicemen died and some of their bodies were dragged through the streets of Mogadishu as a spectacle.

The Legacy of Somalia

The fiasco in Mogadishu was a shock to the Clinton administration and significantly affected his subsequent foreign policy.[53] Clinton announced the withdrawal of US forces from Somalia by 31 March 1994. To help the withdrawal there was a brief surge of US forces in Somalia. The biggest casualty of Black Hawk Down was Aspin, who was soon fired. One of the biggest criticisms leveled against him was a shortage of air power. After Somalia, Clinton would rely increasingly on air power for 1995 and 1999 operations in the Balkans and Desert Fox in Iraq in December 1998. Further, because of Somalia, Clinton was unwilling to use force to stop the genocide in Rwanda in 1994. The feeble coalition that was UNOSOM II quickly unraveled and there was a loss of international support. The final UN exodus was in 1995 and Somalia descended into chaos, with terrorist control, a foreign (Ethiopian) invasion and piracy on the horizon.

Conclusion

The debacle in Somalia in 1993 had numerous lessons for military professionals. It demonstrated how difficult coalition warfare was in a Chapter VII peacekeeping operation. It was clear the United Nations had little grasp on how to use air power and that the United States used air power in a political vacuum. The misuse of offensive air power by the United States was one of the key reasons Aideed turned against the United Nations/United States. In fact, the failure of the forces to understand the appropriate application of air power was the principal cause for the failure of UNOSOM II. The Somalia operation failed because the objectives were changed without congruent military force.

Failure of C2 was a major source of ineffectiveness on the ground and in the air. The various US elements on the ground failed to communicate or coordinate with each other, much less with the UN's forces. American commanders in Somalia were confused about the policies of the Clinton administration and there was poor coordination between Boutros-Ghali and the White House. The battle for Mogadishu showed problems inside JSOC with the intense rivalry between the Rangers and Delta Force and this impacted on their use of air power.

This was the first battle between the United States and al-Qaeda and the United States faced a much more adept and serious foe in the Battle for Mogadishu than has been previously presented. Black Hawk Down was just a foreshadowing of warfare in the twenty-first century, for in this century urban combat between

---

53   On this, see Sale, *Clinton's Secret Wars*, 87–8.

Western forces and insurgents has become quite common, from Iraq to Chechnya. The United States is still involved in fighting armed groups and insurgents in Somalia and has built a military command around this at Combined Joint Task Force Horn of Africa in Djibouti. The United States is still conducting air operations in Somalia with coalition partners. But because of the Black Hawk Down syndrome, the United States (like the United Nations) refuses to put ground forces into Somalia and operates with surrogates, like Ethiopia in 2006. Due to the quagmire in Iraq, the 2006 Quadrennial Defense Review called for surrogate warfare and this has had a great influence on US operations in Somalia. In fact, the future of US military operations will be conducted with a small footprint of Special Forces engaging in foreign internal defense. Large-scale counter-insurgency is dead and operations in Somalia in the twenty-first century will be lessons for the rest of the world.

Since counter-insurgency is unlikely, there will probably be a greater role for peacekeeping and the United Nations. This will mean that the United Nations will have to be more adept at Chapter VII operations. It will have to be more robust in its use of air assets. Of course the United Nations has learned to use kinetic air assets, as is exemplified by its use of Mi-24 Hind helicopter gunships in central Africa, shown by A. Walter Dorn in Chapter 14 in this volume. With extensive US interagency experience in Iraq and Afghanistan, perhaps coordination between the United States and the United Nations will be more effective in the twenty-first century. With yet further famines in Somalia, there is, of course, a new role for the United Nations. This time they will have to work with African Union soldiers. The Salafist Shahab militia has alienated the people of southern Somalia and perhaps there is only a little hope for this ravaged and desperate part of the world.

Chapter 13
Operation Deliberate Force in Bosnia, 1995: Humanitarian Constraints in Aerospace Warfare

Robert C. Owen

On the surface, the notions of "humanitarian war" and humanitarian constraint in war appear conflicted. The first concept, that of going to war in pursuit of humanitarian objectives, can be viewed as a product of moral self-deception at best, or as self-serving propaganda at worst. Opinions on the notion of constraining the employment of military force in deference to humane values vary widely, depending on whom one asks.

However, the humanitarian values exemplified by these two notions have deep roots in Western warfare, as attested by the various articulations of just war theories and by the Geneva Conventions. Even Carl von Clausewitz, who still reigns preeminent among the philosophers of war, recognized that various factors restrain warfare, including moral values.

Certainly in contemporary conflicts, military commanders from democratic states are expected to conduct operations in ways that respect international law and sensibilities, show respect for human life, and do not poison the peace through real or apparently injudicious and/or uncaring applications of military force. Humanitarian objectives and values are natural and integral elements to any broad discussion of modern warfare, in particular warfare by liberal democracies predicated on those very values.

Air power theory and practice have always been infused with humanitarian considerations to a degree probably exceptional in the military. Early air power theorists, such as Guilio Douhet and William Mitchell, articulated city- and economy-busting air strategies that had such obvious implications for non-combatant casualties that both sought humanitarian justification by arguing that such attacks would shorten and/or mitigate wars and minimize suffering.[1] Later, the practice of strategic bombardment during World War II, which basically amounted to the leveling of whole districts in order to strike specific military or economic

---

1   Douhet, G. *The Command of the Air* (New York: Coward-McCann, 1942/ Washington, DC: Office of Air Force History, 1983), 10, 61; Mitchell, W. *Winged Defense: The Development and Possibilities of Modern Air Power, Economic and Military* (New York: G.P. Putnam, 1925/New York: Dover, 1988), 16, 136–8.

targets, forced military and civilian leaders to justify such widespread destruction as either a legitimate objective of total war or an unfortunate consequence of the limitations of the aircraft and weapons of the time. At least implicitly, these justifications also underpinned nuclear warfighting theories during the Cold War.

Since the Gulf War, some discussion of the humanitarian aspects of air warfare have assumed that a vastly superior air force armed with precision weapons is obliged to be sensitive to humanitarian concerns because new technology means that air power has the capability to be selective and precise in its targeting. It should not be surprising, then, that both the advocates and critics of the pre-eminence of aerospace forces in American military endeavors over the past decade often supported their cases with normative suppositions about air warfare's effectiveness and fundamental humanity.[2]

Given the current primacy of aerospace operations in American warfare, this is a good time to re-examine and update our understanding of the connection between air warfare and humanitarian objectives. This examination presents at least two fundamental questions. First, is effective air warfare possible under humanitarian constraints? Second, regardless of the answer to the first question, *should* air warfare be fought under humanitarian constraints? For air strategists, these questions are key to the tactical, operational, and strategic planning of modern warfare.

This chapter examines Operation Deliberate Force, the 1995 North Atlantic Treaty Organization (NATO) air campaign against the Bosnian Serbs, as a case study to evaluate the implementation and impact of humanitarian constraints in air warfare. The Deliberate Force experience provides unique insights into the two questions raised above. On the whole, humanitarian constraints did not debilitate the tactical execution of Deliberate Force; and the humanitarian conduct of the campaign was a vital underpinning of its strategic success.

Operation Deliberate Force

Deliberate Force was conducted between 30 August and 14 September 1995. During that period, actual bombing operations occurred only on 12 days due to periods of poor weather and a brief operational pause in the first week of September.[3] Nations contributing combat and support aircraft included the

---

2   There exists an extensive literature of published direct and indirect discussions of the relationships of precision air warfare and international law and morality. Two summary discussions offer a good start on grasping the salient issues: Schmitt, M.N. "Precision attack and international humanitarian law", *International Review of the Red Cross*, no. 859 (September, 2005), 445–66; Murray, S.R. *The Moral and Ethical Implications of Precision-Guided Munitions* (Maxwell Air Force Base, AL: Air University Press, 2007).

3   Virtually all of the following details regarding Deliberate Force are extracted from Owen, R. (ed.) *Deliberate Force: A Case Study in Effective Air Campaigning* (Maxwell Air

United States, Great Britain, France, Germany, the Netherlands, Italy, Spain, and Turkey. On a typical day, NATO air forces launched about 300 strike and support sorties. Of these, perhaps 200 went "feet dry" – moving across the coast to perform missions over Bosnia. Perhaps 70 of these "feet-dry" sorties might actually deliver weapons, while the others conducted combat air patrols, electronic escort, reconnaissance, and other support operations. The aircraft staying "feet wet" over the Adriatic performed similar missions, with the addition of air refueling. In all, NATO aircraft released 1,070 heavy bombs and missiles against Serb targets for a total ordnance tonnage of about 500. For perspective, this level of effort was a fraction of the 2,000-plus sorties per day and 70,000-t overall effort expended by allied air forces during the 43 days of the 2003 Gulf War.

NATO focused its attacks on a list of targets categorized as "Options 1, 2, and 3".[4] Option 1 targets mainly consisted of Serb artillery, mortar, and other combat systems directly involved in attacks on Bosnian cities declared "safe areas" by the United Nations. NATO planners presumed that these targets could be attacked with minimal risk of collateral damage to non-combatants and their property. Option 2 targets consisted of other heavy weapons, munitions storage sites, and air defense systems in the vicinity of the safe areas and presenting only "medium" risk of collateral damage if attacked. Option 3 targets were dispersed throughout Bosnia-Herzegovina, including the full array of Serb munitions and fuel depots, and their anti-aircraft (AA) and communications systems. These options were described in NATO planning documents as campaign phases to bring increasing pressure against the Serbs. In the actual event, NATO commanders focused their attacks on Option 1 and 2 targets, with some overlap into Option 3, and on some bridge and road targets added to rob the Serbs of their mobility advantage over Bosnian Federation forces.

Within the history of air warfare, Deliberate Force has several distinguishing features. First and foremost, the NATO air forces quickly surmounted and suppressed Bosnian Serb air warfare capabilities. The Serbs had no air force of consequence in the face of an air assault, and their ground-based air defenses consisted of a net of air defense radars, command and control systems, a few medium surface-to-air missile batteries, and a ubiquitous scattering of light AA guns and man-portable air defense missiles (MANPADs). NATO aircraft largely suppressed the missile batteries in the first day of the campaign, kept them

---

Force Base, AL: Air University Press, 2000). From that volume see particularly: Chapter 4, "The Deliberate Force Air Campaign Plan" by Christopher M. Campbell; Chapter 10, "Deliberate Force Targeting" by Richard Sargeant; and Chapter 16, "Summary" by Robert Owen.

4   In actuality, NATO air planners blended together target lists drawn from several pre-conflict plans to build the one that was applied to Deliberate Force. See Owen, *Deliberate Force*, Chapter 10, "Deliberate Force Targeting", by Richard Sargeant, 279–87. See also Chapter 11 in this volume.

ineffective throughout operations, and countered the light AA and MANPADs by operating generally above 15,000 ft.[5]

Deliberate Force was also distinguished by an unprecedented reliance on precision weapons, which comprised 708 of the 1,070 heavy weapons delivered against the Serb Republic. Thus, 69 percent of the weapons dropped during Deliberate Force were precision, compared to 8 percent during the 2003 Gulf War. The precision air weapons used in Bosnia primarily were free-falling, laser-guided bombs, but also included some Tomahawk Land Attack Missiles and Standoff Land Attack Missiles. The air campaign was concurrent to, but not coordinated in detail with, surface operations by the Croatian and Bosnian Federation armies.

Lastly, and perhaps most importantly, Deliberate Force resulted in few casualties on either side. Only two allied aviators were shot down and captured, the crew of a French Mirage fighter aircraft. None were killed. Casualties among the Serb military and non-combatant civilians are not precisely known, but the latter were less than 30, or about 1 for every 30 to 40 heavy weapons dropped by NATO aircraft. This is a notably low ratio given that many of the targets were in joint use, such as bridges, or located in or very near civilian dwellings, such as radio (microwave) relay towers and barracks.

It was a war fought, albeit reluctantly, in defense of humanitarian values in the face of undeniable Serb brutality against military prisoners and non-combatant civilians. For three years before launching the air campaign, the United Nations had pursued a Fabian strategy of public moralizing, diplomacy, and inter-positioning peacekeeping troops between warring factions (that is, the United Nations Protection Force, UNPROFOR). NATO air units patrolled the skies over Bosnia during this period, mainly providing surveillance in support of UN no-fly and safe-zone resolutions. But they did conduct small-scale punitive air strikes against the Serbs in November 1994 and May 1995.[6]

Throughout 1994 and 1995, the confrontation between the Serbs and outside interventionists was a stalemate in which the only real movements seemed to be the tally of civilian dead. The Serbs responded to all UN overtures and half-hearted air attacks with intransigence, arrogance, hostage-taking, and counterattacks.

---

5   NATO pilots breached the 15,000 ft "floor" when required to improve target identification or to use unguided weapons more effectively. Generally, however, releasing precision-guided weapons at or above 15,000 ft improved their accuracy by giving their guidance systems more time to make course corrections and then stabilize weapon trajectories.

6   For a review of the United Nations and the use of force in the former Yugoslavia, see: Woodward, S.L., "The SC and the Wars in the Former Yugoslavia", in *The United Nations Security Council and War: The Evolution of Thought and Practice since 1945*, ed. Vaughan Lowe et al. (New York: Oxford University Press, 2008), 406–41; Smith, R., "The Security Council and the Bosnian Conflict: A Practitioner's View", in *The United Nations Security Council and War* (Oxford: Oxford University Press), 442–51. See also Chapter 11 in this volume for a discussion of the UN military observer mission in and around Sarajevo.

Convinced that United Nations and NATO vacillations meant an unwillingness to risk conflict, the Serb Republic army launched a systematic campaign to conquer the remaining cities under Bosnian Federation control in the spring of 1995 in an attempt to destroy the Federation. Even then the United Nations and NATO held back until news came in July that the Serbs had systematically murdered over 6,000 unarmed Muslim men in the captured city of Srebrenica.

Then, at a London conference in July 1995, NATO ministers committed (some more reluctantly than others) to an air campaign to force the Serbs to halt their advances and adhere to UN directives protecting Bosnian cities from further attacks. As a parallel objective, most foreign ministries hoped that the bombing would force the Serbs to be more cooperative in the peace process. This was the case particularly for the five countries involved in the so-called Contact Group (the United States, the United Kingdom, France, Germany, and Russia), which had been carrying on negotiations with the Serbs for nearly a year. By then, most, including the press, recognized that effective action in Bosnia had become a litmus test for the future of UN peacekeeping and perhaps even for the survival of NATO.[7]

Humanitarian concerns pervaded the planning and execution of the campaign. Perceiving acute public sensitivity to NATO military and Serb civilian casualties, the NATO air commander, United States Air Force (USAF) Lieutenant General Michael Ryan, imposed strict rules of engagement and close personal control over air operations. He insisted that his targeting and tactics planners make every effort to avoid collateral damage and casualties. Targets located immediately adjacent to civilian-occupied sites were not struck unless planners could come up with a combination of weapons and tactics that virtually precluded an errant weapon from causing unintended harm. For example, because smart bombs that went "stupid" generally struck long or short of their targets, Ryan directed that bridge attack runs be made along the rivers they crossed, even though this tactic theoretically placed the crews at risk of AA weapons arrayed along the banks. At one barracks facility, Ryan allowed the bombing of an outer row of munitions bunkers, but not an inner row, to minimize the risk of damage to potentially inhabited buildings.

At some increased risk to crewmen, NATO leaders adjusted standard procedures to further reduce the possibility of collateral damage. Reversing normal weapon selection doctrine, they often employed the smallest weapons capable of taking out targets rather than the largest available weapons their aircraft could carry. In outstanding examples of this approach, A-10 fighters[8] flew into the threat envelopes of Serbian MANPADs and light AA in order to use cannons, rather than bombs, to cut down a microwave relay tower and to destroy the contents of a warehouse, both of which were located near civilian dwellings. While normal procedure called for attacking aircraft to minimize their exposure to enemy defensive systems

---

7   Holbrooke, Richard, *To End a War* (New York: Random House, 1998), 74.

8   Commonly known as the "A-10" or by its nickname, "Warthog", the Fairchild Republic A-10 Thunderbolt II is a fixed-wing, close air support and ground-attack aircraft, flown exclusively by American forces.

by dropping all of their weapons in single passes, General Ryan required many aircraft over Bosnia to make multiple passes, dropping only one weapon at a time and only after the dust from previous weapons had cleared. These tactics exposed crews to the potential of ground defenders improving their aim with practice, but they also assured that all bombs were released as accurately as possible and in no greater number than was required to destroy a target. In other instances, targets were hit late at night to minimize the likelihood that civilians and even military personnel would be in or on them.[9]

Finally, NATO pilots had overarching guidance to bring their bombs back home if they had any doubts about the identity or the presence of non-combatants in or too near the objects they had in their sights. As a consequence, almost 10 percent of the precision weapons sent against the Serb Republic were dumped into the sea, mainly by carrier-borne aircraft without the "take back" capacity to make deck landings with weapons hanging under their wings.

In terms of diplomatic effectiveness, it is hard to argue with the success of Deliberate Force. While other forces were important in coercing Bosnian Serb leaders to comply with UN resolutions – principally the war fatigue of all combatants, compromise negotiating terms worked out at Geneva during the bombing, and the Croatian and Bosnian Federation land offensives – it is clear that the bombing had an immediate and compelling effect on decision-making. Both in post-action interviews and his memoir, Ambassador Richard Holbrooke declared that the bombing's effectiveness and the invulnerability of NATO air power unnerved Slobodan Milošević, President of Serbia, and Radovan Karadžić, President of the Bosnian Serb Republic, and forced them to cooperate.[10] Had they persisted in their intransigence they faced the real danger that the air attacks would strip them of their military superiority over their enemies, a nightmarish thing to contemplate in the Balkans. Moreover, Secretary of State Warren Christopher not-too-subtly exploited the lingering emotional impact of the bombing by holding the initial dinner of the Dayton Peace Talks on the floor of the USAF Museum, amidst the very aircraft and weapons that had done the Serbs so much harm in such a short time.

Assessing the factors that led to Serb cooperation, the USAF "Balkans Air Campaign Study" concluded that, while Geneva diplomacy and ground advances unquestionably sent a message to the Serbs that the time for compromise was near, it was the bombing that put the Contact Group, NATO, and the United Nations in control of the pacing and ultimate shape of the Dayton peace conference and political events in Bosnia.[11]

---

9   For details of the NATO rules of engagement for bombing, see Reed, R. "Chariots of Fire: Rules of Engagement in Operation Deliberate Force", in Owen, *Deliberate Force*, 381–429.

10   Holbrooke, *To End a War*, 148–52.

11   Owen, R. "Summary", in Owen, *Deliberate*, 514–15.

The positive, humanitarian consequences of the air campaign are just as clear as its diplomatic impact. The war resulted in the deaths of about 30 civilians and a still undetermined number of Serbian soldiers. Regrettable as those deaths are, they should be viewed in comparison to the far greater numbers of civilian casualties in Bosnia in the weeks prior to Deliberate Force, especially in Srebrenica. The air campaign also jump-started a process of political settlement that led to political restabilization and the eventual re-emergence of a relatively well-off, multicultural state.

The Implications of Deliberate Force

The experience of Deliberate Force contains several distinct implications for the practicality and necessity of conducting aerospace warfare under humanitarian constraint. In general, the first implication is that air commanders, equipped with air dominance and precision munitions, can conduct effective air operations at the tactical, operational, and strategic levels of combat. A second implication follows from the first: that strategic success in peace enforcement operations depends on the imposition of humanitarian constraint on military operations. Unconstrained air attacks in peace operations are counter-productive. Whatever advantage might be gained from them in the short-run is likely to be lost in the long. Thus, in conjunction with the basic moral imperative for military restraint, there is a fundamental military logic to "fighting" peace operations in as humane a manner as possible under their particular circumstances.

At the tactical level, the level at which forces fight and take or destroy their objectives, the Deliberate Force case suggests that humanitarian constraints did not undermine the ability of NATO air commanders to perform their duties without undue risk to their subordinates. The campaign's record of quick success, small numbers of friendly and collateral casualties, and the elegant execution of operations provide strong evidence that effective aerospace warfare under humanitarian constraints is a practical proposition.

At the operational level, the level at which individual tactical events are planned and linked to achieve strategic objectives, humanitarian constraints were pivotal to the successful execution of Deliberate Force. At the time, General Ryan and other key leaders were certain that a major incident of civilian casualties would fatally weaken NATO's political cohesion and resolve to stay the course. They knew that domestic political support in Europe was vacillating over intervention in Bosnia. "Every bomb is a political bomb", as General Ryan said. A collapse of domestic support in any NATO member state could have brought the air campaign to an abrupt halt since any member could have blocked positive action within the North Atlantic Council. While the North Atlantic Council's unhesitant decision to allow for the recommencement of bombing after the early September pause indicates that support for the campaign was more robust than some commanders assumed, we still do not know what would have happened if a particularly bad or avoidable incident of civilian casualties had occurred, since the leaders and their aircrews successfully prevented such incidents.

Long-range perspective is important in understanding the effects of humanitarian constraints on the outcome of Deliberate Force at the strategic level, the level at which national and alliance objectives are set and achieved. As described above, the campaign was immensely successful in achieving its immediate objectives. From its start, Deliberate Force suppressed attacks on the UN-declared safe areas and Bosnian Serb ethnic cleansing operations. Arguably, the campaign saved thousands of civilian lives. In conjunction with other military and diplomatic events, Deliberate Force pushed the Serbs back to the conference table.

Looking beyond Deliberate Force, but still within the Balkans region, it seems that the operational and humanitarian success of the NATO allies in 1995 set them up for intervention and some tough surprises in Kosovo in 1999. NATO's successful blending of precision air power and humanitarian credibility during Deliberate Force led its leaders to expect a reprise in their confrontation of Serb misrule during Allied Force, the air campaign to protect Kosovo from Serbian repression. The Alliance's confidence that Kosovo would replay Bosnia extended to restricting military planners to preparing for a two- or three-day war only, with air attacks restricted to military installations unlikely to produce collateral civilian casualties. Pre-war planning for a longer war, Alliance leaders generally believed, would undermine their domestic political support and be unnecessary, given their expectation that Serbian leader Slobodan Milošević would fold quickly, as he had in 1995. Milošević did not fold in two or three days, of course, since Kosovo was much more central to his political power base and destiny than had been the fates of Bosnia and the Bosnian Serb Republic. Fortunately for his enemies, however, Milošević proved even more inept strategically than NATO. Apparently willing to gamble on NATO's risk aversion, he launched a major campaign of ethnic cleansing against the Kosovar Albanians, even as the initial air attacks were under way. Presenting NATO with such a blatant and brutal act of misrule virtually forced NATO to stay the war's course, even after its expectations of quick victory proved empty. Further, Milošević's heavy-handed action reinforced the domestic and international political support for the Alliance, even when its attacks against Serb military forces and economic targets began to produce hundreds of civilian casualties.[12]

Deliberate Force has great strategic implications for the enduring, core interests of the United States and perhaps of the whole community of democratic and humanitarian states. As exemplified by Operation Deliberate Force and in 1999 Operation Allied Force, this community faces something of a dilemma. Even though democratic and humanitarian in its foundational values, it asserts the right

---

12   Lambeth, B.S. *NATO's Air War for Kosovo: A Strategic and Operational Assessment* (Los Angeles, CA: RAND, 2001), 182–4.

to intervene with military force in the affairs of other sovereign states or regions not adhering to, or in violation of, those values. Thus the international community saw the crimes against Bosnian Muslims and Croats committed by the Bosnian Serb Republic as repugnant and unacceptable in the context of humanitarian values, and concerned states decided to act.

In the course of asserting moral hegemony in the Balkans, NATO killed a number of civilians whose only "crime" was that of being citizens of the Serb Republic. In the eyes of international law, these deaths are an allowable consequence of war, so long as the state that caused their deaths has taken reasonable and proportional efforts to avoid them. For many citizens within the community of democratic states, however, these deaths are morally wrong and the troubling products of actions that seem to violate humanitarian principles, even as they are undertaken for their protection. If these citizens are galvanized by these contradictions inherent in warfare, they pressure their governments to put an end to the "humanitarian" wars, regardless of merit. Well-meaning states potentially could lose their ability to intervene in humanitarian disasters.

Thus the practice of humanitarian constraint in situations like Bosnia takes on deep, political importance. We can infer that wars of humanitarian intervention must be conducted under humanitarian constraints if intervening powers are to retain their ability to intervene at all. "Fighting dirty" in defense of lofty democratic and humanitarian values would undermine those values in the eyes of the citizens of intervening states and of the world at large. In sum, NATO's modest two-week victory in Bosnia was far more influential than a resounding victory, won more quickly, and with less admirable restraint, could have been. The experience of Deliberate Force suggests that aerospace wars can be fought effectively under humanitarian constraints and that humanitarian concerns actually are essential prerequisites of meaningful strategic victory. Inhumane victory is an oxymoron, at least for states, coalitions, and societies professing to fight in defense of humanitarian values.

As a final point regarding the practicality of humane restraint in warfare, Deliberate Force grants us insight into the profound importance of military superiority as the agent that spares commanders the painful choice between assuming greater risk of failure and shedding restraint in the conduct of military operations. By exploiting their profound superiority in aerospace power, NATO commanders held the Serbs helplessly at arm's length while knocking the daylights out of them. The Allies, therefore, experienced no counterblows or risks that might have driven them to escalate their objectives or reduce the care they were exercising in their attacks. The Serbs, in contrast, could only watch their power and long-term security steadily erode. Once they saw all of their avenues for diplomatic leverage closed at the Geneva negotiations and by the refusal of the Russians to intervene on their behalf, the Serbs gave in. How differently might this story have unfolded, had the Serbs been in a position to inflict significant and embittering "pain" on NATO military forces?

Conclusion

To summarize then, the experience of Deliberate Force offers at least three important insights to those interested in the conduct of warfare under humanitarian restraint. First, at the present time, aerospace forces provide a pre-eminent tool for shaping conflicts in ways that permit the imposition of humanitarian restraint with minimal increased risk or cost. Second, beyond the immediate utility of aerospace power in the realm of peace operations, Deliberate Force also highlights the more general importance of military superiority in the hands of humane powers as the agent that mitigates the violence of war. Military superiority in the hands of the Bosnian Serbs would have fostered a humanitarian disaster. In the hands of NATO, it fostered restraint and peace in the region. There *was* a moral difference between the contestants. Last, governments founded on humanitarian principles must fight under humanitarian constraint if they are to hope for strategic success in any circumstances short of immediate survival. Certainly in peace operations, which are fought in the defense of humane values and, quite likely, other political interests, humanitarian restraint is crucial to long-term success. Fortunately for the United States and its allies, aerospace power provides them with the ability to exploit all of these lessons; by enabling them to fight wars under humanitarian constraint, they can assist beleaguered peacekeeping missions and reignite peace processes. In Bosnia, air power allowed them to come through successfully, with their skins, morale, and treasuries in good enough shape to consider going in again somewhere else.

Chapter 14

Combat Air Power in the Congo, 2003–

A. Walter Dorn[1]

After the United Nations had its first experience of robust peacekeeping in the Congo in the early 1960s (a baptism by fire – see Part I of this volume) the Security Council did not launch another peacekeeping operation in that region or anywhere else in Africa until the end of the Cold War (Namibia mission in 1989). A decade later, peacekeeping returned to the Congo. In 1999, the Security Council created a new mission in the country that by then had been renamed the Democratic Republic of Congo (DRC). In the 1990s, that country had experienced the direct effects of genocide in neighbouring Rwanda and two ravaging civil wars, the second of which could be called a "continental" war, since many African countries sent troops to fight on opposing sides. To help end the second civil war, at the end of November 1999 the Council created the Mission de l'Organisation des Nations unies en République démocratique du Congo (MONUC). At its founding MONUC was designed as a small, non-kinetic mission tasked with assisting the implementation of a peace agreement and the liaising between conflict parties, as well as some basic planning and reporting functions; it was authorized to deploy 500 military observers.[2] However, within three months it was expanded by a factor of more than 10 and given a robust mandate to "take the necessary action" under a Chapter VII[3] mandate to engage in protection operations not only for UN personnel but also for "civilians under imminent threat of physical violence".[4]

In 2010, MONUC was renamed the Mission de l'Organisation des Nations Unies pour la stabilisation en République démocratique du Congo (MONUSCO). It commanded a budget of US$1.35 billion and emplyed some 25,000 personnel, including nearly 21,000 in uniform (military and police).[5] As in the earlier 1960s

---

1 This chapter borrows material from the author's book *Keeping Watch: Monitoring, Technology and Innovation in UN Peace Operations* (Tokyo: United Nations University Press, 2011), Chapter 7. Several sections reproduced with permission.

2 United Nations Security Council, United Nations Security Council Resolution 1279 (1999), 1999, 3.

3 Chapter VII of the United Nations Charter, a chapter that deals with enforcement.

4 United Nations Security Council, United Nations Security Council Resolution 1291 (2000), 2000.

5 United Nations Department of Peacekeeping Operations, "MONUC Facts and Figures", United Nations Organization Mission in the Democratic Republic of the Congo, 1 July 2010. Available at: http://www.un.org/en/peacekeeping/missions/past/monuc/facts.shtml [accessed 7 May 2014].

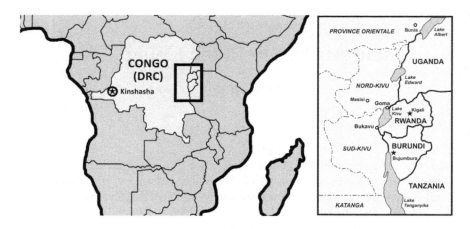

Figure 14.1 Maps showing the position in Africa of the Democratic Republic of the Congo, the Kivu provinces and neighbouring countries of eastern DRC

Source: Author, with graphic design by R. Lang and H. Chilas.

mission the United Nations began small and became more robust as the situation increasingly demanded. One lesson it relearned from the 1960s was the need for substantial armed forces and combat air power.

Any military operation in the DRC is challenged by the vastness of the land, the conditions in the deep jungle and the minimal infrastructure. The lack of a responsible and able national military and government compounds the problem. Operations must cover a forbidding terrain in a country with few paved roads – for example, there are less than 500 km of paved roads in a country the size of western Europe, and most of those paved roads are concentrated in the capital, Kinshasa (Figure 14.1). Most of the UN's effort in the DRC has been in its eastern provinces: the Kivus and Province Orientale. These areas are virtually ungovernable, with various tribes and foreign powers vying for power, revenge, and the precious mineral wealth. It was dangerous for peacekeepers and potentially explosive, especially in 2003.

Experiences of a European Force[6]

As the United Nations sought stability in Province Orientale in 2003, it sought help from the European Union (EU), which sent in a French-led force under the codename "Operation Artemis". The Security Council authorized this new Interim Emergency Multinational Force (IEMF) in Resolution 1484 (2003) of 30 May. The IEMF, under separate command from MONUC, was designed to be

---

6   The section was first drafted by Ryan Cross. Its contribution is gratefully acknowledged.

a short-term deployment: only three months. But the Security Council gave it strong authorization "to take all necessary measures to fulfil its mandate ... to contribute to the safety of the civilian population, United Nations personnel and the humanitarian presence" in the town of Bunia in the Ituri district.[7] The IEMF benefitted from robust air power.

The leading echelon of the IEMF deployment occurred on 6 June 2003 as tactical transport aircraft deployed Special Forces into Bunia's airport.[8] Mirage F-1s flew above to make sure that the forces were inserted safely. The day after the initial arrival of forces, air power was utilized repeatedly, including laser-guided bombs fired from Mirage 2000D jets.[9] Air strikes required that the ground forces include highly trained Joint Tactical Air Controllers to help designate the targets for precision munitions. In close air support (CAS) operations, friendly and opposing ground forces need to be clearly distinguished.[10]

The air component was tasked to provide a day and night deterrent presence over Ituri, gather intelligence, and ultimately provide CAS to threatened ground troops if needed.[11] This required that the expensive planes and helicopters operate out of French military bases in Chad and Gabon, as well as out of Entebbe Airport (Uganda), which functioned as the logistics hub for the operation. With such large distances, "a typical mission would involve multiple air refuelings (sic) and last up to seven hours, with two hours on station over Ituri".[12] As flight paths to Ituri also took the fighter aircraft over large swathes of the DRC, any downed aircraft would involve complex and time-intensive search and rescue operations (although none occurred). The French Air Force:

> came up with a plan that used French Army Cougar helicopters based in Gabon and Cameroon and several pre-positioned stocks of fuel in DRC ... providing an adequate response to any flight emergency remained an important issue.[13]

The operational aviation elements also provided CAS, tactical and strategic airlift, military helicopter operations – including attack, logistics, and surveillance – and presence and overflight missions. Satellite imagery was also available. In addition, ground forces conducted "cordon-and-search operations and vehicle patrols

---

7   United Nations Security Council, United Nations Security Council Resolution 1484 (2003), of 30 May 2003.

8   Laborie, G.J. *The Diplomacy of the Jaguar: French Airpower in Postcolonial African Conflicts*. The Wright Flyer Papers (Maxwell Air Force Base: Air Command and Staff College, Air University, March 2009), 22.

9   Laborie, G.J. "The Diplomacy of the Jaguar: French Airpower in Post-Colonial African Conflicts". Research Report. (Air University, Maxwell Air Force Base, 2008), 26.

10   Laborie, *The Diplomacy of the Jaguar*, 2009, 26.

11   Laborie, "The Diplomacy of the Jaguar", 2008, 24.

12   Ibid., 25.

13   Ibid.

through the region", applying strong rules of engagement to engage when fired on.[14] These actions bought the United Nations time and space to bring in more resources to augment MONUC.[15]

The air power was not only French. The mission was not strictly formed from the current EU member states; future EU member states contributed, as did nations from Africa, Asia, and the Americas.[16] A transport aircraft group comprised aircraft from Belgium, Brazil, Canada, and the United Kingdom,[17] in addition to France.[18] For example, the Canadian Forces provided logistics support in the form of two Hercules transport planes and some 50 personnel in what was codenamed "Operation Caravan".[19]

As there existed a complete security vacuum on the ground, IEMF's task was to halt open warfare between the tribes (Hema and Lendu mostly) and to protect the population at risk; all reports indicate that this protection was provided in

---

14   Marks, J. "The Pitfalls of Action and Inaction: Civilian Protection in MONUC's Peacekeeping Operations", *African Security Review* 16(3) (2007), 73.

15   Sow, A. "Achievements of the Interim Emergency Multinational Force and Future Scenarios", in *Challenges of Peace Implementation: The UN Mission in the Democratic Republic of the Congo*, ed. Mark Malan and João Gomes Porto (Pretoria: Institute for Security Studies, 2004), 211; Hendrickson, R.C., Strand, J.R. and Raney, K.L. "Operation Artemis and Javier Solana: EU Prospects for a Stronger Common Foreign and Security Policy", 39; Homan, Kees, "Operation Artemis in the DRC", in Ricci, Andrea and Kytoemaa, Eero. *Faster and More United? The Debate about Europe's Crisis Response Capacity*. European Communities Commission, Directorate General for External Relations, 2007, 153; International Crisis Group, *Maintaining Momentum in the Congo*. Africa Report No. 4, 26 August 2004; Laborie, "The Diplomacy of the Jaguar", 2008, 24–6; Laborie, *The Diplomacy of the Jaguar*, March 2009, 21–3.

16   Spokesperson of the Secretary General, High Representative for CFSP, "Summary of the Address by Mr Javier Solana, EU High Representative for Common Foreign and Security Policy to the European Parliament", Statement to the European Parliament (Brussels, 18 June 2003). Available at: http://www.eu-un.europa.eu/articles/en/article_2441_en.htm [accessed 7 May 2014].

17   Wodka-Galien, P. "The Tricolor Aloft", *Journal of Electronic Defense* 27(3) (March 2004), 57.

18   Interestingly France was forced to lease strategic lift planes in the form of AN-124s to complete the mission, given American reluctance to provide strategic lift, which would echo similar strategic lift issues for France following 2013 Mali operations. See: Laborie, "The Diplomacy of the Jaguar", 2008, 24. On Mali, see, for example Barrie, D. Hackett, J. and Boyd, H. "Behind the Mali Headlines, an Issue of Airlift", International Institute for Strategic Studies, *IISS Voices*, 30 January 2013. Available at: http://iissvoices blog.wordpress.com/2013/01/30/behind-the-mali-headlines-an-issue-of-airlift/ [accessed 7 May 2014]. Jennings, G. "Analysis: Mali Intervention Highlights France's Strategic Airlift Gap", *Defense and Security Intelligence and Analysis: IHS Jane's*, January 21, 2013. Available at: http://www.janes.com/products/janes/defence-security-report.aspx?id=1065975360

19   Canadian Joint Operations Command Public Affairs, "Operation CARAVAN", Summary of Past CJOC Operations (Ottawa, July 2003). Available at: http://www. forces.gc.ca/en/operations-abroad-past/op-caravan.page [accessed 7 May 2014].

limited areas using the robust force. Likewise, obsessive force did not seem to occur – although it was certainly a concern.[20] For example, it was known at the time that man-portable air-defence systems were in the region where Artemis was operating. Therefore, extensive defence and self-protection equipment was carried by most aircraft, plus a combat-search-and-rescue helicopter and a commando group were deployed into the region to prepare for the loss of aircraft and the necessity to extract downed crews. Ultimately, though, the most predominant threat reported was small-calibre weapons fire.[21]

Air assets were able to monitor improvised airstrips being used for the infiltration of weapons to militias in the region by air. Artemis was able to "disrupt the flow of arms into Ituri through the use of helicopter, fixed wing and other surveillance assets".[22]

Artemis provided critical support to MONUC and the broader region by providing the force necessary to insure a diplomatic breakthrough "against the forces of chaos and violence".[23] The International Crisis Group, a reputable international non-governmental organization (NGO), likewise found that Operation Artemis "largely achieved its stated mandate" through its willingness to use force against "those who interfered with the operation".[24] The French force commander declared that the area of operations, the city of Bunia, would be "'sans [without] armes' and his troops acted quickly – sometimes with deadly force – against those who refused to comply" although this was more likely "a case of Bunia without visible arms".[25]

Artemis created an excellent demonstration of air power and the options it brings for the progressive use of force.

> What could start as a dedicated ISR [Intelligence, Surveillance and Reconnaissance] sortie could evolve into a show of force using non-kinetic means such as a high speed pass and possibly culminate in a strike ranging from gun strafing to the use of LGBs [laser-guided bombs].[26]

With Artemis, an after-action assessment of the French Air Force role argued that air power could provide "effects at the cost of a very small footprint on the ground and little collateral damage".[27]

---

20   Laborie, "The Diplomacy of the Jaguar", 2008, 25.
21   Galien, "The Tricolor Aloft", 57.
22   Homan, "Operation Artemis in the DRC", 153; Sow, "Achievements of the Interim Emergency Multinational Force and Future Scenarios", 211; International Crisis Group, *Maintaining Momentum in the Congo*, 3.
23   Ibid., 220.
24   International Crisis Group, *Maintaining Momentum in the Congo*, 3.
25   Ibid., n. 14.
26   Laborie, "The Diplomacy of the Jaguar", 2008, 26.
27   Ibid.

A UN "lessons learned" study of Operation Artemis concluded with recommendations, including the importance of "the use of air surveillance assets to monitor movements of troops, vehicles or aircraft by MONUC".

After IEMF handed over to MONUC, the United Nations equipped the mission with a powerful asset: armed helicopters. These came in especially useful in the Kivu region of the DRC.

MONUC's Eastern Division

As MONUC took over responsibility from the European force in September 2003, it managed to acquire observation and attack helicopter units from India that immediately proved their worth, though they were initially not permitted to fly at night for safety reasons and were too few in number to cover the vast territory of the eastern DRC effectively. The infiltration routes for arms and fighters from neighbouring countries were still not monitored, although this reconnaissance and surveillance had been mandated. While some rebel leaders were apprehended and sent to the International Criminal Court after 2005, many others were still roving the land with their bands.[28] The United Nations was unable to keep track of their movements or prevent them from pillaging and committing human rights abuses against the general population. MONUC itself was subjected to attacks, kidnappings, and fatalities. Many cordon and search operations proved fruitless. Over time and under necessity, the mission began increasingly robust operations within its Chapter VII mandate.

MONUC created its Eastern Division with Security Council support in 2005 to bring more law and order to the Congo's "wild East", especially in the Kivu provinces (Figure 14.2). It was the first time a peacekeeping operation had included a division-sized component, though in the 1960s the United Nations Operation in the Congo (ONUC) had two brigades in Katanga. The plan was to bring illegal armed groups, both local and foreign, under control through "Disarmament, Demobilization and Reintegration" programmes and if all else failed to confront them forcefully. MONUC's new robust rules of engagement permitted combat action to prevent militia attacks on civilians. But a number of hard-line militia leaders, supported by breakaway factions of the DRC army, continued their abuses and illegal mining activities. They intimidated the local population, attacked villages, and clashed among themselves and with the troops of the country's armed forces (the Forces armées de la République démocratique du Congo, the FARDC). These government troops were themselves frequent perpetrators of human rights violations. Despite having 13,000 UN troops in the East, MONUC's monitoring and reaction capacity was far from satisfactory in the vast and volatile territory. The leaders began to call for more sophisticated technical means, beginning in 2005.

---

28   BBC News, "DR Congo Militia Chief Arrested", World Service, 22 March 2005. Web edition, sec. Africa. Available at: http://news.bbc.co.uk/2/hi/africa/4370843.stm

Neither the Congolese government nor MONUC had resources to track aircraft, let alone control them, in the country's airspace. Commercial aircraft travel in the east depended on the limited air traffic control provided mostly from neighbouring countries. To complicate matters, hundreds of landing strips, built in the era of Congo's dictator Mobutu Sese Seko, were available for arms smuggling with little chance of detection – the United Nations could not afford to place UN military observers at such a large number of landing strips. A Joint Assessment Mission (JAM) was sent by UN Headquarters to the DRC to identify "the exact nature of the surveillance assets". The JAM recommended the acquisition of three mobile surveillance radars, with an effective range of 150–250 km each, "to provide timely warning to enable airborne operations against smugglers".[29]

For aerial surveillance, the JAM noted that:

> With the exception of one flight of Indian Alouette III helicopters, MONUC has no dedicated aerial surveillance capability. It has no airborne imaging capability at all, and no night surveillance capability.[30]

It observed that "the provision of day and night aerial surveillance assets would have an early and positive impact", and specifically recommended unmanned aerial vehicles (UAVs) for local surveillance and overwatch of operations. UAVs had been deployed temporarily in 2006 in western DRC by the European Union Force (EUFOR) during the elections. A UAV contractor bidding process was aborted in 2010 but a new one was launched in 2013. For airspace surveillance, the JAM also noted: "MONUC needs a capability to monitor/control the airspace in eastern DRC". It recommended that MONUC "deploy three mobile air surveillance radars on wheels for temporary surveillance of selected airspace". To accentuate the problem, MONUC suffered numerous fatalities. For instance, in February 2005, a Nepalese officer engaged in providing protection to human rights investigators was fatally shot as he tried to board a departing helicopter. A subsequent investigation showed that MONUC lacked even a basic awareness of the militia's position, strength, equipment, mobility, logistical resources, commanders, organization, and intent.

Engaged in a robust peace operation without the full complement of tools, MONUC's Eastern Division commander strongly supported the conclusions of the JAM. In June 2005, Major General Patrick Cammaert, a senior MONUC military commander, declared a "critical shortfall in dedicated surveillance and intelligence-gathering assets with sufficient reach to provide commanders with accurate, timely and comprehensive intelligence". He identified an urgent requirement for "an aerial surveillance platform with the ability of near real-time

---

29   DPKO, "Report of the Joint Assessment Mission on Intelligence Assets Requirements of MONUC (11 to 19 April 2005)" (JAM Report), DPKO internal document, 2005.

30   Ibid.

enhanced video, geo-coordinated reference data, thermal imagers, and compatible downlink for communications down to the tactical level".[31] In response, UN Headquarters approved a US$5.8 million budget item for aerial surveillance and initiated a bidding process.[32] But to the frustration of the mission leaders, UN Headquarters could identify no compliant or suitable bids from industry.[33] The story became worse after several attempts failed (until finally succeeding in 2013) to contract UAVs for the mission, as noted above.

Despite the setbacks, MONUC has enjoyed more capacity and some remarkable success. It has engaged in extensive cordon and search operations and has employed mobile operating bases and surgical operations using Special Forces equipped with night-vision goggles. With enhanced capabilities for night flying, its attack helicopters were able to support many ground initiatives to prevent militia atrocities. In November 2006, it was able to halt an attack on the town of Goma. Also in 2006, MONUC supervised the largest and most complex elections ever overseen by the United Nations, allowing millions of voters to go to the ballot boxes in relative peace. The EUFOR provided UAVs (Belgian B-Hunters) to assist the UN mission during the tense time. In the DRC, monitoring technology was making a difference and field commanders continued to call for more.

The Mission's Mi-35 Attack Helicopters

The Mi-35 attack helicopter has become a symbol of robust UN peace operations (Figure 14.2). A powerful surveillance, troop transport, and weapons platform, this helicopter was originally designated the "Hind" by the North Atlantic Treaty Organization (NATO) during the Cold War.[34] Designed to fight NATO armoured forces on the central plains of Europe, it was deployed by the Soviet military

---

31   Major-General Patrick Cammaert, "Headquarters Eastern Division Requirement", message from Eastern Division Commander Cammaert to MONUC Force Commander, 24 June 2005, MONUC files.

32   The United Nations budgeted US$5.83 million for an "airborne surveillance system" for MONUC for 2006–2007. The request was advertised by the UN Procurement Service United Nations. Source: MONUC, "2006–7 Acquisition Plan – UN Mission in the Democratic Republic of Congo", 2006. Available at: http://www.un.org/Depts/ptd/2007_monuc.htm

33   MONUC leaders felt the firm Airscan, which had earlier approached them to provide such a service, would have been satisfactory, but the firm was deemed non-compliant in New York because some of its services had been used by governments in South America and Africa to commit human rights abuses. See: International Labor Rights Forum. "Lawsuit Filed Against Occidental Petroleum for Involvement in Colombian Massacre", 24 April 2003. Available at: http://www.laborrights.org/end-violence-against-trade-unions/colombia/news/11403 [accessed 7 May 2014]. Also see O'Brien, K.A. (1998), "Military–Advisory Groups and African Security: Privatised Peacekeeping?", *International Peacekeeping* 5(3), 78–105.

34   Center for Army Lessons Learned, "Hind", *Center for Army Lessons Learned Thesaurus* (Fort Leavenworth, Kansas: United States Army Combined Arms Center,

Figure 14.2   The Mi-35 helicopter gunship used in robust peacekeeping
Source: UN Photo #200146, 17 September 2008, C. Herwig.

in Afghanistan and by several African dictators, including Mobutu, President of the Congo (then called Zaire), to suppress their populations. So it was ironic that this instrument of oppression became an instrument for peace when deployed by Indian forces assigned to MONUC and flown under the UN flag.

Used by MONUC since 2004, the four attack helicopters of the Indian Aviation Contingent, based in Goma, are equipped with state-of-the-art surveillance systems. Though the sensors are designed for target identification and engagement, they are also used extensively for area reconnaissance in support of ground troops in the eastern Congo.

The helicopter's value in the Congo has been demonstrated many times, especially when the rebel group known as the CNDP (Congrès national pour la défense du peuple, or National Congress for the Defence of the People) attempted to attack Goma in 2006 and in 2008. In both cases, the Mi-35 helicopters aided the ground troops of MONUC and the Congolese army (the FARDC) by determining the exact locations of the rebels and, when necessary, aiming rockets or machine-gun fire directly at them.

The CNDP's first major advance on Goma in November 2006 brought the rebels to a town called Sake, some 20 km west of Goma. At this critical juncture,

17 September 2008). Available at: http://usacac.army.mil/cac2/call/thesaurus/toc.asp?id= 36442

the small fleet of UN attack helicopters was used to maintain an overwatch and continuously update MONUC forces on the positions of friendly forces and militia in the area. In one prominent case, the CNDP established a camp near the cell phone tower (Celtel) on a ridge west of Sake. The attack helicopter's onboard sensors were used to scan the Celtel Tower Ridge and 60–100 renegade troops were found at the upper camp, while the FARDC were at the lower camp. It was observed that the forces were exchanging fire with the FARDC troops using machine guns and rocket-propelled grenades.[35] The onboard sensors enabled the crew to relay information about "tubular" and "tripod-mounted" structures that appeared to be rocket launchers and mortars respectively in the CNDP-held area.[36]

On other flights the helicopter crews observed rebel militia clearing areas of growth and engaging in construction. They also reported on deserted villages and civilians fleeing violence.[37] The crews could inform MONUC about the presence or absence of rebel movements along important roads, especially those used in the rebel advance towards Goma.[38] The helicopters were usually not on offensive missions, so the militia were not much deterred from their activities and even ignored the presence of helicopters.[39] But during the intense periods, when the United Nations had warned the CNDP not to advance, the militia would often disperse after spotting or hearing the approaching attack helicopters. During ground battles, on-scene UN commanders observed that rebel firing would usually stop after the arrival of an Mi-35, though not always.

In addition to a colour television camera, the helicopters had fourth-generation, forward- looking infrared (FLIR) cameras and the crew were equipped with special goggles for night flying, which was permitted in special circumstances. The night flights detected hidden militia camps operating with the intent of overwhelming and threatening Goma. Since the militia often moved forward at night to prepare for dawn attacks, the FLIR provided crucial intelligence on developing threats. For instance, on 26 November 2006 an attack helicopter detected a vehicle plying the Sake–Goma road with its headlights off. Closer tracking revealed that this vehicle was shuttling between two towns, stopping on the road as large numbers of armed personnel emerged from their jungle cover at the roadside to meet the occupants. The helicopter concluded that renegade militia were hiding off the Sake–Goma road in order to group for an assault towards Goma. The Indian battalion patrols in the vicinity were advised accordingly and they were able to confirm the deduction by making contact. This vital information could then be passed to the brigade headquarters located in Goma in order to mount joint operations to repel the

---

35   MONUC After Mission Reports ("AMRs") were provided to the author by the mission with permission of the Chief of Staff Forward. See: MONUC, "After Mission Report" (AMR), UNO–888, 26 November 2006 (0612–0748 hrs).

36   MONUC, "After Mission Report", UNO–887, 26 November 2006 (0910–1041 hrs).

37   MONUC, "After Mission Report", UNO–886, 26 November 2006 (0945–1116 hrs).

38   MONUC, "After Mission Report", UNO–888, 26 November 2006 (1410–1530 hrs).

39   MONUC, "After Mission Report", UNO–886, 26 November 2006 (1256–1438 hrs).

threat.[40] The Mi-35 helicopters provided "area domination and surveillance" on the Sake–Goma road and helped to halt militia advances towards Goma in Fall 2006.

The CNDP once again threatened Goma in the period September to November 2008 and, once again, the Mi-35 provided early warning and a potent means to repel the rebel advance. Local UN ground commanders sometimes called for helicopter backup after being attacked. Such was the case on 19 September 2008 when both FARDC and MONUC positions were assaulted near the town of Masisi, some 70 km northwest of Goma. The attack helicopter quickly made radio contact with the local MONUC commander of the Contingency Operating Base (COB), who relayed the supposed position of the rebels on the Kahungole ridge. The nearby FARDC identified their own positions using smoke and white flags. The rebel positions were confirmed by the helicopter crew using visual observation and Mi-35 sensors. The helicopter carried out dummy dives to warn and deter the CNDP elements. After the COB commander reported that CNDP cadres were continuing to threaten UN forces, the helicopter fired a warning shot. When rebel firing continued, salvos of rockets were launched on the CNDP position. This finally caused the CNDP to pull back and stop shooting. The mission was accomplished without any collateral damage and fratricide thanks to the accurate firing from the attack helicopters.

Despite UN warnings and defensive actions, several thousand rebel troops attempted for over two months to seize Goma in 2008. On 28 October 2008, as the rebel offensive continued, an Mi-35 crew was briefed by senior MONUC officers, including the Indian Brigade commander and the Deputy Chief of Staff (DCOS) Forward, Colonel James Cunliffe. The officials shared intelligence on CNDP cadres concentrating in the jungles near the Nyiragongo volcano for an attack on Goma in the night. The attack helicopter arrived in the general area and established radio contact with a MONUC Forward Air Controller (FAC). The DCOS was the on-scene commander. The FAC directed the helicopter towards the location of the "negative elements", as they were called. The helicopter also received information from FARDC troops on the CNDP positions, though the communications with FARDC troops proved technically problematic due to incompatible radio sets.[41] Nonetheless, the attack helicopter identified the ground target and carried out a dummy dive as a warning. The FAC delineated the Forward Line of Own Troops and gave explicit details on the disposition of UN ground troops. He also confirmed the absence of friendly troops and civilians in the vicinity of the target area. The attack helicopters assessed the appropriate attack direction, having to keep clear of the line of fire of a FARDC tank and two army vehicles fitted with heavy-

---

40   MONUC, "After Mission Report", UNO–887, 26 November 2006 (1705–1831 hrs).

41   The Mi-35 crew later suggested that the ground troops be provided with sets for direct communication with the helicopter since this is a mandatory requirement for the use of attack helicopters when providing fire support to ground forces. In another sortie, the attack helicopters had to communicate with ground forces via an UN Lama helicopter that was also in the area.

calibre automatic weapons that were sporadically engaging the rebel target. After receiving confirmation from the FAC, the attack helicopter fired warning shots at the rebel positions. The FAC confirmed that the target was correctly identified. The helicopter then engaged the target during two more passes. The accuracy of the fire was confirmed by the FAC after each pass and the helicopter orbited the target area to carry out a damage assessment.

The helicopter fired again as the government ground troops commenced their assault on the target. This fire had to be accurate because of the forward movement of the FARDC troops. The helicopter carried out a final live pass, engaging the target with four rockets. Henceforth, the proximity of FARDC troops to the target meant no more helicopter attacks could be mounted. Approaching the end of its 1.5-hr flight endurance, the helicopter was replaced on station by another Mi-35. The helicopter crew remarked in their After Mission Report:

> The operation was successful in stopping CNDP advance and stopping their concentration, preparatory to attack on Goma. The attack helicopter support was decisive in stopping the FARDC from falling back, boosting their morale and thus encouraging them to advance and attack the CNDP positions and reclaim lost ground. This was possible due to the co-location of the ground FAC and FARDC officers [so] the operation and the attack helicopter support could be coordinated.[42]

The helicopter and ground actions achieved this tactical success, but the CNDP continued its advance from other directions. The next day, an Mi-35 was dispatched along the Goma–Rutshuru road. About 10 km north of Goma, the attack helicopter observed Congolese troops and army vehicles, including tanks and BMPs,[43] moving in retreat towards Goma. The on-scene commander, again DCOS Cunliffe, informed the Mi-35 crew by radio that the army was withdrawing after a battle with the rebels. Furthermore, the CNDP rebels were advancing in company strength along the road towards Goma. Both UN and FARDC troops were being fired upon with small arms and mortars from about 2–3 km north of Cunliffe's position, which also marked the Forward Line of Own Troops. Colonel Cunliffe approved a helicopter engagement with the CNDP rebel cadre north of his position. The attack helicopter pilots identified the positions from which the rebels were firing. After ascertaining that there were no civilians in the area, the attack helicopters engaged them with four 57 mm rockets. The mission report did not give a damage or casualty assessment. The attack helicopter then reconnoitred the area north using the on-board scanners, but could not spot any movement. The

---

42   MONUC, "After Mission Report", UNO–889, 28 October 2008.
43   The BMP ("Boyevaya Mashina Pekhoty") is a Russian-designed infantry fighting vehicle, combining the features of an armoured personnel carrier (APC) and a light tank.

DCOS asked for a scan of the Rwandan border for possible military elements. No such elements were located.[44]

The limits of joint and combined jungle warfare were shown when an Mi-35 sought to engage CNDP elements near Kibumba at the base of the Mount Nyiragongo volcano on 29 October 2008. After hearing reports of fire on FARDC troops, the crew spent 30 minutes scanning the target area with its TV camera, seeking to spot any movement or arms fire. Finally it found 7 or 8 men approximately 3 km west of the FARDC location moving towards the forest at the base of the volcano. Before engaging, the attack helicopter needed to obtain reassurance that there were no FARDC soldiers in the area. Since the FARDC commander took 7–8 minutes to confirm that these were of the CNDP rebel cadre, the men were able to disappear in the jungle and the attack helicopter lost its ability to track and target them.[45]

The attack helicopter had other limitations as a sensor and weapons platform. Helicopters could typically remain on site for only 1.5 hrs before returning to refuel. They were also limited by poor weather conditions, which sometimes forced them to return early. Nevertheless, in the crucial test of September–November 2008 they proved to be a key enabler to repel aggression. The rebel attack on Goma was thwarted. The United Nations had protected a major population centre, something it had failed to do in other missions. This success served as a lesson on the utility of robust peacekeeping. When India decided to withdraw the Mi-35 aircraft in 2011, citing needs back home, the United Nations made it a priority to find a replacement and Ukraine stepped in to provide the Mi-35 service.

The Mi-35 attack helicopter (or its variants) has been used successfully not only in the DRC but also in Côte d'Ivoire, Liberia and (much earlier) in Croatia. It is a highly mobile and powerful platform for peace enforcement. The United Nations has made progress in the twenty-first century to incorporate such forceful means into some of its operations. There are many more cases to be studied and lessons to be learned about how to use attack helicopters. The use of force for peacekeeping is an irony and a dilemma with tremendous importance for the peace of the world.

---

44   MONUC, "After Mission Report", UNO–888, 29 October 2008 (0855h–1027hrs.).

45   Even though it had lost sight of the confirmed CNDP fighters, it fired in their general area repeatedly with 28 rockets. The success of these shots could not be ascertained due to thick vegetation in the area. The crew remarked in the after mission report: "A golden opportunity to engage CNDP cadre in the open and thus helping stem their advance was lost due to the long channel of communication between on-scene Cdr [commander] and Attack Helicopter". It also recommended that as far as possible, the commander should be on-scene "to provide accurate and timely intelligence and guidance to Attack Helicopter" (MONUC, "After Mission Report", UNO–888, 29 October 2008 (0855h–1027hrs)).

Chapter 15
Allied Air Power over Libya

Christian F. Anrig[1]

In a private meeting during the Libya crisis summit at the Elysée Palace in Paris, French President Nicolas Sarkozy informed US Secretary of State Hillary Clinton and British Prime Minister David Cameron that French combat aircraft were en route to the Libyan coast to enforce United Nations Security Council Resolution (SCR) 1973,[2] which had been adopted on 17 March 2011. With none of them objecting, the French Air Force opened the allied campaign in the afternoon of 19 March.[3] In these opening strikes, Rafale and Mirage fighter–bombers destroyed several armoured vehicles at the outskirts of Benghazi, the rebel stronghold in eastern Libya.

The initial strikes highlighted specific characteristics of the air operations over Libya. In contrast to the practice found in conventional Western air power doctrine, the campaign did not begin with offensive counter-air strikes to take down the Libyan integrated air defence system (IADS) but sought to produce an immediate impact on the ground. It is also the first allied air campaign of the post-Cold War era in which selected European air forces shouldered a significant portion.

---

1   This chapter is a revised version of Christian F. Anrig's article "Allied Air Power over Libya: A Preliminary Assessment", *Air and Space Power Journal* XXV(4) (Winter 2011), 89–109. Permission from *Air and Space Power Journal* to use the paper is gratefully acknowledged.

2   United Nations Security Council Resolution 1973 mandated the implementation of a no-fly zone in Libya's airspace and the use of force to protect civilians under the threat of attack without deploying an occupation force. From 19 March 2011 until the end of that month, an American-led coalition of the willing enforced SCR 1973. The operation was codenamed Operation Odyssey Dawn (though various countries had their own codenames) and involved the United States, other selected NATO alliance members and countries from the Middle East. NATO-led Operation Unified Protector began as an arms embargo operation on 23 March 2011 to enforce the earlier SCR 1970 of 26 February 2011. The enforcement of SCR 1970 mostly involved alliance navies. On 25 March 2011, Operation Unified Protector was extended to include a no-fly zone component. Finally, on 31 March, Operation Unified Protector became the framework for all miltiary operations to address the humanitarian crisis in Libya and supplanted Operation Odyssey Dawn. Sweden joined this operation as a non-NATO member.

3   Tran, P. and Chuter, A. "U.K., France Vault to Center of Euro Defense", *Defense News* 26(14) (11 April 2011), 1.

One can argue that French and British decision makers diplomatically and militarily confronted their counterparts with a fait accompli before reaching consensus. From a French and British perspective, the situation on the ground dictated the pace, which required immediate action that only air power could deliver. Finally, on 31 March 2011, 12 days after the initial air strikes, the North Atlantic Treaty Organization (NATO) took over the allied air operations.

The Opening Diplomatic Moves

In the run-up to the air strikes against Colonel Muammar Gadhafi's military machine, which was violently oppressing the domestic anti-government movement, France and the United Kingdom forced the diplomatic pace. In late February 2011, Cameron unambiguously stated:

> We do not in any way rule out the use of military assets, we must not tolerate this regime using military force against its own people. In that context I have asked the Ministry of Defence and the Chief of the Defence Staff to work with our allies on plans for a military no-fly zone.[4]

For his part, Sarkozy was the first Western leader to acknowledge the Libyan National Transitional Council on 10 March 2011, 21 days after the popular uprising began in Benghazi on 17 February 2011.

Although the United Kingdom and France displayed unusual unanimity, the European Union (EU)'s view on tackling the crisis in Libya was far from homogeneous. An EU summit in early March ended without support for military intervention. On the diplomatic front, a crucial turning point was the Arab League's endorsement of a no-fly zone (NFZ) over Libya on Saturday, 12 March 2011. Amr Moussa, Secretary-General of the Arab League, indicated after a six-hour-long meeting that "the Arab League has officially requested the United Nations Security Council to impose a no-fly zone against any military action against the Libyan people".[5] Reportedly, Algeria, Sudan, Syria, and Yemen opposed the Arab League's vote for a NFZ.

While diplomatic support for a NFZ gradually grew, the disorganized Libyan rebel forces continued to lose ground to the superior firepower of Gadhafi's forces, which, after the initial shock of the revolution, started to reorganize and seize the initiative. Besides heavy tanks and artillery, Gadhafi's forces had a decisive

---

4   "Cameron: UK Working on 'No-Fly Zone' Plan for Libya", BBC News, 28 February 2011. Available at: http://www.bbc.co.uk/news/uk-politics-12598674

5   Freedman, C., Meo, N. and Hennessy, P. "Libya: Arab League Calls for United Nations No-Fly Zone", *The Daily Telegraph*, 12 March 2011. Available at: http://www.telegraph.co.uk/news/worldnews/africaandindianocean/libya/8378392/Libya-Arab-League-calls-for-United-Nations-no-fly-zone.html [accessed 7 May 2014].

advantage in air- and ship-borne firepower. On 12 March, when the Arab League declared its support for a NFZ, forces loyal to Gadhafi reconquered the oil port of Ras Lanuf, in eastern Libya, at the gates to the rebel stronghold Benghazi. As a consequence, the situation for the Libyan opposition movement became drastically serious. Gadhafi's son Saif al-Islam confidently predicted that loyalist forces would soon thwart the revolution, announcing no negotiations with the rebels but a war to the end.[6]

Support for a NFZ by Arab nations and the deteriorating situation of the anti-Gadhafi forces on the ground encouraged the United Kingdom and France to step up their diplomatic efforts. Along with Lebanon, the two permanent members of the UN Security Council came up with a draft resolution, increasing the pressure for military intervention.[7] The Obama administration, originally sceptical of a military intervention, as is examined below, suddenly changed course on 15 March. In fact, it not only changed course but also produced a new draft resolution going beyond a NFZ and providing any intervening force with sufficient leeway to decisively shape events on the ground.[8] Finally, in the evening of 17 March 2011, the Security Council adopted Resolution 1973 by a vote of ten in favour, with five abstentions (Brazil, China, Germany, India, and Russia). SCR 1973 authorized member states, that:

> acting nationally or through regional organizations or arrangements, to take all necessary measures to protect civilians under threat of attack in the country, including Benghazi, while excluding a foreign occupation force of any form on any part of Libyan territory.[9]

Hence, SCR 1973 relegated any potential military intervention to the predominant use of air power, avoiding the presence of Western militaries on the ground of yet another Arab nation. The key passage "all necessary measures", endorsed by the Obama administration and giving SCR 1973 substantial teeth, was instrumental in mounting an effective air campaign. Yet the resolution did not explicitly include regime change and remained vague in desired strategic end-states – a prerequisite for the resolution to be passed.

Two days after the Security Council adopted Resolution 1973, Sarkozy ordered fighter–bombers to take off towards hard-pressed Benghazi. Critics of the French

---

6   Ibid.

7   "Libya: UK No-Fly Zone Proposal to Enter UN Talks", BBC News, 16 March 2011. Available at: http://www.bbc.co.uk/news/uk-politics-12755896 [accessed 7 May 2014].

8   Chivvis, C.S. "Libya and the Future of Liberal Intervention", *Survival* 54(6) (December 2012–January 2013), 70.

9   Security Council, "Security Council Approves 'No-Fly Zone' over Libya, Authorizing 'All Necessary Measures' to Protect Civilians, by Vote of 10 in Favour with 5 Abstentions", SC/10200, 6498th Meeting (night) (New York: United Nations, Department of Public Information, News and Media Division, 17 March 2011). Available at: http://www.un.org/News/Press/docs/2011/sc10200.doc.htm [accessed 7 May 2014].

president argue that he primarily acted for domestic reasons. Whatever Sarkozy's motivations, the threat of a massacre in Benghazi was imminent in the second half of March 2011 and required immediate military action.

In contrast to the British and French, former US Secretary of Defense Robert M. Gates used cautious rhetoric at a press conference on 1 March 2011:

> All of the options beyond humanitarian assistance and evacuations are complex. … We also have to think about, frankly, the use of the U.S. military in another country in the Middle East.[10]

Gates's words unambiguously signalled scepticism within the Obama administration about military intervention in Libya. Admiral Mike Mullen, Chairman of the Joint Chiefs of Staff, and General James N. Mattis, head of US Central Command, publicly shared his concerns. Accordingly, the Secretary of Defense might primarily have had humanitarian assistance and evacuation operations in mind when he ordered the two amphibious assault ships USS *Kearsarge* and USS *Ponce* from the Red Sea into the Mediterranean. The focus on evacuation operations and humanitarian relief is underlined by the absence of a carrier strike group and by the fact that 400 additional Marines deployed from the United States to the *Kearsarge* while the 1,400 Marines assigned to the ship were fighting in Afghanistan.[11] In short, Gates questioned the wisdom of military intervention in yet another Muslim country.

According to Washington-based commentators, the Obama administration's passive stance in the opening diplomatic moves partly stemmed from a concern that Arab leaders would have difficulty sanctioning an American-led operation, not to mention the spectre of another protracted military involvement.[12] Yet realities unfolding in Libya seem to have brought about a drastic change within the Obama administration on 15 March 2011.

A Common European Defence Identity?

The intervention in March put into concrete action what American, British and French leaders had deliberated in the preceding months. In particular, a new entente cordiale was emerging in 2010. In November, for instance, the United Kingdom

---

10   "Gates: Libyan No-Fly Zone Would Require Attack", *CBSNews.com*, 2 March 2011. Available at: http://www.cbsnews.com/stories/2011/03/02/501364/main20038352.shtml

11   Whitlock, C. "Pentagon Hesitant on No-Fly Zone over Libya", *Washington Post*, 2 March 2011. Available at: http://www.washingtonpost.com/wp-dyn/content/article/2011/03/01/AR2011030105317.html [accessed 7 May 2014].

12   Leiby, R. and Wilson, S. "Arab League's Backing of No-Fly Zone over Libya Ramps Up Pressure on West", *Washington Post*, 13 March 2011. Available at: http://www.washingtonpost.com/wp-dyn/content/article/2011/03/12/AR2011031205484.html

and France signed treaties foreseeing military cooperation in various areas such as common support of A400M airlifters, cross-deck operations of aircraft carriers (no longer an option after the United Kingdom's U-turn in its decision to purchase F-35B, instead of F-35C aircraft), or maintenance of nuclear warheads. This rapprochement was underlined by increased cooperation between the Royal Air Force (RAF) Eurofighter Typhoons and the French Air Force Rafales.[13] According to Liam Fox, the UK's Secretary of State for Defence, cooperation with France was desirable because it met two key criteria: its willingness to deploy and its willingness to spend on defence.[14]

Unlike his predecessor Jacques Chirac, Sarkozy wished to reinforce French ties with his Anglo-Saxon counterparts. For example, under his presidency, France returned to NATO's integrated military command structure in 2009. Yet against the backdrop of the Libya campaign, he preferred a coalition of the willing framework and only reluctantly accepted NATO command. The changed French attitude was also seen on an air force level. The United States Air Force (USAF), the RAF and the French Air Force established strategic studies groups staffed by officers from each organization. According to General Norton Schwartz, the USAF Chief of Staff, this exchange of ideas concerns "how the best air forces in the world mix and match their capabilities for the best defense".[15] These ties were borne out during the campaign itself. In particular, the French and British exchanged and mixed aircrews on the dual-seat Tornado GR4 and Mirage 2000D fighter–bombers. Accordingly, General Jean-Paul Palomeros, Chief of Staff of the French Air Force, argued in June, "I can tell you the level of confidence with the Royal Air Force is very, very high".[16]

One month after the start of operations, the troika became especially apparent again in a letter signed by US President Barack Obama, British Prime Minister Cameron, and French President Sarkozy. Leading newspapers of the three countries published the letter with the intention of demonstrating continued resolve and a united front against Colonel Gadhafi. It even went beyond SCR 1973, stating unambiguously that "it is impossible to imagine a future for Libya with Gaddafi in power".[17] The letter appeared after the US military officially ceded its leading role and pulled all combat aircraft from operations in early April. Consequently, doubts emerged, particularly in the United States, about whether NATO air

---

13   Tran, P. and Chuter, A. "French, U.K. Air Forces Boost Cooperation", *Defense News* 25(43) (15 November 2010), 8.

14   Tran, P. and Chuter, A. "Cross-Channel Cooperation", *Defense News* 25(33) (6 September 2010), 1.

15   Reed, J. and Tran, P. "Leading Air Forces Cooperate: France, U.K., U.S. Work Strategy, Doctrine, Systems", *Defense News* 25(35) (20 September 2010), 1.

16   Tran, P. "[Interview with] Gen. Jean-Paul Palomeros, Chief of Staff, French Air Force", *Defense News* 26(25) (27 June 2011), 22.

17   "Libya: Obama, Cameron and Sarkozy Vow Gaddafi Must Go", BBC News, 15 April 2011. Available at: http://www.bbc.co.uk/news/world-africa-13089758

strikes could succeed with US aircraft such as the A-10 Warthog or the AC-130 gunships grounded.[18]

Although the United Kingdom and France were willing to make substantial contributions, the situation in NATO and Europe remained very heterogeneous. With regard to Libya, one finds basically three categories of NATO countries: those that conduct offensive air operations; those that relegate their actions to air policing, effectively a non-combat role; and those that fail to appear at all. As of mid-April, only six alliance countries, including France, the United Kingdom, Canada, Belgium, Denmark, and Norway, were conducting strike missions, directly influencing events on the ground.[19] Canadian forces undertook a particularly swift overseas deployment when seven CF-188 (informally referred to as the CF-18 Hornet) and two CC-150T Polaris tanker aircraft departed from Canada to Trapani Air Base, Sicily, on 18 March. Canadian aircraft began combat operations on 21 March.[20]

Interestingly, the Royal Netherlands Air Force, formerly at the vanguard during the Balkan air campaigns and a significant participant in operations over Afghanistan, was restricted to imposing the NFZ. Since early 2010, a marked shift seems to have occurred in Dutch policy, which also led to The Netherlands armed forces pulling out of Afghanistan. In contrast, Belgian aircraft operated across the spectrum of military force. Usually, the role of the two countries had been reversed, The Netherlands military taking a more proactive stance. Belgium's proactive involvement and the active lobbying for an air campaign by Guy Verhofstadt, the liberals' leader in the European Parliament, put into question remarks made by a prominent British defence scholar in 2004 – that Belgium is the most conspicuous example of a European tendency to use military force only reluctantly.[21]

Italy initially offered lukewarm support for the campaign. Though it provided seven air bases, its active military contribution to the air campaign was limited – particularly in the opening stages. Having maintained extensive economic ties with Libya, Italy felt uneasy about resorting to military force. Only from late April did the Italian Air Force become involved in offensive strike missions, but then used almost its complete inventory of precision-guided munitions (PGM). After the Italian Air Force's MQ-9 Reaper medium-altitude, long-endurance unmanned aerial vehicles (MALE UAV) had achieved initial operational capability, Italy

---

18   See, for instance, Dwyer, D. "Doubts about NATO in Libya as US Takes Backseat", *abcNews*, 1 April 2011. Available at: http://abcnews.go.com/m/story?id=13274607&sid=77

19   "Libya: Obama, Cameron and Sarkozy Vow Gaddafi Gaddafi Must Go", BBC News.

20   Joyce, Brigadier-General D. (Commander TF Libeccio), "End of Tour Report – Task Force Libeccio", (unclassified document provided by the Canadian Aerospace Warfare Centre), document 1630–31, November 2011, 1.

21   Freedman, L. "Can the EU Develop an Effective Military Doctrine?", in *A European Way of War*, ed. Steven Everts et al. (London: Centre for European Reform, 2004), 16. Available at: http://www.cer.org.uk/sites/default/files/publications/attachments/pdf/2011/p548_way_ofwar-4464.pdf [accessed 7 May 2014].

found itself in a position to provide a special capability to the campaign.[22] Yet the global financial downturn had a severe effect upon Italy's budget. As a cost-saving measure, Italy removed its aircraft carrier *Giuseppe Garibaldi* from the operational theatre in July. Earlier, in late June, Italian decision makers called for a ceasefire, manifesting Italy's ambiguous position towards the allied campaign.[23] Since the Italians could not afford not to shape Libya's future, they were literally forced to participate in the operations. Doing so rather reluctantly, they attempted to mitigate military operations in addition to hosting various forces on Italian territory.

It is also interesting to look at the European non-contributors, Germany foremost among them. A dilemma between its strong emphasis upon NATO as the bedrock for German security and the country's reluctance to employ its armed forces across the spectrum of military force – a prerequisite for making credible contributions to alliance operations – will likely persist. Germany's historical legacy still exerts tremendous inertia upon a proactive defence policy. For the foreseeable future, the use of military force will remain a sensitive issue for the German constituency. Nevertheless, the German military has developed into balanced forces in the post-Cold War era, particularly in the last decade. Consequently, Germany has evolved as a key player in several air and space dimensions, including synthetic-aperture radar satellite reconnaissance/surveillance, theatre ballistic missile defence, and deep strike by acquiring an impressive number of indigenous air-launched cruise missiles. Moreover, it has retained niche capabilities such as a very sophisticated and proven suppression of enemy air defences (SEAD) capability. In 1999 a lean German Air Force SEAD component, including 10 Tornado Electronic Combat/ Reconnaissance (ECR) aircraft, released approximately one-third of all High-speed Anti-Radiation Missiles (HARM™) expended during Operation Allied Force over Yugoslavia.[24] By opting out of military operations against Gadhafi, Germany missed a further opportunity to translate the German Air Force's new potential into effective operational output.

Equally interesting was the absence of the new NATO countries – the former Warsaw Pact nations, in particular Poland, which operates an advanced F-16 attack force. One might speculate on three reasons for their absence: lack of operational preparedness; lack of funding for deployed fighter operations; or lack of political willingness to contribute – the latter due perhaps to Gates's (and therefore American) lukewarm support for operations against Gadhafi. Eastern

---

22   Luca Peruzzi, "Italian Air Force Chief Details Libyan Operations", *Flight International*, 17 June 2011. Available at: http://www.flightglobal.com/articles/2011/06/17/358044/italian-air-force-chief-details-libyan-operations.html [accessed 7 May 2014].

23   Brunnstrom, D. "Analysis: French Arms Move Shows Libya Pressure on West", Reuters, 30 June 2011. Available at: http://af.reuters.com/article/libyaNews/idAFLDE75T18L20110630 [accessed 7 May 2014].

24   Anrig, C.F. "The Quest for Relevant Air Power – Continental Europe", *Royal Air Force Air Power Review* 12(1) (Spring 2009), 84. Available at: http://www.raf.mod.uk/raf cms/mediafiles/CA6EA006_1143_EC82_2E796A6FC58BA05B.pdf

European nations, particularly Poland, put a premium upon staying in line with American goals – hence their support in 2003 for Operation Iraqi Freedom. With the United States ceding its leading role in Operation Unified Protector to NATO, Poland might have felt less inclined to get involved.

Besides the NATO allies, Sweden, Qatar, the United Arab Emirates, and Jordan have taken part in the operations. For Sweden – as is examined below – participating in the Libya campaign was a first in the post-Cold War era. On 1 May, Mirage 2000-9s of the United Arab Emirates, up to that time restricted to air policing, reportedly were carrying PGMs and targeting pods. Actual strikes, however, could not be confirmed at the time.[25] For its part, Qatar deployed six Mirage 2000-5s to Crete and flew that country's first air-policing sorties on 25 March alongside French Mirage 2000-5s, marking the first combat mission of an Arab League nation against the backdrop of operations over Libya.[26]

To conclude, Europe's defence political fragmentation persisted and Libya has offered the latest examples of this political reality. Historical national experiences are too different when it comes to the use of military force. Yet as the Libya campaign aptly highlights, no carved-in-stone patterns about particular national behaviours exist. Who could have foreseen the reversed roles between Belgium and The Netherlands or, even more tellingly, the "renewal" of the entente cordiale between Britain and France, particularly after the fierce debates against the backdrop of the invasion of Iraq? In early 2003, Donald Rumsfeld, the US Secretary of Defense, divided Europe into the new and old. Establishing such fixed patterns, however, does not adequately address the problem. National historical experiences as well as the context of a particular campaign, regarding both domestic and foreign policies, will likely determine European contributions and the resulting European force mix. It is therefore also highly unlikely that Europe as a whole will ever bring to bear its full military potential for a specific political purpose.

Accordingly, the author argued in an article published in 2009 that, although one cannot expect all European alliance partners to contribute to a particular operation, it is realistic to assume that any two of the larger European air forces, combined with a number of smaller air forces, will commit themselves. Hence it is vital that the RAF, the French Air Force or the German Air Force retain a balanced core of air power capabilities that the smaller European air forces can augment.[27] Provision of this European core of air power capabilities by the RAF and the French Air Force could successfully sustain the air operations over Libya. Yet as this article further analyses below, a significant imbalance exists between combat air assets and force enablers such as air-to-air refuelling. This disequilibrium

---

25   Hewson, R. "United Arab Emirates Signals Shift to Attack Role", *Jane's Defence Weekly* 48(19) (11 May 2011), 6.

26   Ministère de la Défense [Ministry of Defence], "Libye: point de situation opération Harmattan no. 7" [Libya: Progress report on Operation Harmattan no. 7], 26 March 2011." Available at: http://www.defense.gouv.fr/content/view/full/112895 [accessed 7 May 2014].

27   Anrig, "The Quest for Relevant Air Power", 86–7.

between the spear and the shaft will likely hamper European operations in the future. In the case of Libya, significant US support in the domain of force enablers and the geographical proximity of Libya mitigated the problem.

The Air Campaign Unfolds

On Saturday, 19 March 2011, French combat aircraft entered Libyan airspace in the early afternoon. Seeking to obtain an immediate impact, the aircraft aimed at armoured vehicles just outside Benghazi.[28] However limited this opening strike was, it proved crucial to stop Gadhafi's forces outside the rebel stronghold; inside the city, it would have been extremely difficult to discriminate between combatants and non-combatants or between the various parties. At night, US Navy ships launched over 100 Tomahawk Land-Attack Missiles (TLAM) against critical nodes of Libya's IADS and fixed-site surface-to-air missile systems. Royal Navy submarine HMS *Triumph* also participated in this effort, which preceded the ensuing fixed-wing aircraft strikes.

During the initial strikes, significant confusion arose about command and control arrangements. According to French official sources, national general staffs commanded their respective assets and the sorties were coordinated among the allies.[29] De facto, U.S. Africa Command's Air Operations Centre located at Ramstein Air Base, Germany, directed coalition operations. Prior to NATO taking over air operations in support of SCR 1973, the United States essentially led the campaign, with the USAF bringing to bear a vast array of capabilities. Of these, units participating in Operation Odyssey Dawn included B-2 stealth bombers from the 509th Bomb Wing at Whiteman Air Force Base, Missouri; F-15Es from RAF Lakenheath, United Kingdom; F-16CJs – dedicated SEAD aircraft – from Spangdahlem Air Base, Germany; and EC-130 Commando Solo psychological operations aircraft from the 193rd Special Operations Wing, Pennsylvania Air National Guard.[30] Although each of these aircraft offered unique capabilities, KC-135 tanker aircraft were about to make the USAF's key contribution for the remainder of the campaign. According to the chief of staff of the French Air

---

28 Ministère de la Défense, "Libye: point de situation opération Harmattan no. 1" [Libya: Progress report on Operation Harmattan no. 1], 25 March 2011. Available at: http://www.defense.gouv.fr/actualites/operations/libye-point-de-situation-operation-harmattan-n-1. Ministère de la Défense, "Libye: point de situation opération Harmattan no. 2" [Libya: Progress report on Operation Harmattan no. 2], 25 March 2011. Available at: http://www.defense.gouv.fr/actualites/operations/libye-point-de-situation-operation-harmattan-n-2

29 Ministère de la Défense, "L'opération Harmattan" [Operation Harmattan], 27 September 2011. Available at: http://www.defense.gouv.fr/actualites/operations/l-operation-harmattan [accessed 7 May 2014].

30 Tirpak, J.A. "Odyssey Dawn Units Identified", *Air Force Magazine*, 22 March 2011. Available at: http://www.airforce-magazine.com/DRArchive/Pages/2011/March%202011/March%2022%202011/OdysseyDawnUnitsIdentified.aspx [accessed 7 May 2014].

Force, they shouldered approximately 70 per cent of NATO's air-to-air refuelling, highlighting the European gap in this important domain of air power.[31] In light of the United Kingdom's expecting its new Airbus tankers, the RAF managed to muster just three of its 1960s-vintage VC10 air refuelling aircraft to support air operations over Libya.[32]

Just prior to the United States pulling out all combat aircraft from operations over Libya in early April, the Department of Defense announced that the A-10 and AC-130 had begun operations over Libya on 26 March.[33] Both aircraft, especially suited for this particular campaign, thus made only brief appearances.

NATO's assumption of operations over Libya on 31 March 2011 coincided with the adaptation of Gadhafi regime forces to the air strikes by shifting to non-conventional tactics. Libyan government forces started to blend in with civilian road traffic and to use civilians as a shield for their advance. On many occasions, they used pick-up trucks and "technicals" (trucks armed with heavy machine guns) instead of main battle tanks and armoured personnel carriers. Moreover, weather conditions deteriorated for a few days. Against this backdrop, Gadhafi's regime forces partly seized the initiative again and recaptured territory in eastern Libya, once more posing a threat to the rebels in Benghazi.[34] At the time, many Western commentators blamed NATO for not dealing with the situation adequately. It can indeed be argued that the transition from Operation Odyssey Dawn (American-led) to Operation Unified Protector (NATO-led) initially had a negative impact on the planning side – in particular, NATO's combined air operations centre in Poggio Renatico, Italy, was not prepared for an operation of this scale. Regardless, the Gadhafi forces' gradual shift to non-conventional tactics at the time mitigated the effectiveness of Western air power.

As a result, allied air power had to adapt to the regime forces' non-conventional tactics – witness the efforts of the French armed forces. From 7 to 14 April, French Air Force and naval aviation flew 20 per cent of the overall NATO sorties and 25 per cent of the offensive sorties, neutralizing slightly more than 20 targets, of which 15 were military vehicles and five artillery pieces, including one multiple

---

31   Tran, "[Interview with] Gen. Jean-Paul Palomeros", 22.

32   Ripley, T. "UK RAF Looks to Slow VC10 Drawdown Rate", *Jane's Defence Weekly* 48(31) (3 August 2011), 12.

33   Gertler, J. CRS Report for Congress: *Operation Odyssey Dawn (Libya): Background and Issues for Congress*, 30 March 2011 – R41725 (Washington, DC: Congressional Research Service, Library of Congress), 8. Available at: http://fpc.state.gov/documents/organization/161350.pdf [accessed 7 May 2014].

34   Ministry of Defence, "NATO Delivers Update on Libya Operations", 7 April 2011. Available at: http://www.mod.uk/DefenceInternet/DefenceNews/Military Operations/NatoDeliversUpdateOnLibyaOperations.htm [accessed 7 May 2014]. Ministry of Defence, "Typhoon Joins Tornado in Libya Ground Attack Operations", 13 April 2011. Available at: http://www.mod.uk/DefenceInternet/DefenceNews/MilitaryOperations/TyphoonJoins tornadoInLibyaGroundAttackOperations.htm [accessed 7 May 2014]

rocket launcher.[35] One and a half months later, from 26 May to 2 June, the French conducted 30 per cent of the overall offensive sorties, enabling them to take out twice as many targets.[36] From 23 June to 1 July, French efforts neutralized approximately 100 targets, of which 60 were military vehicles, including tanks and armoured personnel carriers, and 10 were artillery positions.[37] Just prior to the pulling out of the French aircraft carrier *Charles de Gaulle*, from 3 to 11 August, targets destroyed by French aviation peaked at 150, among them 100 military vehicles and 20 artillery pieces, including multiple rocket launchers.[38] By the end of September, a month prior to the formal closure of Operation Unified Protector, French fighter–bombers released more than 1,140 PGMs, including air-launched cruise missiles.[39]

On 20 October, a French Mirage fighter–bomber and a USAF MALE UAV spotted and fired on a convoy attempting an escape out of Gadhafi's home town of Sirte. After the convoy had been disrupted by the air strikes, the former Libyan leader was quickly captured by the anti-regime forces.[40] In the initial strikes, French combat aircraft operated from the French mainland and from Corsica. To save transit time, those aircraft gradually forward-deployed to Souda Bay, Crete, and later to Sigonella, Sicily.[41] The composition of the French contingent changed over time. In mid-August, after pulling out the *Charles de Gaulle*, France had eight Mirage 2000D, four Mirage 2000N, and four Mirage F1 strike aircraft at Souda Bay. Five Rafale multirole aircraft were stationed at Sigonella.[42] According to official French sources, with these aircraft in place at forward-deployed bases,

35 Ministère de la Défense, "Libye: point de situation hebdomadaire no. 21" [Weekly progress report no. 21], 18 April 2011. Available at: http://www.defense.gouv.fr/actualites/operations/libye-point-de-situation-hebdomadaire-n-21 [accessed 7 May 2014].

36 Ministère de la Défense, "Libye: point de situation no. 28" [Libya: Progress report no. 28], 6 June 2011. Available at: http://www.defense.gouv.fr/content/view/full/121771 [accessed 7 May 2014].

37 Ministère de la Défense, "Libye: point de situation no. 32" [Libya: Progress report no. 32], 7 July 2011. Available at: http://www.defense.gouv.fr/actualites/international/libye-point-de-situation-n-32#.U2p76bLDwzU [accessed 7 May 2014].

38 Ministère de la Défense, "Libye: point de situation no. 38" [Libya: Progress report no. 38], 11 August 2011. Available at: http://www.defense.gouv.fr/content/view/full/130650

39 "UK, France Detail Sorties Mounted, Ordnance Expended", *Jane's Defence Weekly* 48(44) (2 November 2011), 5.

40 Hebert, A.J. "Libya: Victory Through Airpower", *Air Force Magazine* 94(12) (December 2011), 4.

41 Ministère de la Défense, "Libye: point sur le dispositif Harmattan" [Libya: Brief on Harmattan deployment], 19 July 2011. Available at: http://www.defense.gouv.fr/operations/actualites/libye-point-sur-le-dispositif-harmattan [accessed 7 May 2014].

42 Ministère de la Défense, "Libye: réorganisation du dispositif militaire français" [Libya: Reorganization of the French military deployment], 17 August 2011. Available at: http://www.defense.gouv.fr/actualites/international/libye-reorganisation-du-dispositif-militaire-francais#.U2p8lVsg7uE [accessed 7 May 2014].

French armed forces continued to conduct one-third of the offensive sorties.[43] The *Charles de Gaulle* supported combat operations from 22 March until 12 August, when it returned to its home port, Toulon in southern France. Counting its previous deployment to support operations in Afghanistan, it operated more than eight months at sea with a brief break at the beginning of March. The carrier's air component included Rafale and Super Etendard Modernisé strike aircraft, E-2C Hawkeyes, and a combat search and rescue component.[44]

Naval gunfire complemented the air strikes, with British and French navy vessels contributing to lifting the siege of Misrata. In the night from 7 to 8 May, for instance, the French Navy frigate *Courbet* detected rocket launchers firing into the city and, after receiving authorization, effectively engaged the targets.[45] Royal Navy vessels supported air strikes by firing illumination rounds, allowing fixed-wing aircraft to engage regime targets accurately and, like their French counterparts, they engaged artillery positions along the shore.[46]

In mid-April, after the United States had ceased its lead in offensive operations against Gadhafi's regime, the *Washington Post* claimed that the US Armed Forces were doing virtually all of the intelligence, surveillance, and reconnaissance (ISR) and "thus are chiefly responsible for targeting".[47] True, the United States continued to make significant contributions to ISR, but the newspaper's claim completely ignored European ISR assets involved in the campaign. Accordingly, the chief of staff of the French Air Force put into perspective American contributions in an interview of June 2011. Although he acknowledged the vital US support in air-to-air refuelling, European reliance upon American ISR was less severe. In particular, he highlighted the French Air Force and the French Navy's role in supplying the coalition with imagery intelligence by means of the Rafale's advanced digital reconnaissance pod.[48] The French Navy also deployed maritime patrol aircraft

---

43 Ministère de la Défense, "Libye: point de situation no. 40" [Libya: Progress report no. 40], 29 August 2011. Available at: http://www.defense.gouv.fr/operations/actualites/libye-point-de-situation-n-40

44 Ministère de la Défense, "Libye: réorganisation du dispositif militaire français" [Libya: Reorganization of the French military deployment]; and Ministère de la Défense, "Libye: qu'est-ce que la Task Force 473?" [Libya: What is Task Force 473?], 29 March 2011. Available at: http://www.defense.gouv.fr/content/view/full/113211#.U2p9ECSeqwc [accessed 7 May 2014].

45 Ministère de la Défense, "Libye: point de situation no. 25" [Libya: Progress report no. 25], 16 May 2011. Available at: http://www.defense.gouv.fr/content/view/full/119212

46 Ministry of Defence, "RAF and Navy Continue to Strike at Gaddafi Regime", 21 July 2011. Available at: http://www.mod.uk/DefenceInternet/DefenceNews/MilitaryOperations/RafAndNavyContinueToStrikeAtGaddafiRegime.htm

47 Cody, E. "French, British Leaders Meet about West's Role in Libyan Uprising", *Washington Post*, 14 April 2011. Available at: http://www.washingtonpost.com/world/france-britain-call-for-summit-on-natos-role-in-libyan-uprising/2011/04/13/AFy71qVD_story.html

48 Tran, "[Interview with] Gen. Jean-Paul Palomeros", 22.

to Souda Bay, those platforms performing surveillance and guiding coalition strike aircraft.[49] Moreover, the Harfang – the French MALE UAV – conducted its first sortie over Libya on 24 August.[50] Finally, one should note that France is the European key player in military satellite ISR.

Within the first 24 hours of Odyssey Dawn, the RAF's Sentinel R1 Airborne Stand-Off Radar aircraft, essentially an equivalent of the E-8 Joint Surveillance Target Attack Radar System, began to conduct wide-area surveillance.[51] Given the size of Libya, it provided NATO with a unique capability. In particular, it proved instrumental in cueing the USAF's MALE UAVs, which then identified targets and cleared them for air strikes.[52] During the siege of Misrata, USAF MQ-9 Reaper MALE UAVs were crucial in identifying regime forces in built-up areas.[53] In the ensuing sensor-to-shooter loop, NATO, USAF, RAF, and French E-3 Airborne Warning and Control System (AWACS) aircraft relayed attack authorizations from the combined air operations centre at Poggio Renatico in northern Italy to NATO's strike aircraft.[54]

According to a statement by Brigadier General Mark van Uhm, Chief of Allied Operations at NATO's Supreme Headquarters Allied Powers, Europe, in late April, only 10 per cent of the daily sorties represented designated targets; dynamic strikes dealt with the remaining targets. In these cases, strike pilots regularly loitered for a couple of hours in search of targets.[55] So-called "strike coordination and reconnaissance" (SCAR) boxes were established over specific areas and main lines of communications.

About a month after NATO had taken charge of the air operations, it claimed to have degraded Gadhafi's military machinery by one-third.[56] In light of an apparent stalemate, these claims seemed to lack credibility. The target sets consisted

---

49   Ministère de la Défense, "Libye: arrivée des Atlantique 2 à la Sude" [Arrival of the Atlantique 2 at Souda], 4 August 2011. Available at: http://www.defense.gouv.fr/actualites/international/libye-arrivee-des-atlantique-2-a-la-sude#.U2p9bMfMpHw [accessed 7 May 2014].

50   Ministère de la Défense, "Libye: premier vol du Harfang au profit d'Harmattan" [Libya: Harfang's first flight for Harmattan], 25 August 2011. Available at: http://www.defense.gouv.fr/operations/autres-operations/operation-harmattan-libye/actualites/libye-premier-vol-du-harfang-au-profit-d-harmattan [accessed 7 May 2014].

51   Ministry of Defence, "RAF ISTAR Squadrons Keep Watch over Libya", 1 April 2011. Available at: http://www.mod.uk/DefenceInternet/DefenceNews/Military Operations/RafIstarSquadronsKeepWatchOverLibya.htm [accessed 7 May 2014].

52   See, for instance, Ripley, T. "AWACS Provides Key Link for NATO Strikes over Libya", *Jane's Defence Weekly* 48(31) (3 August 2011), 7.

53   Ripley, T. "USAF Predators Direct Tornado Strikes", *Jane's Defence Weekly* 48(20) (18 May 2011), 6.

54   Ripley, "AWACS Provides Key Link", 7.

55   Tigner, Brooks. "NATO Cripples Ghadaffi Forces, C2 Capabilities", *Jane's Defence Weekly* 48(17) (27 April 2011), 8.

56   Ibid.; and Tigner, B. "'More Work to Be Done' in Libya, Reports NATO", *Jane's Defence Weekly* 48(19) (11 May 2011), 6.

of: military headquarters; communications nodes; ammunition bunkers; defence radar sites; artillery pieces, including multiple rocket launchers; tanks; armoured personnel carriers; armed vehicles; and other military assets. The French effort, as is examined above, concentrated on fielded forces that immediately threatened the civil population. This focus, however, did not preclude taking out operational- and strategic-level headquarters. Unlike Operation Allied Force in Kosovo, this operation included no dispute about the most effective centres of gravity. In 1999 some military leaders were not inclined to emphasize the destruction of Serb forces in the field.[57] Despite NATO's continued focus on fielded forces, better-armed regime troops forestalled rebel advances. As of late June, the Western Mountains south of Tripoli represented the only front where the rebels had steadily advanced.[58] Though this front initially received the least attention by allied air power, it finally proved decisive in overcoming the stalemate on the ground – reportedly French special forces played a crucial role in establishing an effective air–land interface.

The extremely fluid situation on the ground in the early stages of the campaign complicated the synchronization of ground manoeuvres and air strikes. Unlike the early phases of Operation Enduring Freedom in Afghanistan in 2001, during which American special operations forces tightly synchronized air strikes with Northern Alliance movements, the political situation dictated that NATO air power should not serve as the immediate air arm of the rebels.[59] Thus NATO air power occasionally hit rebel forces, particularly when they used captured tanks, though this might have been by accident.[60] Synchronization also proved difficult because the rebel forces lacked effective organization. By early June, coordination of air and ground manoeuvres had reportedly improved.[61] Yet one might attribute this to the fact that the front lines had become less fluid and more rigid. Coalition aircraft also minimized collateral damage by using only PGMs, a landmark for Western air power.[62]

Like its French counterpart, the RAF shouldered a heavy burden of the air attacks and proved its effectiveness once more. Over the weekend of 9 to 10 April, for instance, NATO reportedly destroyed 61 armoured vehicles and air defence assets, the RAF engaging one-third of the targets.[63] In the second half of May, RAF attack

---

57   Clark, General W.K. *Waging Modern War: Bosnia, Kosovo, and the Future of Combat* (New York: PublicAffairs, 2002), 303.

58   Mili, A. "Libya Rebels Prepare for Fight over Strategic Town", Reuters, 30 June 2011. Available at: http://af.reuters.com/article/libyaNews/idAFLDE75T1S220110630

59   See Biddle, S. *Afghanistan and the Future of Warfare: Implications for Army and Defense Policy* (Carlisle Barracks, PA: Strategic Studies Institute, US Army War College, 2002), 38–39.

60   See Tigner, B. "NATO Blames 'Fluid' Situation for Allied Airstrikes on Rebels", *Jane's Defence Weekly* 48(15) (13 April 2011), 7.

61   Gelfand, L. "NATO Extends Operation in Libya for 90 Days", *Jane's Defence Weekly* 48(23) (8 June 2011), 4.

62   Chuter, A. and Hale, J. "NATO's Libyan Air War Uses Only Precision Bombs", *Defense News* 26(28) (25 July 2011), 3.

63   Ministry of Defence, "Typhoon Joins Tornado".

aircraft also engaged Gadhafi's navy. On 19 May, they destroyed two corvettes at the naval base at Al Khums, the nearest military harbour to the port of Misrata, as well as a facility in the dockyard that constructed fast, inflatable boats by means of which, regime forces intended to mine the harbour of Misrata and attack nearby vessels.[64] The RAF particularly excelled through demanding targeting. According to sources in the United Kingdom, the RAF had flown approximately 90 per cent of its combat missions against dynamic targets, which are more demanding than pre-planned static objectives.[65] As of 24 August 2011, UK forces had destroyed over 890 former regime targets, including several hundred tanks, artillery pieces, and armed vehicles.[66] When the street fighting started in Tripoli, RAF aircraft maintained a presence over the city, destroying military intelligence facilities in a pre-dawn strike on 21 August or engaging heavy weapons such as main battle tanks on the outskirts of Tripoli.[67] Interestingly, British attack aircraft staged a mini Scud hunt on 24 August, destroying three Scud-support vehicles near Sirte, a site from which former regime forces launched Scud ballistic missiles against the city of Misrata.[68] British forces flew more than 3,000 sorties, including more than 2,100 strike sorties. The latter corresponds to approximately 22 per cent of the coalition's strike sortie total. By 24 October, a week before the formal cessation of operations, RAF fighter–bombers released approximately 1,400 PGMs, including air-launched cruise missiles. These were supplemented by Royal Navy TLAM strikes in the early stages of the campaign.[69]

As in the case of the French Air Force, the RAF contingent changed over time. Originally, the RAF force consisted of 10 Typhoons in the air defence role and eight Tornado GR4s in the attack role. Libya was a first for the Eurofighter Typhoon. Two days after the start of the air campaign, on 21 March 2011, RAF Typhoons patrolled the Libyan NFZ, their first-ever combat mission. However, the air-to-air component gradually decreased in favour of the ground attack component. In early April, two Typhoons returned to the United Kingdom, while the addition of four aircraft boosted the Tornado GR4 component to a total of 12. Simultaneously, four

---

64   Ministry of Defence, "RAF Strikes Gaddafi's Navy", 20 May 2011. Available at: http://www.mod.uk/DefenceInternet/DefenceNews/MilitaryOperations/RafStrikes GaddafisNavy.htm [accessed 7 May 2014].

65   Jennings, G. "Royal Air Force Downplays Carrier Aviation", *Jane's Defence Weekly* 48(30) (27 July 2011), 12.

66   Ministry of Defence, "NATO Mission Not over Yet in Libya", 24 August 2011. Available at: http://www.mod.uk/DefenceInternet/DefenceNews/Military Operations/NatoMissionNotOverYetInLibya.htm [accessed 7 May 2014].

67   Ministry of Defence, "RAF Ready for Further Action over Tripoli", 23 August 2011. Available at: http://www.mod.uk/DefenceInternet/DefenceNews/Military Operations/RafReadyForFurtherActionOverTripoli.htm [accessed 7 May 2014].

68   Ministry of Defence, "RAF Strikes Targets Near Tripoli and Sirte", 25 August 2011. Available at: http://www.mod.uk/DefenceInternet/DefenceNews/ MilitaryOperations/RafStrikesTargetsNearTripoliAndSirte.htm [accessed 7 May 2014].

69   "UK, France Detail Sorties Mounted", *Jane's Defence Weekly*, 5.

**Figure 15.1   A Royal Air Force Tornado GR4 takes off from a United
Kingdom airfield for service over Libya**
Source: UK Ministry of Defence, Photo 45152525.

of the remaining eight Typhoons had shifted from air defence to ground attack.
The resulting 16 ground-attack aircraft allowed the RAF to provide a quarter of
NATO's ground-attack assets.[70] In the second half of July, the RAF once more
boosted its attack and reconnaissance capabilities by deploying another four
Tornado GR4s, one of them equipped with a reconnaissance pod. Henceforth,
the RAF operated 16 Tornado GR4s and six Eurofighter Typhoons from Gioia
del Colle Air Base in southern Italy.[71] Notably, the combat-proven Tornado GR4
(Figure 15.1) remained the RAF's preferred aircraft.

70   Ministry of Defence, "RAF Typhoons Patrol Libyan No-Fly Zone", 22 March
2011. Available at: http://www.mod.uk/DefenceInternet/DefenceNews/MilitaryOperations/
RafTyphoonsPatrolLibyanNoflyZone.htm; Ministry of Defence, "PM Visits RAF
Crews Deployed on Libya Operations", 5 April 2011. Available at: http://www.mod.
uk/DefenceInternet/DefenceNews/MilitaryOperations/PmVisitsRafCrewsDeployed
OnLibyaOperations.htm; Ministry of Defence, "Action in Libya Has Prevented
'Humanitarian Catastrophe'", 6 April 2011. Available at: http://www.mod.uk/Defence
Internet/DefenceNews/MilitaryOperations/ActionInLibyaHasPreventedhumanitarian
Catastrophe.htm; and Ministry of Defence, "Typhoon Joins Tornado".
71   Ministry of Defence, "Additional RAF Tornado Jets Arrive in Italy", 20 July
2011. Available at: http://www.mod.uk/DefenceInternet/DefenceNews/MilitaryOperations
/AdditionalRafTornadoJetsArriveInItaly.htm.

Other Nations' Contributions and the Importance of a UN Mandate

Canada put a particular premium on a robust UN mandate authorizing the use of military force. In his 18 March statement, just one day prior to the beginning of combat operations, Canadian Prime Minister Stephen Harper established an explicit linkage between SCR 1973 and Canada's military commitment.[72] With a Canadian, Lieutenant-General Charles Bouchard, in charge of the NATO mission, the Royal Canadian Air Force (RCAF) provided – besides its fighter–bomber deployment – some sought-after capabilities such as air-to-air refuelling. Against the backdrop of a scarcity in ISR assets, the two deployed Canadian CP-140 Aurora maritime patrol aircraft (a derivative of the Lockheed P-3C Orion) also played a significant role throughout the campaign.[73]

The European F-16 operators – Denmark, Norway, and Belgium – once more proved that smaller air forces with the right equipment, training, and attitude can punch above their weight. Though the Royal Netherlands Air Force has proven time and again that it fulfils the criteria above, it was politically hamstrung in displaying its full potential and as such was not authorized to carry out air-to-ground attacks. Without a UN mandate, one could hardly have expected such significant contributions from Denmark and especially from Belgium and Norway.

What made Libya different – according to Danish scholars – was the perceived need to act swiftly to prevent genocide and the fact that ground forces were ruled out from the start. Libya thus presented a perfect opportunity for doing good with UN authorization in a way that presented few risks to Danish personnel.[74] The Royal Danish Air Force (RDAF) was at the vanguard of operations against the Gadhafi regime. Its six F-16 fighter–bombers – two of them kept in reserve – released in excess of 900 PGMs.[75] Given the limited size of Denmark, the number of PGMs expended is impressive and comes close to UK and French PGM volumes released over Libya. The RDAF's outstanding performance was also fully embraced by Danish political decision makers. Lene Espersen, Denmark's foreign minister, stated:

> We went into this operation in Libya with open eyes and knew that it could cost us. ... The important thing is that Denmark has been at the forefront, and helped to keep civilians safe and ensure that the UN resolution is carried out.[76]

---

72   Canada, Government of, Statement by the Prime Minister of Canada on the situation in Libya, 18 March 2011. Available at: http://pm.gc.ca/eng/media.asp?id=4048

73   Joyce, D. "TF LIB Operations; Chronologies–Acronyms", Annex A in Joyce, "End of Tour Report", 2, November 2011.

74   Jakobsen P.V. and Møller, K.J. "Good News: Libya and the Danish Way of War", in Nanna Hvidt and Hans Mouritzen (eds), *Danish Foreign Policy Yearbook 2012* (Copenhagen: Danish Institute for International Studies, 2012), 109–112.

75   Haynes, D. "Denmark's Top Guns Trump RAF in Libya", *The Times*, 29 September 2011, 13.

76   O'Dwyer, G. "Libya Operations Threaten Nordic Budgets", *Defense News* 26(24) (20 June 2011), 8.

Yet Denmark's outstanding contribution to the air campaign also proved challenging. In June, the Danish government was reported to be in talks with a number of NATO allies, particularly the United States, to get its PGM stocks topped up.[77]

On 23 March, the Norwegian Prime Minister adopted a royal decree authorizing the Royal Norwegian Air Force to contribute to the implementation of SCR 1973 and participate in the American-led Operation Odyssey Dawn.[78] The decree explicitly highlighted the legal foundations for Norway's participation in the Libya campaign, referring not only to SCR 1973 but also to the Arab League's 12 March decision to request the UN Security Council to establish a NFZ and safe havens to protect the civilian population. Yet Norwegian decision makers viewed SCR 1973, based on Chapter VII of the UN Charter, as the ultimate legal authorization for the use and necessity of military force.[79]

The Royal Norwegian Air Force (RNoAF) deployed its aircraft to Souda Bay in Crete. By the end of July, when Norway formally ceased its combat aircraft contribution to Operation Unified Protector, Norwegian F-16 fighter–bombers had dropped 588 PGM – again an impressive volume in relation to the size of the Norwegian armed forces. The RNoAF engaged a variety of targets, ranging from tanks and armoured personnel carriers to Scud-related facilities.[80] Yet Libya operations apparently also represented an unsustainable burden for Norway, hence the premature redeployment of the Norwegian F-16 detachment from Souda Bay. The Norwegian Defence Minister, Grete Faremo, stated on 13 June that:

> It's important that Norway continues to contribute, but we must expect understanding from our allies that having such a small air force means we cannot maintain such a large fighter contribution over a prolonged period.[81]

There is also speculation that – with the Libya operation going beyond the protection-of-civilians mission towards regime change – there was no longer

---

77   Kington, T. and O'Dwyer, G., "Small Bombs Loom Big as Libya War Grinds On", *Defense News* 26(25) (27 June 2011), 1.

78   Statsministerens Kontor [Prime Minister's Office], "Norge med i operasjoner i Libya" [Norway participates in operations in Libya], *Pressemelding* [press release], 23 March 2011. Available at: http://www.regjeringen.no/nb/dep/smk/pressesenter/presse/meldinger/2011/norge-med-i-operasjoner-i-libya.html?id=636399 [accessed 7 May 2014].

79   Minister of Defense Grete Faremo, "Fullmakt til deltakelse med norske militære bidrag i operasjoner til gjennomføring av FNs sikkerhetsrådsresolusjon 1973 (2011)" [Authorization for the participation of a Norwegian military contribution in operations for the implementation of UN Security Council Resolution 1973 (2011)], *Kongelig Resolusjon* [Royal Decree], 23 March 2011, Sections 1, 2. Available at: http://www.regjeringen.no/upload/FD/Temadokumenter/Libya-deltakelse_kgl-res-23-3-2011.pdf [accessed 7 May 2014].

80   Forsvaret [Defence (Norway)], "Sluttrapport Libya" [Final Report on Libya Operations], 2 December 2012. Available at: http://forsvaret.no/operasjoner/rapporter/sluttrapporter/Sider/Sluttrapport-Libya.aspx

81   O'Dwyer, "Libya Operations Threaten Nordic Budgets", 1.

sufficient consensus within Norway's government to wholeheartedly back the RNoAF's fighter–bomber commitment.

According to Belgian scholars, the Belgian Air Force's participation in Libya air operations was made possible primarily by three factors: SCR 1973, which was widely regarded as a solid foundation for action; the wide media coverage, which created a sense of necessity; and the public antipathy towards Gadhafi. Like the RDAF, the Belgian Air Force was among the first to contribute to Operation Odyssey Dawn. The Belgian detachment, based at Araxos air base, Greece, conducted its first combat air patrol to enforce the NFZ on 21 March, only two days after the initial strikes were flown by French fighter–bombers.[82] The first air-to-ground strikes followed suit on 27 March. These were offensive counter-air missions.[83] Shortly before the formal closure of Operation Unified Protector on 20 October, the Belgian Minister of Defence, Pieter de Crem, stated at a press conference that the Belgian detachment at Araxos airbase, consisting of six F-16 fighter-bombers, had accumulated approximately 2,500 flying hours and conducted 473 weapon engagements.[84]

The Royal Netherlands Air Force (RNLAF) for its part was restricted from conducting air-to-ground strikes and so had to focus on the air-to-air role. Nevertheless, a by-product of the air-defence sorties was intelligence gathering. In this area, Royal Netherlands Air Force F-16 fighter-bombers could make a valuable contribution to the campaign that went beyond simply imposing the NFZ. In total, the RNLAF conducted 591 sorties and accumulated 2,845.5 flying hours. For a brief period, at the beginning of the campaign, one of the RNLAF's two KDC-10 tanker aircraft also provided air refuelling to both Netherlands and alliance F-16s.[85]

Undoubtedly, the European F-16 operators punched above their weight and their performance was by any standards remarkable. Yet it also needs to be pointed out that, with the exception of the brief appearance of the Netherlands KDC-10 tanker, they primarily contributed to the offensive efforts and were completely dependent on their alliance partners, particularly the United States, when it came to force-enabling aspects, such as air-to-air refuelling.

For Sweden, the Libya crisis resulted in the first deployment of combat aircraft to a real operation since the early 1960s, when Swedish fighter-bombers supported

---

82 La Défense [Defence (Belgium)], "Aperçu hebdomadaire des opérations extérieures" [Weekly Report on Deployed Operations], 17 March 2011–23 March 2011. Available at: http://www.mil.be/def/doc/viewdoc.asp?LAN=fr&FILE=&ID=2452

83 La Défense [Defence (Belgium)], "Aperçu hebdomadaire des opérations extérieures" [Weekly Report on Deployed Operations], 24 March 2011–30 March 2011. Available at: http://www.mil.be/def/doc/viewdoc.asp?LAN=fr&FILE=&ID=2457

84 La Défense [Defence (Belgium)], "Khadafi et son régime neutralisés" [Gadhafi and His Regime Neutralized], 20 October 2011. Available at: http://www.mil.be/def/news/index.asp?LAN=fr&ID=3155

85 Erwin van Loo, Netherlands Military History Institute, to the author, email, 21 February 2012.

UN operations in the former Belgian Congo.[86] Initially, this Nordic country with a legacy of neutrality deployed eight JAS 39 Gripen aircraft supported by a Swedish Air Force C-130 tanker on 2 April. The deployment took place only 23 hours after a Swedish parliamentary decision to help enforce the NFZ over Libya. National rules of engagement were tight. This meant that the Swedish government relegated missions to implementing the NFZ and conducting counter-air-oriented reconnaissance missions, so that Swedish Air Force intelligence-gathering was basically restricted to airfields and ground-based air defence systems. These restrictions were in place despite the first Swedish Air Force detachment's aircraft and pilots being multirole-capable. After the first Swedish detachment had been redeployed and relieved by a reduced force consisting of five Gripen combat aircraft, national rules of engagement were relaxed. As a consequence, the successor detachment conducted a variety of reconnaissance missions. Equipped with dedicated reconnaissance and Litening III targeting pods, the Swedish detachment delivered 250,000 images. The total amounted to 650 sorties and 2,000 flying hours.[87]

Task Force Hawk Coming of Age

During the course of Allied Force in 1999 over Kosovo and Serbia, General Wesley Clark – Supreme Allied Commander, Europe – assembled Task Force Hawk in Albania, intending to bring more pressure to bear against Slobodan Milošević, then President of the former Federal Republic of Yugoslavia. Task Force Hawk's main manoeuvre element was its Apache combat helicopter component. After Clark's attempts to request permission to employ the Apaches, Washington finally turned him down. The Joint Chiefs of Staff had severe concerns about risking sophisticated combat helicopters to attack tactical forces. According to Clark, though, the Apaches could identify targets from across the border that fixed-wing aircraft had not struck.[88]

Twelve years later, in May 2011, the resolve to deploy combat helicopters gradually grew both in the United Kingdom and France in order to further restrain the ground manoeuvres of Gadhafi's forces. In the night from 3 to 4 June, French and British combat helicopters for the first time engaged ground targets. British Army Apache helicopters, launched from helicopter carrier HMS *Ocean*, operated in the area of Brega, helping to soften the front deadlock in eastern Libya. They reportedly faced incoming fire.[89] Despite the threat, *Ocean* again launched its

---

86　*Editor's note*: see Part I of this volume on the UN's 1960s Congo mission; in particular see Chapter 2 in this volume.

87　Pengelley, R. "Swedish Gripens over Libya", *Jane's International Defence Review* 45 (January 2012), 34–5.

88　Clark, *Waging Modern War*, 303, 337 and 367.

89　"Libya: UK Apache Helicopters Used in NATO Attacks", BBC News, 4 June 2011. Available at: http://www.bbc.co.uk/news/uk-13651736 [accessed 7 May 2014].

combat helicopters the next night to engage multiple-launch rocket systems.[90] French and British combat helicopters operated in close cooperation with fixed-wing aircraft, the latter gathering intelligence both to select targets and to provide assessments of potential surface-to-air-missile threats. They also remained on standby to launch complementary strikes.[91] On a raid in early June, British Army Apache helicopters first destroyed high-speed inflatable boats attacking the harbour of Misrata and then opened fire on a ZSU-23-4 self-propelled anti-aircraft gun near Zlitan, as well as a number of armed vehicles, displaying the flexibility of helicopter operations in this particular theatre.[92]

Launched from France's amphibious assault ship *Tonnerre* in the night from 3 to 4 June, Tigre and Gazelle combat helicopters engaged approximately 20 ground targets.[93] Like their British counterparts, the French Army combat helicopters reportedly faced incoming fire by man-portable air defence systems. In the first week of French helicopter operations, the number of destroyed Libyan military vehicles increased. Among the 70 targets destroyed by French forces from 2 to 9 June, approximately 40 were military vehicles, two-thirds of them destroyed by helicopters.[94] In mid-August, French attack helicopters, launching from the amphibious assault ship *Mistral*, conducted a major interdiction strike. Ten of them struck at two choke points along the lines of communications west of the front deadlock at Brega, destroying several vehicles, surveillance radars, and defensive positions.[95] According to rebel commanders, sustained helicopter strikes were crucial in turning the table at the Brega front. French attack helicopters carried out

90   Ministry of Defence, "HMS Bangor En Route to Libya", 7 June 2011. Available at: http://www.mod.uk/DefenceInternet/DefenceNews/MilitaryOperations/HmsBangorEn RouteToLibya.htm [accessed 7 May 2014].

91   Ministry of Defence, "Good Progress Seen in Libya Operations", 24 June 2011. Available at: http://www.mod.uk/DefenceInternet/DefenceNews/MilitaryOperations/Good ProgressSeenInLibyaOperations.htm [accessed 7 May 2014].

92   Ministry of Defence, "RAF Destroys Nine of Gaddafi's Underground Storage Bunkers in Libya", 13 June 2011. Available at: http://www.mod.uk/DefenceInternet /DefenceNews/MilitaryOperations/RafDestroysNineOfGaddafisUndergroundStorage BunkersInLibya.htm [accessed 7 May 2014].

93   Ministère de la Défense, "Libye: engagement des hélicoptères français depuis le BPC Tonnerre" [Libya: Use of French helicopters from BPC Tonnerre], 4 June 2011. Available at: http://www.defense.gouv.fr/actualites/operations/libye-engagement-des-heli copteres-francais-depuis-le-bpc-tonnerre [accessed 7 May 2014].

94   Ministère de la Défense, "Libye: point de situation no. 29" [Libya: Progress report no. 29], 11 June 2011. Available at: http://www.defense.gouv.fr/actualites/operations/libye-point-de-situation-n-29 [accessed 7 May 2014].

95   Ministère de la Défense, "Libye: les hélicoptères français en opération dans le secteur de Brega" [Libya: French helicopters operating in the area of Brega], 25 August 2011. Available at: http://www.defense.gouv.fr/actualites/international/libye-les-helicopteres-francais-en-operation-dans-le-secteur-de-brega#.U2p9-89hKq8 [accessed 7 May 2014].

the majority of these strikes, launching in excess of 430 HOT[96] anti-tank missiles and an unspecified number of cannon rounds and rockets.[97] Still, helicopter strikes against the backdrop of Operation Unified Protector remain a controversial issue. In particular, many Western airmen believe that their employment was tying down too many fixed-wing aircraft, which were needed to provide cover and could have done the same job as effectively.

Drawing upon Comparative Advantages: General Observations

In his book *The Causes of Wars*, renowned British scholar Sir Michael Howard outlined four dimensions of strategy: the social, operational, logistical, and technological. In his view, "no successful strategy could be formulated that did not take account of them all, but under different circumstances, one or another of these dimensions might dominate".[98] The German Wehrmacht of World War II, for instance, is a prime example of an armed force that attempted to exploit the operational dimension. On most occasions outgunned and outnumbered, it nevertheless remained confident of achieving victory by virtue of superior skills in the operational dimension. Yet as the logistical dimension started to dominate, superior allied resources in equipment and manpower undermined this German strategy. The technological dimension very much shaped the battle of the Atlantic. The British achievement in breaking the Enigma code, combined with US and British advances in anti-submarine warfare, gave the Western allies the decisive advantage to secure a safe passage across the Atlantic and to mitigate the German U-boat threat to a "tolerable" level. Counter-insurgency campaigns, such as the involvement of France or the United States in Vietnam, are by their very nature dominated by the social dimension while one strives for success in the operational dimension. As recent campaigns in Iraq and Afghanistan have borne witness, winning hearts and minds is extremely difficult. Can Western armed forces effectively bring across their benign intentions in a culturally alien environment?

Hinging upon air and naval power, the Western alliance could confine its intervention to the operational and technological dimensions as the predominant ones with regard to Libya, the wider Arab community, and their domestic constituencies. Support for the campaign in France and the United Kingdom did not wane. The zero own-casualty toll, enabled by the superior technology of air

---

96   The HOT missile is a joint French–German creation, a second-generation, long-range anti-tank missile that can be launched from ground vehicles or helicopters. HOT stands for *Haut subsonique Optiquement Téléguidé Tiré d'un Tube*, or High Subsonic Optical Remote-Guided, Tube-Launched.

97   "UK, France Detail Sorties Mounted", *Jane's Defence Weekly*, 5.

98   Howard, M. "The Forgotten Dimensions of Strategy", in *The Causes of Wars and Other Essays*, 2nd ed., enlarged, ed. Michael Howard (Cambridge, MA: Harvard University Press, 1983), 105.

power, might have significantly contributed to this public backing. In the absence of ground troops in Libya, France disclosed on 29 June that it had airdropped weapons to rebel fighters in the Western Mountains south of Tripoli – the first time that a Western country acknowledged arming the rebels.[99] Qatar, for its part, supported the rebels by funnelling arms into Benghazi, from where they were further distributed to the various fronts, also by air.[100] In addition, various allied countries sent military-liaison advisory teams to support the National Transitional Council, and Western alliance special forces evidently offered immediate advice to rebel front-line forces. All of these measures fall short of deploying regular ground forces with a large footprint into the theatre.

By staging successive offensives, Western forces have repeatedly attempted to turn the Afghan conflict into a situation dominated by the operational dimension. Though most of these offensives have been militarily successful, the conflict remains dominated by the social dimension, making it nearly impossible for the West to effect decisive results at the strategic level, even after 10 years of continuous deployments.

In the 1970s and the 1980s, the United States confined its military involvement in the Persian Gulf to carrier strike groups and naval air power without a single boot on the Arabian Peninsula. "Offshore balancing" allowed the United States to secure its oil interests effectively at the lowest price. In the context of Michael Howard's theory of the dimensions of strategy, the reason for this becomes obvious. By concentrating on the maritime and air environments, the United States could draw upon comparative advantages, at the same time managing to avoid becoming an occupying force and arousing grievances in the local populations. This was no longer the case in the 1990s. Al-Qaeda founder Osama bin Laden's speeches and sermons drew attention to the massive Western, particularly American, military presence on the Arabian Peninsula. In this regard, the American scholar Robert Pape, author of *Bombing to Win: Air Power and Coercion in War* and the more recent book *Dying to Win: The Strategic Logic of Suicide Terrorism*, argues that the presence of American ground troops in Muslim countries is the main factor driving suicide terrorism.[101] According to this logic, Islamic fundamentalism is not the principal driving factor of suicide terrorism against the United States' interests. This explains the absence of al-Qaeda terrorists from Iran or Sudan, which harboured bin Laden in the 1990s. Suicide attacks aimed against the West, however, surged in Iraq after Western forces with a different religious background occupied that country. This difference in religion between the occupier and the occupied community is – according to Pape – the key reason for suicide

---

99   Mili, "Libya Rebels Prepare for Fight".

100   *Editor's note*: supplying arms to any group in Libya was a violation of the arms embargo imposed by UNSCR 1970 (2011).

101   See: Pape, R.A. *Bombing to Win: Air Power and Coercion in War*. Cornell Studies in Security Affairs (Ithaca, NY: Cornell University Press, 1996); Pape, R.A. *Dying to Win: The Strategic Logic of Suicide Terrorism* (Random House, 2006).

attacks. Prior to Iraqi Freedom, Iraq reportedly had never experienced a suicide terrorist attack.[102]

From this vantage point, arguments made by various commentators, for example, retired general Henning von Ondarza, former commanding officer of Allied Forces Central Europe, which called for ground troops to control the situation in Libya, do not take account of all dimensions of strategy.[103] Although such an approach might have delivered swift military results in the operational dimension, "infidels" on the ground scoring decisive victories and "occupying yet another Muslim country" might have led to strategic backlashes, with the great potential for the social dimension to predominate. Western boots on the ground, also not backed by the Arab League, would likely have caused massive grievances, including suicide terrorism. The very fact that the Western alliance refrained from deploying ground units helped retain the intervention in a situation that placed the operational and technological dimensions at the forefront, despite concerns about collateral damage and international objections to issues such as airdrops of weapons violating the UN arms embargo.

Most interestingly, making sure that the operational and technological dimensions remain predominant helps to prevent significant strains in the logistical dimension of strategy. According to the UK Defence Committee's fifth report, of 19 July 2011, estimates of additional costs of operations in Afghanistan during that year amounted to just over £4 billion (approximately US$6.3 billion). Yet the report admitted that the total costs of operations in Afghanistan remained unknown.[104] In contrast, Secretary of State for Defence Fox estimated the costs of six months of military operations in the framework of Operation Ellamy, the United Kingdom's contribution to the allied effort in support of SCR 1973, at £260 million (approximately US$410 million). This figure includes the cost of replenishing munitions.[105] Accordingly, one can estimate an entire year at approximately £520 million (approximately US$820 million). Although they are very rough estimates, these figures by no means fail to reveal the large discrepancy between the costs of UK operations in Afghanistan and Operation Ellamy in Libya.

---

102   Pape, R. "The Logic of Suicide Terrorism: It's the Occupation, Not the Fundamentalism", *American Conservative*, 18 July 2005. Available at: http://www. theamericanconservative.com/article/2005/jul/18/00017/ [accessed 7 May 2014].

103   Solms-Laubach, F. and Reichelt, J. "Warum packt die NATO Gaddafi nicht?" [Why doesn't NATO seize Gaddafi?], *Bild.de*, 30 July 2011. Available at: http://www. bild.de/politik/ausland/muammar-gaddafi/warum-packt-die-nato-ihn-nicht-19129868.bild. html [accessed 7 May 2014].

104   Defence Committee, *Fifth Report: Ministry of Defence Main Estimates* (London: Defence Committee, 19 July 2011). Available at: http://www.publications.parliament.uk/ pa/cm201012/cmselect/cmdfence/1373/137302.htm [accessed 7 May 2014].

105   Ministry of Defence, "Cost of Libya Operations", 23 June 2011. Available at: http://www.mod.uk/DefenceInternet/DefenceNews/DefencePolicyAndBusiness/CostOf LibyaOperations.htm [accessed 7 May 2014].

To put the UK costs involved into perspective, the RAF was providing about a quarter of the ground-attack assets as of mid-April.[106] Given the estimated yearly UK costs of US$820 million and its estimated 25 per cent share of the offensive air campaign, about US$3.3 billion would have theoretically covered the costs of an entire operation at that pace for a year's duration. Particularly expensive were TLAMs launched from US Navy ships to shut down Libya's IADS and other strategic key targets at the onset of the campaign. The approximate cost of missiles and other American munitions expended from 19 to 28 March came to US$340 million.[107] The above figures combined would be significantly less than the United Kingdom's estimated additional costs of operations in Afghanistan during 2011.

Towards the end of operations Northern and Southern Watch over Iraq before March 2003, General John P. Jumper, the USAF Chief of Staff, argued that the air blockades caused his service to fly some aircraft longer than the average amount of time. However, he was not certain whether doing so would actually result in more wear and tear on the fleet, since the majority of missions did not involve violent manoeuvring.[108] The degree to which European air forces in Libya will feel the effects of increased wear and tear and additional costs involved remains to be seen. Based upon Jumper's comments on the USAF's experience in Iraq, though, these additional costs are unlikely to be excessive.

Not only are costs in treasure significantly lower in comparison to those associated with operations in Afghanistan but also – and even more importantly – the human cost is dramatically reduced. For instance, in the first half of 2011 the British armed forces suffered 27 fatalities in Afghanistan, not to mention the number of wounded and maimed. The 108 fatalities in 2009 and 103 fatalities in 2010 made the two previous years the bloodiest for British troops in Afghanistan.[109] Throughout the 2011 Libya campaign, however, the allies had suffered no fatalities in Libya. Unlike the situation in Afghanistan, the allies could draw fully upon their asymmetric advantages in the technological dimension of strategy, significantly improving force protection.

This article does not contend that the use of ground forces is too costly in modern warfare. In fact, joint manoeuvre warfare, as conducted by the West's most advanced forces, has proven extremely effective and powerful in conventional campaigns, sweeping away conventional resistance. Yet in stabilization operations, Western allies should shape their involvement in ways that allow them to effectively draw upon the comparative advantages in the operational and

---

106   Ministry of Defence, "Typhoon Joins Tornado".

107   Gertler, *Operation Odyssey Dawn (Libya)*, 21–2.

108   Tirpak, J.A. "Legacy of the Air Blockades", *Air Force Magazine* 86 (2) (February 2003), 50. Available at: http://www.airforce-magazine.com/MagazineArchive/ Documents/2003/February%202003/02legacy03.pdf [accessed 7 May 2014].

109   "British Dead and Wounded in Afghanistan, Month by Month", *Guardian. co.uk*, 2011. Available at: http://www.guardian.co.uk/news/datablog/2009/sep/17/afghanistan -casualties-dead-wounded-british-data [accessed 7 May 2014].

technological dimensions. In contrast, winning hearts and minds is excessively difficult, highlighting the extreme challenges for Western intervention forces in the social dimension.

As a rule, warfare does not lend itself to a recipe and the weight and characteristics of each dimension of strategy depend upon its context. In Bosnia in 1995, deployment of a heavy multinational brigade in the United Nations Protection Force did not undermine the West's standing in the social dimension. Together with air power, it produced synergistic joint effects against the Bosnian Serbs' ground manoeuvres, thereby providing the significant combined-arms leverage that Allied Force lacked in 1999. Hence, ground forces strengthened the operational dimension of strategy during Operation Deliberate Force (see Chapter 13 in this volume), which led to the Dayton Peace Accords in late 1995. Due to the specific circumstances, however, the West made air power its weapon of choice against Gadhafi. However protracted the campaign seemed, it proved significantly cheaper in both resources and lives than current or recent stabilization operations in Iraq and Afghanistan, which demanded a great influx of ground forces.

Conclusion

The Libyan campaign stands as a successful example of how Western air power shifted the balance of power in favour of a resistance movement against superior armed regime forces. Essentially, it levelled the playing field. Nevertheless, the Libyans themselves made the final decision. Without intervention from the West's air power, forces loyal to Gadhafi could have inflicted tremendous carnage on both Benghazi and Misrata. Gadhafi's siege of Misrata was terrible, but without air power it most certainly would have become another dark chapter in history. With the United States relegating its major contribution to force enablers, there was a need for offensive contributions by smaller NATO countries. As such, SCR 1973 – viewed by many as the ultimate legal authorization – was a prerequisite to muster sufficient NATO air power.

During the course of the campaign, renowned commentators made various claims. Against the backdrop of the air campaign's becoming protracted, one of them argued that the West should have better armed and trained the rebels before intervening militarily. Aside from political concerns, this proposed course of action completely ignores the time-sensitivity of this operation. The overrunning of the rebel strongholds in late March would have left no time for such arming and training. Other commentators downplayed the intervention as a rather small campaign. Yet assessing a campaign by assets involved is not the most sophisticated approach. At the end of the day, the effect is important. Probably the most frequently raised criticism involved the need for ground forces to effectively turn the tables in Libya. Granted, this strategy might have produced swift military effect, but at the strategic level of warfare it might have caused backlashes – allowing the social dimension of strategy to dominate the conflict.

Moreover, commentators raised concerns about a "protracted" air campaign, implicitly referring to the excessive costs involved. Both the Iraqi NFZs and the Libya campaign, however, bear witness to the fact that relegating an intervention to air power – if circumstances permit – is far less costly than, for instance, ongoing operations in Afghanistan. For some unjustified reason, interventions by air power attract criticism that they consume vast amounts of treasure. Yet air power, combined with its ability to reduce collateral damage significantly, helps keep an intervention in the operational and technological dimensions of strategy, where the West can draw upon its comparative advantages. In particular, the technological dimension yields an asymmetric advantage in force protection that can reduce allied fatalities to a minimum. Short of deploying ground troops, the British and French deployed combat helicopters. After their first missions in the night of 3 to 4 June, commentators expected casualties. These daring attacks undoubtedly and visibly demonstrated NATO's resolve and thereby generated additional coercive leverage.

Other critics charged that, instead of conducting a shock-and-awe campaign, the West used air power only gradually, thus dissipating its true value. Even if the coalition had staged massive air strikes, who could have actually exploited their effects in the early phase of the conflict? This campaign was as much about protecting civilians as about a contest of will between Gadhafi's regime and NATO, whose willingness and ability to conduct a protracted air campaign slowly ground down the dictator's forces and denied him the use of superior conventional weapons on the ground. As it proved, NATO occupied a position from which to do this. The French Air Force's contingent on Crete, for instance, contained about one-tenth of the entire French Mirage 2000D and 2000N fleets, a ratio perfectly suited for a prolonged air campaign.

However, the campaign once more revealed the European imbalance between spear and shaft (or "tooth and tail"), the effects of which could be mitigated only through significant American support and Libya's geographical position. This imbalance will likely persist – witness the RAF's and the French Air Force's acquisition of or plans to acquire 12 to 14 modern multirole transport tanker aircraft each and the remainder of Europe placing even less emphasis on air-to-air refuelling, a situation that will hamper Europe's reach and mobility in the future. Luckily, Europe's only true aircraft carrier, the *Charles de Gaulle*, was immediately ready for action, but France had to pull it out of operations on 12 August after more than eight months of almost continuous service. Clearly, the West could have waged the Libyan campaign without naval air power, but the geographical position of the next contingency might require the availability of more seaborne flight decks.

The campaign has also shown the limits of force specialization within Europe. With countries such as Germany opting out, or others, such as Italy, offering only hesitant support, the campaign kicked off without vital European capabilities (both Germany and Italy operate the most advanced European SEAD forces). To secure political discretion, the larger European countries need to retain balanced air forces. Smaller European air forces that are willing to deploy could punch above their weight by reinforcing Europe's force enablers. A willingness to take risks could also make up for the absence of certain capabilities. Thus French fighter–bombers opened the campaign on 19 March with no dedicated SEAD aircraft, and the employment of combat helicopters effectively compensated for limited numbers of fixed-wing aircraft.

The campaign is likely to reshape European force transformation. For example, the authors of the United Kingdom's *Strategic Defence and Security Review* of late 2010 undoubtedly wrote that document against the backdrop of ongoing operations in Afghanistan.[110] The RAF earmarked such assets as the Sentinel wide-area surveillance aircraft, which saw only limited use in Afghanistan but proved extremely valuable in Libya, for phasing out in the coming years. Consequently, decision makers might need to reconsider certain plans. At the least, the RAF deferred retiring its last Nimrod R1 signals intelligence aircraft by three months, extending its service to support Operation Ellamy – the United Kingdom's contribution to NATO's air campaign. Overall, even though the military gap across the Atlantic undoubtedly remains, the Libyan campaign demonstrated that the gap had narrowed, not only in terms of quality of equipment but also in terms of willingness to intervene.

---

110   HM Government, *Securing Britain in an Age of Uncertainty: The Strategic Defence and Security Review*, presented to Parliament by the Prime Minister by Command of Her Majesty, October 2010 (London: The Stationery Office, 2010). Available at: http://www.direct.gov.uk/prod_consum_dg/groups/dg_digitalassets/@dg/@en/documents/digitalasset/dg_191634.pdf [accessed 7 May 2014].

PART VI
Evolving Capabilities

This volume demonstrates that peacekeeping is much more than soldiers in blue helmets and white vehicles patrolling on the ground. Peacekeeping also means air action, co-ordinating with the soldiers, sailors, police, and civilians below. As has been shown, UN aircraft have fired guns, launched missiles, and even dropped bombs. This volume has also offered new information and insights into the staple of UN aviation: transportation of personnel and supplies. Several million peacekeeper–passengers have been carried to the field and within missions along with their equipment and weapons, including, on occasion, tanks, helicopters, and other heavy-lift items. Described in this volume, using selected case studies, is the UN experience with air power for each of the three core air capabilities: transport, observation, and firepower. The authors have shown in the most comprehensive manner to date how the United Nations has made use of the "third dimension" of space.

Currently, the United Nations relies on 200 to 300 aircraft – all on loan or lease – to provide air support for some 15 peacekeeping missions around the world. It spends about US$1 billion annually on aviation. The organization pays about a dozen of its member nations for the loan of their aircraft and crew, but the majority (two-thirds) of aircraft and crews are leased from commercial contractors. As has been covered, in addition to the aviation work of the departments of Peacekeeping Operations and Field Support, another major contributor is the United Nations Humanitarian Air Service run by the World Food Programme, flying relief workers and supplies using about 50 leased aircraft to the world's emergency zones.

Even with the huge (fivefold) expansion in peacekeeping since the turn of the century, the United Nations is still under-equipped and under-resourced, in air power as in many other areas, and remains unable to fully meet the ambitious mandates given by the UN Security Council. The world organization faces a perennial shortage of helicopters and has few combat aircraft at its disposal: only about a dozen armed helicopters and no jet fighters. It is therefore remarkable that the United Nations can do so much with so little in so many places of the conflict-ridden world. A comparison with the operations of the North Atlantic Treaty Organization (NATO) in Bosnia in the second half of the 1990s shows how much more advanced were the NATO aircraft fleet, which included Joint Surveillance, Targeting Attack Radar System (JSTARS) and Airborne Reconnaissance Low (ARL); and unmanned aerial vehicles (UAVs) like Lofty View, Predator, and Pioneer. Even more advanced was NATO's air fleet for the UN-authorized

Operation Unified Protector in Libya in 2011. The United Nations conducts so-called "poor man's aviation", with too little in the way of high technology. Despite this, the UN has largely been successful in keeping flights safe, cost effective, and green (environmentally, through a recent fuel-efficiency initiative). There are also hopeful signs of advancement in UN aviation technology: in 2013, the United Nations flew its own imaging UAV for the first time – a small but important step forward. However, a tremendous capability gap remains between the peacekeeping UN and the made-for-warfighting NATO.

Fortunately, there is vision and commitment both inside and outside the United Nations for the world organization to advance its aviation capabilities. To give a view of recent progress and where the United Nations is headed, the Chief of Aviation Projects at UN headquarters, Kevin Shelton-Smith, offers a highly useful Chapter 16, full of practical insights. In Chapter 17, Robert David Steele presents a grand vision with many creative and novel elements that represent "big sky" and "out-of-the-box" thinking on the UN air power of the future, while being quite critical of current US support.

Chapter 16

Advances in Aviation for UN Peacekeeping: A View from UN Headquarters

Kevin Shelton-Smith

The aviation fleet used in UN peacekeeping has changed significantly since the turn of the century. It has significantly expanded in size, fleet composition, utilization, route complexity and support. The UN aviation fleet is about to undergo another far-reaching change. This chapter describes the recent changes and looks at what can be expected in the future for the peacekeeping fleet. It does not include any discussion of the United Nations Humanitarian Air Service provided by the World Food Programme (this is the subject of Chapter 6) and the service of World Food Programme's own cargo fleet, or African Union aircraft, or any national air assets supporting UN mandates that are not under direct UN operational tasking and hence not carrying "UN" markings.

UN aviation is required to provide logistics support for the UN's peacekeeping and political missions. Such tasks include Very Important Person (VIP) liaison between political centres, which is often necessary due to the absence of reliable, convenient or safe local commercial air networks. Within mission areas, UN aviation transports food and materiel to UN troops and provides transport for UN staff for duty tasks. Operationally, aircraft provide aerial patrol, observation and monitoring, armed security protection force attack helicopters, search and rescue and, crucially, 24/7 aeromedical evacuation. In addition, UN aviation has increasingly conducted troop rotations between mission areas and home nations, which are otherwise flown using short-term contracts for airliners. In both cases, ground support in mission areas is invariably provided by the United Nations.

In 1999, the United Nations had 47 aircraft worldwide on long-term contracts. The fleet consisted predominantly of King Air B-200, L-100 Hercules, An-26 fixed-wing (FW) aircraft and Mi-8T helicopters with a number of military light and medium helicopters. Since then, peacekeeping missions such as the United Nations Mission in Liberia (UNMIL), the United Nations Mission in Sierra Leone (UNAMSIL) and the United Nations Transitional Administration in East Timor (UNTAET), have each had added dozens of aircraft, while large missions such as the United Nations Organization Mission in the Democratic Republic of the Congo (MONUC) – see Chapter 14 – THE United Nations Mission in Sudan (UNMIS; from 2005) and the African Union–United Nations Mission in Darfur (UNAMID) in the Darfur region of Sudan (from 2007) have each operated more aircraft than the entire UN fleet of 1999. In 2011 the UN aviation fleet size reached 289 aircraft, comprising 111 FW, 108 civilian rotary wing (RW) aircraft and 70 military RW aircraft.

Peacekeeping missions have grown significantly in troop size and geographical area. Whereas aircraft usually operated largely out of capital cities with reasonably developed airport infrastructure, no matter how weakened by the ravages of war, missions are now found with major bases in remote areas and with virtually no infrastructure and poor runways.

The United Nations does not own any of the 200+ aircraft in its fleet. Rather, they are chartered. The provision of aviation services has always remained under the operational control of the air carriers chartered by the United Nations for long-term contracts. The quality of the service has often reflected the level of oversight available. The UN's non-aviation management with a non-aviation culture was in no position to fully appreciate the risks inherent in aviation. Simply knowing that aircraft accidents must be avoided did little to help managers recognize the causes of accidents and the subtle and less subtle impact of external influences. Some crews may have put themselves under pressure to carry out operations in unsatisfactory meteorological conditions, fly long hours, or operate from short runways under the perceived threat that their contracts were at risk if the customer was not fully served. For some crews this may even have carried over from their operational experience in their home countries, with such practices more prevalent among crews from countries with poorer safety standards. The highest standards were not seen as being necessary, applicable or affordable in war-torn regions in which peacekeeping operations occurred. Many UN staff with military backgrounds compared UN operations comfortably with their past experience, while ignoring the absence of training and awareness possessed by civilian passengers from many cultures, who spoke many languages and included the old, the frail and the casual. Some carriers even flew troop rotations for the United Nations using Ilyushin 62 jet airliners – a 1960s-era Soviet plane – having no oxygen supply and argued that it was acceptable. The United Nations routinely flew passengers on cargo aircraft, even with cargo on board, hindering exit routes, supposedly approved by the air operators' civil aviation authorities.

The tipping point came in 1998–1999 when UN air accidents reputedly caused more loss of life to UN personnel and troops than hostile action, even with a small fleet. There were many examples. An executive jet carrying a Special Representative crashed on approach in West Africa.[1] At the end of 1998, two L-100 aircraft were shot down while supporting the United Nations Observer Mission in Angola (MONUA). In 1999 a helicopter attached to the United Nations Civilian Police Mission in Haiti (MIPONUH) flew into the ground in Haiti, killing 14. Senior management from the Secretary-General, Kofi Annan, down were determined to bring about change. No new resources were immediately provided, but the change of attitude had begun. Aviation officers were finding

---

1   In June 1998, the UN Secretary-General's Special Representative to Angola, Mr Alioune Blondin Beye, was killed in a plane crash in Côte d'Ivoire while on mission.

support in seeking higher standards. As the MONUC mission started, contracts for airport services now included firefighting, meteorological forecasting, air traffic control, passenger-handling and cargo-handling. The use of cargo aircraft to carry passengers was stopped, with the limited exception of the L-100 Hercules, in which strict compliance with Civil Aviation Authority requirements for forward-facing seating, oxygen, escape signage, lighting, emergency briefings and additional exits were applied, and passengers were never carried with cargo on board. Antonov 24 passenger aircraft and a Boeing 727 airliner were introduced and more types such as Boeing 757, CRJ, MD-83, Let 410, ATR-42, DHC7 and DHC8 have since joined the fleet.

Crews were happy to accept the mantra "safety first". They were assured of the freedom to cancel a flight for safety reasons. Rules on crew duty periods became strictly applied. Alcohol abuse was not tolerated and crews were repatriated without hesitation if deemed guilty of infringing this rule. English-language standards of crews were significantly tightened.

Over time, aircraft equipment standards were also raised to include colour weather radars, flight data and cockpit recorders, ground proximity warning systems, impact-operated G-switches, digital 406 MHz emergency locator transmitters, aircraft collision avoidance systems and satellite tracking of aircraft.

All this has led to an improved perspective of the operating environment and the safety culture has risen with it, among managers, crews and passengers alike. There is much more to be done, since advances in infrastructure support and investment have not kept pace with changes in the aircraft operations. Staffing levels have improved in some areas, but always against a resistance to provide resources in good time ahead of operations commencing, making development of the aviation environment before the aircraft arrive difficult if not impossible. By way of example, MONUC commenced operations with one aviation officer and three aircraft. In 2005, UNMIS had only 32 personnel to support 51 aircraft flying seven days a week, with 10 of those staff required to cover 11 airfields where aircraft and customers were based. Staff levels inevitably increased over the following years. However, in 2008, the United Nations Mission in the Central African Republic (MINURCAT), in Chad, was funded for 96 staff to support 31 aircraft at five airfields from the outset. There was still a long way to go, but the improvement was remarkable.

Unlike most military forces around the world, the United Nations is simply not funded to retain standby capability for missions yet to exist. However, the United Nations is regularly called upon to react to crisis needs at very short notice. Many countries fund the provision and regular training of professional armed forces with no specific conflict in mind and at far greater cost than that of missions of the United Nations. Despite this, the nations of the world direct the United Nations to conduct over 15 peacekeeping and political missions worldwide. However, there is some progress. Following the Brahimi Report of 2000, it was recognized that significant equipment took many months to deliver, even once the lengthy procurement process was complete. Airfield equipment such as loaders and fire trucks are generally not put into production without firm orders. Thus, a "Strategic Deployment Stock" was established for such items at the United Nations Logistics Base in Brindisi, Italy. The United Nations now possesses a number of items such as mobile air traffic control towers, refuelling bowsers, main deck loaders, airport fire trucks, illuminated windsocks, ground power units and emergency runway lighting kits, ready to be deployed at all times.

Cargo Aircraft

The growth in the size of peacekeeping from 15,000 to over 100,000 personnel since 1999 has outstripped the ability of suitable cargo aircraft to provide support. Airfields with little or no infrastructure oblige the United Nations to use military-style cargo aircraft such as the L-100 Hercules, which is able to land on rough runways and has rear ramps, enabling easy loading and unloading. Unfortunately, very few suitable civilian types exist. The IL-76, with its ability to load sea containers directly onto flatbed trucks, is also becoming unloved in a greening environment and with rising fuel costs due to its enormous Jet A1 fuel burn rate of 9,500–11,000 l/hr. UN Headquarters in New York has been busy establishing the availability of ground-handling equipment able to support conventional freighters such as Boeing 777 and Airbus A-300 aircraft at fully surfaced airfields, in order to release L-100s to greater focus on semi-prepared strips. For the longer term, the United Nations is showing interest in very large airships and hybrid airships such as Sky Cat,[2] which can deliver heavy loads over long- and short-range distances with almost no requirement for ground support or runways. More immediately, the United Nations is also currently engaged in attracting Short Takeoff and Landing (STOL) FW cargo aircraft such as the Alenia C-27 and the Airbus C-235 and C-295 aircraft through the active encouragement and support of operators by the manufacturers. Such aircraft could suitably carry troops on journeys currently operated by helicopters. The increased speed may facilitate quick reaction forces arriving faster or from more centralized locations while increasing UN capability at lower cost.

---

2   The SkyCat (short for "Sky Catamaran") is a hybrid aircraft proposed for heavy lift and ultra-heavy lift. The aircraft derives its lift from helium buoyancy and aerodynamic shaping. The hover cushion technology allows it to take off and land almost anywhere, including in remote locations. It is also alleged to be impervious to rifle and mortar fire. See for example: *Aerospace Technology Magazine*. "CargoLifter CL160". Project Data Sheet. London, 2012. Available at: http://www.aerospace-technology.com/projects/cargolifter/ [accessed 30 March 2013]. Dillon, R.M. "High-tech cargo airship undergoing tests". *The Associated Press*. 30 January 2013. Available at: http://www.militarytimes.com/news/2013/01/ap-high-tech-cargo-ship-being-built-california-013013/ [accessed 30 March 2013].

Medium Helicopters

The UN requirement has grown so much that the demand has outstripped the availability of medium helicopters. Provision of STOL FW aircraft will go some way to alleviate this, particularly with emphasis on improving runways as the United Nations Mission in South Sudan (UNMISS) proposes. New sources of helicopters will contribute to the solution, as well as increasing operational capability and potentially reducing cost. The United Nations has been supported by S-61, Puma and Super Puma helicopters in the past, but currently all large medium helicopters are Mi-8MTV helicopters, favoured for their single-compartment capacity, load capability and long-range performance. However, these aircraft are slow to start up, slow in flight, lose valuable seats when carrying luggage, cannot be loaded by forklift trucks, are uncomfortable, noisy, fuel-inefficient and have a significant downwash hazard.[3] While these aircraft are safe enough, they could certainly be safer.

The United Nations recognizes that its contract terms may not be attractive to operators of other helicopter types that are in high demand for oil and gas companies, which require high standards of safety, equipment and service. However, the contracts the United Nations offers are long enough to make investment in new, efficient helicopters worthwhile over the long term. A review of past UN operators' views made it clear that large medium helicopters were generally not available for lease. Operators had to make US$20–$25 million investments in new assets such as the S-92, which would be delivered in two years. Hoping to satisfy an unknown UN demand made no financial sense, given that the demand would only be revealed in an "Invitation to Bid" with three months' notice (including a three-week bidding period and eight to 10-week UN decision-making process, leaving successful bidders with two to three weeks' notice to position), which the contractor might not win and which could be cancelled at 30 days' notice. Any contract won would be on an uneven playing field with Mi-8 operators who knew the UN business well and were paid per flying hour, including lengthy start-up periods and additional hours to be paid for due to slow flying speeds. The procurement process took little account of value or past performance and was generally for a two-year contract with an option for a single-year extension. The resultant near-monopoly of Mi-8 helicopters in UN operations has left other manufacturers unaware of the significant business opportunity that the United Nations presents.

The United Nations has engaged in active "enlightenment" of manufacturers such as Sikorsky, Bell, AgustaWestland and Eurocopter. It hopes to ensure that future UN contractors will be encouraged by original equipment manufacturers providing training, regional spares and technical support. Briefings to leasing companies at various aviation exhibitions and air shows are leading to lessors

---

3   *Editor's note*: Rotating blades produce air vortices that can, under certain circumstances, reduce the stability of an aircraft, especially for a helicopter near uneven ground or in walled-in areas.

considering ways in which they can make assets available to operators in the future and thus de-risk the competition for UN contracts by operators.

Meanwhile, the United Nations is looking to tackle its problems holistically: it has considered virtually every aspect it can influence; has examined its own contract arrangements; and is considering the introduction of contracts lasting five to seven years. Most significantly, the United Nations is changing from its Initiation to Bid (ITB) process to seeking Request for Proposals (RFP). The ITB simply identifies mandatory parameters and specifications that aircraft and operators have to satisfy. All those bids that meet the minimum acceptable standard are then reviewed for price and the lowest acceptable bid is selected. Under the RFP system, the United Nations will be able to take account of useful operational benefits over and above the minimum necessary. It will be able to compare offers in terms of the overall cost of achieving the task and allow vendors to offer a range of solutions to a task-based requirement. For instance, operators could offer a greater number of smaller assets such as the EC-145 (Eurocopter twin-engine light utility helicopter) to better match daily variance in demand, or offer faster helicopters such as the AW-139 (AgustaWestland medium-sized, twin-engine helicopter), or FW aircraft (or even a mix of the two) to reduce the number of assets required, or simply add operational benefit. Maintenance standards and reliability would be considered, which may also impact on the number of aircraft required. Fuel costs and the cost to the United Nations of housing crews can also be considered, along with past performance and back-office support. Early positioning may also be an important consideration, but for the first time, the United Nations will be able to recognize the value of small fast helicopters for emergency medical response. This is a major change for the United Nations and, while labour-intensive to implement, offers huge benefits and increased competition.

Operational demand in the United Nations Assistance Mission for Iraq (UNAMI) has forced the United Nations to introduce civil helicopters fitted with Civil Anti-Missile Protection Systems and Kevlar-protected floors. These have been purchased by the United Nations and installed on a pair of chartered EC-117 light medium helicopters. Operational techniques have also been revised for this service, with the EC-117s flying as a pair at low level over threat areas, with each able to accept passengers "cross-decking" from the other aircraft in the event that one should make an unplanned and undesired landing. While this capability might well be deemed valuable in many missions, financial constraints on all budgets have yet to make this advantage sufficiently attractive to be deemed essential and there is the potential for this capability to be lost in the near future.

Aerial Delivery

The United Nations is also looking to support its peacekeeping and political missions in new and novel ways as technological advances become more established elsewhere. These include the introduction of global positioning system (GPS)-

guided parafoils (rectangular sports parachutes) to deliver supplies. Being able to supply troops is an important role for the United Nations. This can include routine resupply or emergency reaction and may not be possible by road. Reasons for this include obstacles such as hostile areas, mines, poor road conditions, urgency, floods and unavailability of ground transport. Hostile acts, poor visibility, night prohibitions and the short range of helicopters may make resupply impossible or unsafe. GPS-guided parafoils are capable of flying autonomously 15–20 mi if dropped from 25,000 ft, thus keeping air assets in a safe environment while the delivery system descends to its landing zone in full cognizance of wind and predetermined ground obstacles.

Unmanned Aerial Systems

Technologically savvy military forces have been renowned for their increasing reliance on unmanned aerial vehicles (UAV) and systems (UAS). The United Nations has repeatedly considered the use of UAVs in its operations since 2005, when MONUC considered the provision of UAVs and included them in its 2006 Force Requirement. Also in 2006, the Special Committee on Peacekeeping of the General Assembly (the "C-34") required the United Nations to "examine all forms of monitoring and surveillance technology, particularly aerial monitoring", and to "ensure the safety and security of peacekeepers". The Department of Peacekeeping Operations conducted a comprehensive study in 2007.[4] The United Nations recognized the following benefits of UAVs:

- able to fly day or night;
- much reduced risk to personnel;
- able to fly in hostile areas;
- view "over the hill" for troops;
- command aerial view of live operations;
- protect against ambush;
- protect civilian populations;
- live video to headquarters from long distances, 24/7;
- able to track targets for hours/days;
- able to detect targets at extreme range;
- covert or deterrent;
- over-horizon communications relay;

---

4   One study prepared for the C-34 and welcomed by it in 2007 was Dorn, A.W. "Tools of the Trade? Monitoring and Surveillance Technologies in UN Peacekeeping", *Peacekeeping Best Practices*, 2007. That study formed the basis for the subsequent book, Dorn, A.W. *Keeping Watch: Monitoring, Technology and Innovation in UN Peace Operations* (Tokyo: United Nations University Press, 2011). See also: http://www. keepingwatch.net [accessed 29 March 2013].

- able to intercept phone and radio messages;
- able to see through foliage (infrared- and radar-sensing);
- guide troops to action early, pointing to targets;
- support disaster relief/humanitarian efforts;
- continuous patrol enables immediate diversion response;
- where aircraft are only option, UAVs are lowest cost option.

The United Nations sought troop-Contributing countries to offer UAVs for MONUC in the Democratic Republic of the Congo (DRC). With no offers forthcoming at the time, the United Nations then looked to contractors to fill the requirement to provide a contractor-owned and contractor-operated surveillance capability in the eastern DRC. Eight bids were received to meet the following requirement:

- two systems comprising three UAVs, each offering day/night video;
- systems to be based in the eastern DRC cities of Goma and Bunia;
- operating range 250 km, pre-programmable to 800 km or using additional relay UAVs or remote viewing terminals;
- night capability to peacekeeping operations;
- able to identify 2 m target at 5 km;
- safety modes;
- 10-hr operations, 5 days/week;
- 72-hr surge capability;
- self-reliant contractor embedded with UN military operations;
- provide view to forward and regional commanders of live operations;
- stealth or deterrence;
- relay for mission voice communications;
- able to relocate by C-130 in 48 hrs.

Political considerations are a reality with the United Nations and a UAV avoided the need to use relay satellites controlled by non-UN entities. A second UAV was therefore to be used as a relay to forward control data from the ground-station pilots to the mission aircraft and retransmit images from the aircraft back to the ground station. This would eliminate line-of-sight range restriction for regular communications and enable the United Nations to operate UAVs at lower altitudes to compensate for weather and civil airspace restrictions, while seeing over the horizon and mountainous terrain of the eastern Congo. The UASs that were offered and considered suitable for MONUC included the Elbit Skylark, the Elbit Hermes 450, the MMist Snow Goose, the Aeronautics Aerostar and the IAI Searcher.

The UAS was to be a great leap forward in UN capability and its adoption overcame the many concerns of governments. Typically, the United Nations can only operate in a country with the consent of the host government and this had been achieved in the DRC. Consideration, and the fears of neighbouring countries, were factors also considered. Other nations might express concerns that more properly reflect their regard for the United Nations later coming to use UAVs in their national areas of interest. While troop-contributing countries could find their troops better protected, some have expressed reservations that greater efficiencies and capabilities might lead to a reduction in the need for their troop contribution – the same might be felt by countries providing helicopters. The desire to provide military capability may result from an interest in exerting influence in the region or at UN Headquarters, or from the value of the experience and the training that the military receives from its peacekeeping operations; some countries may seek to contribute to the peacekeeping effort, others to receive monetary income from the United Nations. In some countries the decision to provide military capability may stem from a combination of one or more of these factors. The validity of such concerns is not proven. Certainly, greater situational awareness through UAVs should enable the United Nations to place troops in the right place at the right time with less fear of exposure to undue risk. This could, in fact, result in the greater use of troops and the far greater impact of UN peacekeepers. Delivery of such troops will inevitably result in increased use of troop-carrying helicopters. Attack helicopters (the second type, in addition to utility helicopters) may also become more productive and be used effectively to deliver air power while not consuming as much time and fuel flying over benign areas. Just as computers have not eliminated or even saved on paper, UASs may be so effective that they actually result in more work and not less, but with far more beneficial results.

Sadly, financial pressure was brought upon the United Nations following the global credit crunch and the fiscal crisis of 2008–2009. This was coupled with a request from the government of DRC for MONUC to reduce its strength and resulted in the project being cancelled just as the procurement process was reaching a conclusion and award of contract was anticipated.

Nevertheless, the situation in the DRC continues to cause concern and the United Nations has once again sought to provide a UAS for the eastern parts of the country. Considerable political effort has been made to garner support from member states. Troop-contributing countries have once again failed to offer suitable UAS support, but contractor interest appears to have grown significantly and in January 2013 the United Nations once again commenced a procurement process for a UAS in the DRC.[5] The winning UAV was the Falco of Selex ES, a division of the Italian company Finmeccanica. The UAS achieved initial operating capacity in December 2013. Figure 16.1 shows one of the UAVs prior to its official launch by the Under-Secretary-General for Peacekeeping Operations, Hervé Ladsous (shown on the left). An Mi-35 helicopter can be seen in the background.

MONUC had not been alone in considering UAVs. In 2007 the United Nations Mission in the Republic of Georgia (UNOMIG) also considered their introduction

---

5   See, for example, Binnie, J. "UN peacekeepers request UAVs". *Defense and Security Intelligence and Analysis: IHS Jane's*. 14 January 2013. Available at: http://www.janes.com/products/janes/defence-security-report.aspx?id=1065975094 [accessed 30 March 2013].

**Figure 16.1   A Falco unmanned aerial vehicle before the official launch
ceremony in Goma, Democratic Republic of the Congo, on
3 December 2013**

Source: UN Photo 572910, S. Liechti.

to support surveillance operations of UN Military Observers in the Kidori
Valley in the breakaway region of Abkazia. In 2008, the UNOMIG was called to
investigate the shooting down of a Georgian Hermes 450 UAV. Although it was
claimed to have been shot down by an Abkazian L-39 light fighter aircraft, after
reviewing wreckage, radar tracks and film footage captured by the UAV itself –
which included the firing of the missile from a MiG-29 and its approach to the
UAV – the UN team investigating the incident concluded that the Hermes 450 had
been shot down by an aircraft of the Russian Federation. In due course, following
the Russian intervention in the South Ossetia region of Georgia, Russia vetoed
renewal of the UN mandate for UNOMIG, preferring to conduct peacekeeping
operations itself.

Meanwhile, other UN missions have expressed interest in UAVs. The United
Nations Operation in Côte d'Ivoire (UNOCI) has considered the suitability of
such systems for its operations, mainly along the border with Liberia. Fighting
in bordering Mali in late 2012 and into 2013 will only add to this potential
requirement. If the United Nations is successful in operating UAVs in the DRC
and the benefits become clear, short-term budgetary considerations that seek
immediate offsets for increases in operational efficiencies may attract less credence
and enable the United Nations to benefit from the great strides in technology that

many armed forces and air operators of the world are already able to enjoy. For more information on UAV's and UN peacekeeping operations see Chapter 9 in this volume.

Summary

UN aviation today is a far cry from that of earlier times. Emphasis on safety and an increase in the scale of UN peacekeeping operations has led to a whole-scale change in the size of operations and the types of aircraft used. Much has been done but much remains to be done. This has been of particular note for passenger aircraft, which have been overhauled. Delivery of air cargo is primed for change, with airports and ground-support equipment being made available for freighter aircraft such as the Boeing 777 and the Airbus A-300 in lieu of IL-76 types. Focus in recent times has shifted to L-100 operations and the multiple benefits that STOL cargo aircraft such as the C-27 and the C-235/295 can offer. Helicopter provision is in the process of a radical change. Helicopters such as the AW-139, the Super Puma, the S-61T and the EC-145 may greatly improve UN operations and at reduced cost.

The United Nations now has experience of providing helicopters with missile defensive suites and is looking at a range of new technologies. The greatest change to UN aviation is likely to come in the form of unmanned aircraft. While this might or might not result in a change to the remainder of the UN aviation fleet, UASs and their UAVs are primed to transform peacekeeping efforts.

Chapter 17

Peace from Above:
Envisioning the Future of UN Air Power

Robert David Steele

The United Nations has never had an air power strategy or an air power campaign plan. UN air power has always been on loan from donor member nations, often as an afterthought, and generally only in relation to land forces. It has rarely been used outside of normal support functions for the UN force (generally for transport), and only recently has it been used for modern intelligence-collection purposes, including imagery and mapping of unmapped territories such as the eastern Congo. It has yet to be used creatively as a primary UN function with a decisive impact, with at least two exceptions.[1] One has the impression that UN staff, despite their leavening of experts from the air staffs, of UN member states, do not really have an appreciation for all that air power might do before, during, and after UN forces are on the ground. This is a critical knowledge gap at the leadership level.

In recent times multinational air power has been used to compel (as in Serbia–Kosovo, 1999) and protect (as in Libya in 2011 – see Chapter 15), but it has generally failed to achieve its objectives. It has been very expensive and it delayed more holistic strategic coalition planning and operations. Air power as force projection and air power as a political tool are greatly over-hyped – it simply does not do all that is promised.[2]

Today's context is much, much scarier than traditional inter-state conflict, civil war, and routine genocide. Today we are looking at over 100 states in various stages of dysfunction and instability; and even very great governments such as those of Russia and the United States are increasingly being seen as "imitation"

---

1   *Editor's note*: Cases of a decisive and deliberate impact of air power in UN peace operations include: the Congo in the 1960s (see Chapter 2) and the Congo in the 2000s (Chapter 14).

2   A good summary specifically in relation to the United Nations and the North Atlantic Treaty Organization (NATO) air operations is provided by Andrew Cockburn. Additional commentary is provided by Chuck Spinney, long-time critic of US military acquisition policies. See, respectively, Cockburn, A. "The Limits of Air Power", *Los Angeles Times*, 3 April 2011. Available at: http://articles.latimes.com/2011/apr/03/opinion/la-oe-cockburn-libya-20110403 [accessed 7 May 2014]. Spinney, C. "The Limits of Air Power", *Journal of Public Intelligence*, 2011.

governments, unable to fulfill the role of a proper state for their peoples and the international community. Consequently, the future of UN air power can be, and should be, centered on "just in time" responsiveness to catastrophic situations and pre-catastrophic "peaceful preventive measures".[3] Both require the ability to deliver massive precision assistance from the air, and the precision multinational, multiagency, multidisciplinary, multidomain intelligence (M4IS2) – generally open source, not secret[4] – to manage air power in context. Transcontinental airlift, the ability to carry out regional air management and cross-decking from big air to small air,[5] precision parachute deliveries, and integrated ground-to-air communications from all possible indigenous sources, as well as "peace jumpers" (explained below, under the heading "UN Air Power – Kinetic Peace from Above"), are all required.

Two recent natural disasters turned into long-term catastrophes for lack of adequate global responsiveness, the first in Haiti (January 2010 to date) and the second in Japan (March 2011 onwards). It is valuable to examine what the United Nations could have done but did not do in such instances, especially relating to air power. A UN air-power strategy and concept of operations are presented in which air power becomes central to the strategic mandate, the operational campaign plan, and the tactical employment of UN forces.

But before looking at case studies and examples in air power, more context is needed, with general prescriptions. The future of the United Nations lies in the coherence of peace intelligence and the coherence of the air-power plan.

Background

Over the course of the past two decades, the United Nations has sought to evolve and mature. Although this process is far from complete, the course has been well set by the 1999 Brahimi Report on peace operations, then the 2004 Report of

---

3   The concept of using open source intelligence to identify needed "peaceful preventive measures" was first brought forward by General Al Gray, US Marine Corps (then Commandant of the US Marine Corps). See Gray, A.M. "Global Intelligence Challenges for the 1990s", *American Intelligence Journal* 11(1) (1989).

4   The full term is "Multinational, Multiagency, Multidisciplinary, Multidomain Information-Sharing and Sense-Making". Originally a Swedish and then a Nordic military concept, (M4IS2) has been adapted to be central to the concept of public and collective intelligence as oriented toward creating a prosperous world at peace.

5   Cross-decking is a standard UN/military term for moving people or materiel from one deck (air, sea, land) to another (air, sea, land). It is a logistics operation that needs planning. Big air refers to C-17's and Boeing 777's and other aircraft that require major (long) runways. Small air refers to C-130's and localized smaller civilian aircraft that can land in out of the way smaller airports.

the High-Level Panel on Threats, Challenge, and Change and, more recently, by the 2006 Report of the Secretary-General's High-level Panel on System-wide Coherence: "Delivering As One".[6]

At the same time that the United Nations has been evolving, I have been a global proponent for Open Source Intelligence (OSINT) and, more recently, for its more sophisticated and expansive replacement, M4IS2.

The above two evolutionary trends suggest that the future of UN air power will be centered on global information management – using information that is unclassified and intelligence decision-support that is unclassified – to identify needs on the ground with precision and then harmonize precision delivery of specific needed items from across a very broad range of actors.

As the world grows in complexity, and particularly in demographic and cultural complexity, there is one word that must become central to UN policy-making, acquisition, and operations: integrity. The Merriam-Webster Dictionary defines it as: "the quality or condition of being whole or undivided; completeness". Integrity is not just about honor and avoiding corruption. It is about wholeness of perspective, openness to diversity of view, and the ability to embrace and apply truth. As Dr Russell Ackoff, one of the leading systems thinkers of our generation, would say, we have to do the right thing, not do (as we do now), the wrong things righter.

Table 17.1 presents a snapshot of the change between global threats of the past and those of the future:

Table 17.1   Global threats then and now

Old (State) Emphasis	New (Hybrid) Emphasis
Inter-state conflict	Poverty
Civil war	Infectious disease
Proliferation	Environmental degradation
Failed states	Inter-state conflict and civil war
Refugees	Genocide and other atrocities
	Proliferation
	Terrorism
	Transnational crime

---

6   See, respectively, "The Secretary-General's High-level Panel Report on System-wide Coherence: 'Delivering as One'", UN Doc. A/61/583 (2006); "The Secretary-General's High-level Panel Report on Threats, Challenges and Change: 'A More Secure World – Our Shared Responsibility'", UN Doc. A/59/565 (2004). "The Secretary-General's High-level Panel Report on United Nations Peace Operations" ['The Brahimi Report'], UN Doc. A/55/305–S/2000/809 (2000).

The new threats are much more human and demand two things the United Nations does not do well now:

1. Hybrid operations with diverse multinational players sharing an operations center.
2. Deep, honest, timely decision-support not available from member nations.

UN operations demand multinational, multiagency, multidisciplinary, multi-domain intelligence. The UN missions that UN air power must support in the future are much more nuanced and much more demanding as well. It is no longer possible for a UN military or observation force to be sent into the field with a simple order to provide transport, observation, or even combat. In the future, UN air power will be essential to all forms of presence, and must excel at intelligence – decision-support. It must never be fielded without its own organic intelligence collection, processing, analysis, and dissemination capability. The Member states cannot be relied upon to meet UN needs for intelligence in out of the way places they do not care about and have no relevant intelligence collection capabilities for.

Table 17.2 illustrates the changes in emphasis for UN air power operations.

Table 17.2   Old versus new emphasis for UN future operations

Old (State)	New (Hybrid)
National security	Human security
Sovereignty	Legitimacy
Borders	Human rights
Governments	Public health
Armed forces	Precision giving
No intelligence	Precision intelligence
Veto possible/decisive	Veto unlikely/irrelevant
Marginal outcomes	Sustainable outcomes

In the face of this mix of nuance and complexity, precision intelligence using all open sources of information in all languages – and the application of multinational multiagency multicultural perspectives, is the new standard. I know of no government that is even remotely close to meeting this new standard.

In the above context, the nature of UN air power changes radically, with a new emphasis (Table 17.3).

Table 17.3 UN air power in the old and new paradigms

Old (State)	New (Hybrid)
Limited fixed wing	Contracted long-haul
– Imagery	Regional management
– Transport	– Big air/boats to small
Limited vertical/short takeoff and landing (VSTOL)	Long loiter surveillance
	Air-enabled M4IS2
– Gunships	VSTOL for all purposes incl.
– Hand-held imagery	gunships and medevac
– VIP transport	Precision cargo drops
– Medevac	Peace jumpers*

Note: * Explained later in this chapter, see: "UN Air Power …", Step 1.

In this new era, and bearing in mind my continued emphasis on hybrid operations with non-governmental organizations and private military corporations being fully engaged, there are two observations that I would make to any UN leader with respect to UN air power:

1. The United Nations will be, at best, a coordinator, not a commander.
2. The United Nations will be most successful if it becomes the central provider of trusted unclassified intelligence (decision-support) to all of the participating agencies.

In other words, regardless of what the UN mission is, regardless of what mix of UN air power is engaged, the primordial role of the United Nations will be as a service of common concern with respect to information and intelligence, and it is that primordial role that must be first in the mind of anyone who is creating a UN air power mandate; acquiring a UN air power force structure; or devising campaign plans for the employment of UN air power.

The primary role of UN air power in the twenty-first century is to serve as the hub – a service of common concern – to hybrid networks requiring intelligence – decision-support – and responding to shared information as a harmonizing influence instead of "command and control".

The Changing Craft of Intelligence

With that background, we can now look at how the world has changed and how the intelligence field should change along with it, as shown in Table 17.4.

Table 17.4   Modern intelligence emphasis

Old (State)	New (Hybrid)
Depend on members	Harness "World Brain"
Depend on secrets	90 per cent open source
Depend on active lies	Lies quickly exposed
Lack 90 per cent of what is needed to carry out mandate, force structure, campaign plan, and day-to-day tactics	

The new model will be possible if leaders will be leaders – 90 percent of what is needed for precision peacekeeping and "peaceful preventive measures" can be obtained from open sources. The future of intelligence as decision-support is not federal, not secret, and not expensive.

A recent book, *No More Secrets: Open Source Information and the Reshaping of U.S. Intelligence,*[7] is helpful in understanding this point. I worked at the Top Secret Codeword level from 1976 until 2006 and have also been a global proponent for open source intelligence for 20 years. The United Nations, and UN air power, can take from me two points on the matter of intelligence:

1. Member states do not have national or military intelligence systems suited to support hybrid networks that require unclassified, holistic intelligence.
2. Of all that a UN commander or any leader of any hybrid element working with the United Nations "needs to know", 90–95 percent is not available from a UN member state, is not secret, and can be obtained from open sources.

There are eight information-sharing groups, or "tribes", that the United Nations must engage: government, military, law enforcement, media, academia, civil society, not-for-profit organizations, and business. Within government we can distinguish between: secret internal, secret shared, and sensitive shared; and open (public) information categories. The information commons available to the United Nations is created by these eight tribes in the aggregate.

During the conference at which this chapter was first presented, Lieutenant-General (retired) Roméo Dallaire touched on the disconnect between the United Nations and everyone else, as well as the paucity of actionable intelligence (decision-support) from member states. I have served in three "country teams" overseas, on multiple assignments in Washington, and did a second graduate thesis on strategic information management. I found that a typical country team collected at best 20 percent of what could have been known and, in the process of sending

---

7   Bean, H. *No More Secrets: Open Source Information and the Reshaping of U.S. Intelligence*. Praeger Security International (Santa Barbara: Praeger, 2011).

that back to Washington spilled 80 percent of it (for example, sending in hard copy to a single agency desk instead of disseminating electronically).[8]

Most governments make decisions based on ideology and very limited real-world information; at the same time they are terrible about sharing what they do know with the United Nations and other international or non-governmental organizations. This pathology is not limited to government. Academic, civil society, commercial – all forms of organization have real difficulty doing external information-sharing and sense-making.

This is a major reason for the United Nations to take the lead and for UN air power commanders and staff to be especially well versed in facilitating M4IS2.

Today, intelligence as a discipline is very immature. If one keeps firmly in mind the Brahimi Report, the Report of the High-Level Panel on Threats, Challenges, and Change, and the Report of the High-Level Panel on Coherence, it is possible to draw two conclusions when observing the state of intelligence (decision-support) among governments today.

1. It is obsessively focused on inter-state conflict and on secret sources and methods.
2. They have, as Norman Cousins summarizes in his book *Pathology of Power*, grown away from the public interest. This is important enough to warrant one summary and one quotation from him.

Summary

The tendency of power to drive intelligence underground; to become a theology, admitting no other gods before it; to distort and damage the traditions and institutions it was designed to protect; to create a language of its own, making other forms of communication incoherent and irrelevant; to spawn imitators, leading to volatile competition; to set the stage for its own use.

Quotation

> Governments are not built to perceive large truths. Only people can perceive great truths. Governments specialize in small and intermediate truths. They have to be instructed by their people in great truths. And the particular truth in which they need to be instructed today is that new means for meeting the largest problems on earth have to be created.[9]

---

8 Hard copy material sent from one Embassy section back to their corresponding desk officer is generally thrown in the trash or filed; it is rarely "exploited" as in digitized, evaluated, distilled, or shared outside the receiving section of a single agency.

9 Cousins, N. *The Pathology of Power* (New York: W.W. Norton & Company Inc., 1987).

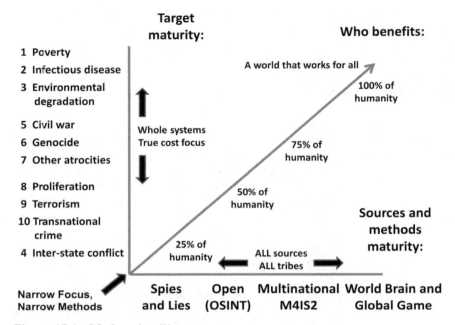

Figure 17.1   Modern intelligence
Source: The author.

In my view, the United Nations and its air power have an infinite opportunity to do well by being the global proponent for mature intelligence (decision-support), as depicted in Figure 17.1.

Opportunities Lost

Strategic Level: Iraq II

Although I and many others tried to speak the truth to the public on Iraq, our cash-up-front full-page newspaper ads in 2003 were rejected by the mainstream media. The fact is that Dick Cheney led us to war on the basis of 935 now-documented lies; and then Secretary of State Colin Powell went along with these lies when he appeared before the UN General Assembly.[10] In the future, without relying on member states, the United Nations must be able to detect lies, expose them, and make the case for peace such that imperial invasions and occupations are rejected by massive informed public opinion.

---

10   For more information, see Rampton, S. and Stauber, J. *Weapons of Mass Deception: The Uses of Propaganda in Bush's War on Iraq* (Tarcher/Penguin, 2003).

Operational Level: Haiti 2010[11]

Haiti is another opportunity lost, one that has enormous significance for the future of UN air power. This nationwide disaster in January 2010 was converted into a catastrophe with no end by a lack of intelligence, a lack of imagination, and very poor decisions across all parties from the US Southern Command (USSOUTHCOM) to the International Committee of the Red Cross and the United Nations, among many others. The tragedy was elevated virtually into crimes against humanity by the blatant corruption visible to informed observers among the charities that collected funds, ostensibly for Haiti, yet delivered from 1 percent to 10 percent with the average being closer to 2 percent, in the months thereafter. Presented to any court, the facts would have led to convictions for misrepresentation, fraud, and failure to perform as promised. Haiti was an intelligence and an imagination failure, with an air–sea lift management failure.

Haiti was an intelligence–imagination failure. Despite the fact that there was an obvious and desperate need for an orchestrated delivery of water, food, and lightweight shelter materials, the best the USSOUTHCOM could do was send in 20,000 troops with their own high-end logistics needs in the aftermath of the disaster.

USSOUTHCOM also refused to heed the many warnings about the impending rainy season (approximately June to October), and did nothing to address the urgent need for both sanitation facilities across the country, and lightweight housing options, among which I recommended light nested plastic geodesic domes.

USSOUTHCOM, regional authorities and the United Nations, including MINUSTAH already located in Haiti, failed to imagine how easily a regional sea–air management plan could be put into place, diverting large air and sea vehicles to major air and sea ports (for example, in Santo Domingo, Guantanamo, Havana, Miami, Tampa and Dover Air Force Base or Caracas), for breakdown and reshipment to the small air and sea ports still working in Haiti.

Put quite simply, the United Nations, the United States, and others remained in the unilateral action mindset, failing to see the advantages of a regional concept of operations and the necessity of orchestrating all incoming materiel and the craft it traveled on via a robust regional sea–air management plan.

MINUSTAH was neither staffed nor of the mindset to orchestrate a regional assistance campaign. What would have been of enormous assistance are two intertwined capabilities. First, an air-mobile M4IS2 element able to operate from in-country, from a fixed base with global communications, and across a distributed network of professionals intimately familiar with all of the logistics categories, all of the participating elements (most of them non-governmental). Second, all of the associated communications – computing, intelligence, and information equipment, frequencies, and analytic protocols corresponding to a global diversity of well-intentioned but "out of control" parties. This is harder than it might appear.

---

11   *Editor's note*: For an alternate interpretation of US relief operations in Haiti in 2010, see Chapter 5 of this volume.

At the strategic and technical levels, it could have been realized that the best way to migrate over a million people away from Port-au-Prince would have been to deliver building materials, sanitation kits, water, and food to the other five small ports in Haiti, via landing craft loaded out of Guantanamo, Havana, and Santo Domingo.

At the strategic and technical levels, it could have been realized that creating several factories to manufacture sanitation, plastic geodesic domes, and other mid-term sustainability packages, would employ people and accelerate the general recovery, with a special focus on having two million people under leak-proof plastic with reliable sanitation before summer rains came (an obvious and known future turn of events).

At the strategic and technical levels, it could have been realized that at least 100 Mobile Army Surgical Hospitals (MASH) were needed. In so many ways, Haiti was and remains a lost opportunity to save and resurrect an entire country through the orchestrated, imaginative infusion of assistance with a coherence that the United Nations is striving to achieve, but cannot – absent the concepts in this chapter.

Technical: Earthquakes, Nuclear Plants, and Tsunamis

I believe the United Nations, because it is not trained, equipped, or organized to "do" global intelligence or "intelligence from above", has been a failure at encouraging proper planning across nuclear and other energy options, and also unable to prevent or retroactively punish covert attacks on nuclear capabilities.

Nuclear power needs to be safe, and it needs to be developed in conjunction with a proper understanding of the true cost over the full life cycle of all possible sources of energy.

The fact is that most nuclear plants are at sea level, and are not only subject to flooding from tsunamis (near term) and global warming (longer term) but are also subject to being closed down by jellyfish. Add to that the very broad range of earthquakes known to be imminent in the next decade, and one can only conclude that if the United Nations does not become a global facilitator for information-sharing and sense-making, it will become even less relevant to the future of humanity than it is now.

In each of the above cases – strategic, operational, tactical, and technical – the role of UN air power will be central. It will be the kinetic counterpart to UN global information and intelligence operations, linked to "information peacekeeping", a term I coined in early 1997 for the United States Institute of Peace, and then developed further for a book edited by my colleagues Doug Dearth and Alan Campen.[12]

---

12   Steele, R.D. "Information Peacekeeping: The Purest Form of War", in *Cyberwar 2.0: Myths, Mysteries and Reality*, ed. Alan D. Campen and Douglas H. Dearth (Fairfax: AFCEA International Press, 1998). Available at: http://www.fas.org/irp/eprint/cyberwar-chapter.htm [accessed 29 April 2013]; Steele, R.D. "Virtual Intelligence: Conflict Avoidance and Resolution Through Information Peacekeeping", *Journal of Conflict*

In the balance of this chapter, I will present my view of how UN intelligence and UN air power must develop together. I will present five ideas for UN intelligence and seven ideas for UN air power. Intelligence without air power is irrelevant – air power without intelligence is noise. They need each other.

UN Intelligence – Non-kinetic Peace from Above

Despite some significant progress during the tenure of Major-General Patrick Cammaert, The Royal Netherlands Navy, as Military Advisor to the Secretary-General, and the proven success of Joint Military Analysis centres (JMACs) and Joint Operations centres (JOCs), today the United Nations remains largely incapable of producing coherent intelligence (decision-support) across all of its needs, inclusive of the specialized agencies. The world can, essentially, be divided into eight tribes, illustrated in Figure 17.2. They currently operate far too much in isolation.

Figure 17.2 The UN and eight tribes of information–intelligence
Source: The author.

Studies 19(1) (1999). Available at: http://journals.hil.unb.ca/index.php/JCS/article/view /4380 [accessed 29 April 2013].

This figure was created during my teaching of a six-agency UN class in Beirut in August 2007, a three-day orientation on "Information Sharing & Analytics".[13] We called it the "Class Before One". Although some excellent multinational information-sharing and intelligence courses are run, notably in the Nordic countries, the United Nations does not yet have a proper organization with standards for intelligence, and such analysts as exist are scattered and too easily corrupted by their supervisors with departmental or agency agendas.

In relation to what can be known, the United Nations is severely ignorant on all fronts.

Harmonized Coherence from Shared Information

Over the next decade I anticipate that the United Nations and various governments will have less money to spend and will achieve their good effects through shared information and multinational intelligence (decision-support). The United Nations cannot control or coordinate its own specialized agencies with any degree of coherence today, in part for lack of authority and in part for lack of the informal authority that a UN intelligence organization would provide.

Coherent integrated intelligence is how the United Nations can influence and lead through shared and generally unclassified or open-source intelligence.

UN Intelligence Must Be Multilevel in Nature

I learned a great deal from Lieutenant-General Roméo Dallaire and Major General Patrick Cammaert. The particular lesson I want to emphasize here is that the United Nations must be able to "do" intelligence (decision-support) at all four levels of thought: strategic, operational, tactical, and technical; it must be able to do integrated intelligence among the civil, military, and environmental domains; and finally, it must be able to do integrated intelligence and counterintelligence across all of the different mission and logistics areas – it does no good to deliver cans without can openers, or food and shelter without sanitation. Holistic intelligence is both an art and a science, something the United Nations desperately needs to learn; a model of this is shown in Figure 17.3.

It is far beyond the limits of this short chapter to introduce the concept of "true cost" in detail, but as others have documented, at least 50 percent and often as much as 70 percent of all "costs" (upon which profits are calculated) are either waste, or corruption, or both. This has been documented across the policy areas and is especially relevant to areas of deep concern to the United Nations and human security: education, energy, family, health, and justice.

---

13   The slides with words in notes format can be found at http://tinyurl.com/UN-Class-1 [accessed 24 April 2014].

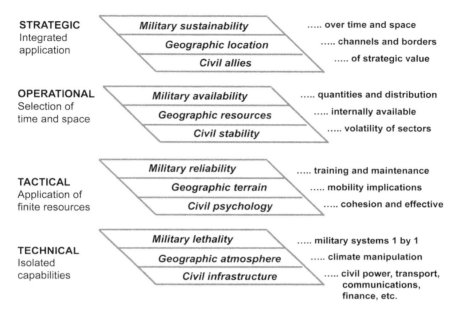

| STRATEGIC
Integrated
application | *Military sustainability*
Geographic location
Civil allies | ..... over time and space
..... channels and borders
..... of strategic value |

Figure 17.3 Integrated multilevel intelligence model
Source: The author.

On the following page, in Table 17.5, is the depiction of what it means when we get a grip on "true costs" and are able to eradicate waste and corruption – we can achieve a prosperous world at peace for one-third the cost of what we spend now on war alone. Add to that savings from eliminating waste across education, energy, health, and justice, and we achieve heaven on earth.

Table 17.5 is based totally on the brilliant lifelong work of Professor Medard Gabel, number two to Buckminster Fuller in the creation of the analog World Game, and today the developer of the EarthGame™ with integrated "true cost" information.[14]

The United Nations must free itself from the monopoly of judgment within the Security Council, whose permanent members are the worst proliferators and practitioners of war, and must instead leverage public intelligence (decision-support) and hybrid networks of activists, to achieve both public diplomacy within the donor states, and coherence or unity of effort as delivered to the beneficiary states. It should do this by becoming a global public intelligence network, one that I have suggested should be called the "United Nations Open Source Decision-Support Information Network (UNODIN)".

---

14   I recommend Professor Gabel's book, Gabel, M. *Designing a World That Works for All: How the Youth of the World are Creating Real-World Solutions for the UN Millennium Development Goals and Beyond* (CreateSpace Independent Publishing Platform, 2010).

Table 17.5   Cost of war versus cost of prosperity and peace for all (annual expenditures)

Total world military expenditures	US$1.3 trillion
versus	US$ billion
Eliminate starvation and malnutrition	20
Clean water for all	12
Health care for all	25
Shelter for all	21
End illiteracy/education for all	10
Clean safe energy for all	45
Save the environment	32
Democracy and diversity	5
Secure world for all	17
Retire debt/credit for all	30
Stabilize population everywhere	11
Total prosperity and peace for all	**228**

UN Strategy – Peace from Above with Intelligence and Air Power

Over a decade ago when I starting thinking about the United Nations, peace, and the role of a mature intelligence professional in fostering peace, I created the graphic in Figure 17.4 opposite. It is implementable today.

All of this has been a necessary preamble – a context and a foundation – for where I believe UN air power must evolve in the twenty-first century.

If intelligence (decision-support) is the non-kinetic foundation for UN relevance and effectiveness in the twenty-first century, then air power is the kinetic foundation. UN air power will deliver precision peace from above.

UN Air Power – Kinetic Peace from Above

I have been a critic since 1988 of how the US government makes decisions, focusing my concerns on the areas that I know best: secret intelligence and defense policy, acquisition, and operations. Watching the US government make a complete mess of Haiti was a deeply troubling experience for me.

Member states and their secret intelligence communities are ignorant. They lack intelligence (decision-support) and they lack integrity in the holistic sense of the word. The United Nations is no better, but the United Nations has an opportunity that I present, centered on air power, in seven steps:

Intelligence-Driven Precision Peace Targets, Space, Cyber, Air

Figure 17.4   UN strategy 21: Intelligence-driven precision peace
Source: The author.

Step 1: Peace Jumpers

Here is an image of what I call a "Peace Jumper". Peace Jumpers, like smokejumpers (who parachute into remote areas to fight wildfires), jump into the fire, but with one big difference. Peace Jumpers are armed with man-portable communications and instant access to real-time translations in all languages. The Peace Jumper program would have the following elements:

1. It would be globally celebrated and exercised regularly, with tangible deliveries to demonstrate to every public in every clime and place that when Peace Jumpers land, they bring with them Peace from Above.
2. Peace Jumpers would be a multinational force, especially trained, a form of "Special Forces for Peace" drawn from the best volunteers of all nations. They would be qualified to speak the language(s) of their region generally, but have instant access to real-time translation across 183 languages and dialects as needed.
3. Peace Jumpers would be on "strip alert" status in each major region (for example, in Africa there would be a cadre each for North, West, East, and South Africa). When a need occurs, as many as 100 Peace Jumpers could

be spread across the area of concern, and begin immediately calling in "Peace Targets", a form of Reverse Time-phased Force Deployment Data.

Step 2: Precision-guided Cargo Parachutes

I skip ahead to precision-guided cargo parachutes (between the parachutes and the Peace Jumpers are Steps Three, Four, and Five) because this is the essence of Peace from Above, and the tangible "deliverable".

There are some very exciting developments in precision airdrops and, while provision must be made for recovering the guidance units this is in my view the single greatest advance in the possibilities of Peace from Above when combined with Peace Jumpers, regional Air–Sea Management, and a Multinational Decision-Support Centre.[15] Flocking of sets of parachutes, active collision avoidance, and multiple locations hit accurately with one string of parachutes, each to a separate village – these are all possible today.

What is not available to the United Nations today is 1:50,000 combat charts for the 90 percent of the world where the United Nations and coalition forces go in harm's way – the Russian government has many of these and could be part of the solution.

Step 3: 1:50,000 Combat Charts and Air Intelligence

The aviation piloting standard for charts is 1:250,000. This is inadequate for close air support, and the coordination of fire support, and also for coordinating complex ground activities by a very wide range of non-governmental actors.

I helped General Cammaert when he was UN Force Commander in the eastern Congo, as part of the UN Stabilization Mission in the Democratic Republic of Congo (MONUSCO), His priorities were for 1:50,000 combat charts, which did not exist at the time. They were eventually created by the Dutch. Laptops do not work with bullet holes in them. I want the world at 1:50,000 on the shelf ready to go.

A very important point was made during the conference by Professor Robert Owen of Embry–Riddle Aeronautical University; bandwidth is more expensive than pilots. I believe the US model of many unmanned aerial vehicles is bad at multiple levels from cost to processing speed to flexibility. Pilots will be the heart of UN air power. Not only will air power be central to all operations but it may often be the major force employed to define and deliver Peace from Above.

---

15   An "Air–Sea Management and Multinational Decision-Support Centre" is not a JMAC. The first is focused on coordinating information shared across the full spectrum of participants in hybrid networks and combines logistics optimization with intelligence support to non-UN elements as well as UN elements. The JMAC is primarily a UN-centric peacekeeping operations capability.

Step 4: Regional Multinational Intelligence Centers

Now we come to the centerpiece of the UN intelligence and UN air power mosaic. As we saw with respect to Haiti, the US government, and USSOUTHCOM in particular, disgraced themselves. The Joint Intelligence Center of USSOUTHCOM was totally without value, without imagination, and without effect. This is what happens when intelligence "professionals" are not held accountable for actually being able to collect, process, analyze, and deliver decision-support as an outcome. The US military is on the verge of a 50 percent cut to its global budget. In my view, the world has relied for much too long on US military support, at the same time that the world has been much too tolerant of unilateral militarism and unwarranted extra-legal interventions by US covert and overt forces. My antidote to US irresponsibility is to create multinational, multiagency, regional intelligence centers designed for Asia and regions of the world, each with their own staff.

In both the United Nations and its member states, it has long been recognized that intelligence focused on a single country is not adequate. From blood diamonds going out to mercenaries and proliferation precursors coming in, only a regional, multinational, multiagency information-sharing and sense-making network stands any prospect of being effective. The United Nations *can* lead.

Step 5: Liberation Technology and Satellites for Peace

The final two elements of UN intelligence and UN air power are both centered on harnessing the distributed intelligence of the Earth's populations, and empowering them with the tools to communicate their needs both to the United Nations and to hybrid networks of individuals and organizations seeking to render assistance; and with those tools to create infinite wealth. I refer here to the three billion extremely poor people for whom access to the Internet should be, as the United Nations recently declared, a human right.[16]

In the twenty-first century, UN peacekeeping and conflict prevention will be centered on discerning and sharing the truth more than on deploying armed forces. The greatest source of information about conditions of instability and preconditions of revolution for the United Nations will be the public, not member states.

I am not happy with the "shadow Internet" the US government claims to be building. I believe it will be underfunded and generally out of touch with what is already available in the way of solar-powered Internet hubs, wireless mesh, and adapters that turn any cell phone into a satellite phone. Free satellite and airborne relay stations should, in my view, be a big part of UN air power. Cheap adapters already exists (US$169). For instance, the SPOT app can connect your smartphone to communication satellites to update social media, send short emails or text messages or send your global positioning system coordinates and

---

16   Report of the UN Special Rapporteur on the Promotion and Protection to the Right to Freedom of Opinion and Expression, UN Doc. A/66/290, 10 August 2011.

emergency messages. When combined with free satellite time subsidized by the hybrid network seeking to gain the information advantage for peace, this turns any citizen into a priceless information source with the added advantage that they are less likely to be detected by those who mean them harm in this era of mass surveillance, especially if they are using throw-away cell phones.

It is important to emphasize that the multiplicity of sources will only be as valuable as the social network that filters and validates the traffic, and the back office machine and human processing that can be applied by a multinational regional intelligence center or an overhead UN air coordination and collaboration aircraft.

Related initiatives in this area can be found by using the terms "Liberation Technology", "Autonomous Internet", "Mesh Network", and "Invisible Communications". When the Egyptian government of Hosni Mubarak cut off the Internet in 2001, its action accelerated global public interest in achieving complete independence from any government or corporation that might wish to interfere with public communications.

Conclusion

Over half the ports of the world are inadequate for sealift due to shallow waters, limited turning radiuses for gray- and black-bottom ships,[17] and poor port and supply-line development. Even if goods can be offloaded, the probability of their being spoiled or stolen before reaching their furthest destination is very high. However, all countries of the world can be serviced by C-130 aircraft, and the beauty of precision parachute operations is that it can provide precision deliveries to the most remote villages, in a manner that is fully transparent and almost devoid of corruption and theft possibilities. This is the "UN intelligence – UN air power advantage". Informed air operations. Precision assistance. The rendering of UN assistance in a coherent fashion – deliver as one – and the influencing of all others (governmental and nongovernmental) so as to mobilize, channel, and leverage a thousand times more value than could possibly be donated to the specialized agencies or delivered by the United Nations acting alone.

UN intelligence – UN air power can lead the way toward a prosperous world at peace where peaceful preventive measures are identified quickly and acted on quickly, at very low costs in comparison to all alternatives.

Since most charities, including the Red Cross, deliver less than 20 percent of the total funds collected – some as little as 10 percent, and one, the Bono Foundation, only 1 percent, – this is a major opportunity for the United Nations to use UN intelligence to capture funding that can be placed in direct action with a perfect accountability trail. It can – it should – change the future of global assistance, of conflict prevention, and of peacekeeping.

---

17   Gray-bottom ships are military ships. Black-bottom ships are merchant ships. These are standard terms of art.

Put directly, I anticipate the day when the United Nations no longer relies on member state funding; comprises representative parliaments from all eight tribes of information–intelligence (academic, civil society, commerce, government, law enforcement, media, military, and nongovernmental); and influences over US$1 trillion a year in planned giving and planned assistance, much of it real-time or near-real time in nature. *To do that, UN intelligence and UN air power will have to evolve in spectacularly innovative, effective, and world-changing ways.*

Acknowledgements: Oh Canada!

Five Canadians have made, in my view, extraordinary contributions to the evolving dialog about where the United Nations should be going with respect to UN intelligence, and I would like to take a moment to single them out, for each has in their own way changed my life, my focus with respect to the potential of the United Nations and, hence, with respect to the future of humanity.

In rank order, for lack of a better rationale for sorting, they are:

Madame Louise Frechette, Deputy-Director-General of the United Nations and former Deputy Minister of Defence for Canada. For seeing the shortfalls in decision-support at the strategic level, and seeking to remedy them.

Lieutenant-General Roméo Dallaire, Force Commander in Rwanda, for his heroic and personally costly sacrifice in illuminating the complete disconnect between UN Headquarters and reality, a disconnect that is by itself a crime against humanity not to be tolerated any more.

Brigadier-General James Cox, Deputy N-2 for NATO, who single-handedly led the leadership toward an appreciation of OSINT and directed the preparation of the *NATO Open Source Intelligence Handbook* and the *NATO Open Source Intelligence Reader*, still standards in the field but overdue for updating and expansion. He personally inspired OSINT units across the 66 member countries at the time of NATO and the Partnership for Peace (PfP).

Dr A. Walter Dorn, Consultant to the United Nations, author, and investigative academic, has been the foremost observer and reporter on UN intelligence from its brilliant days in the 1960s in the Congo to its lesser decades; and now as it has evolved in the 1990s. It is not possible to reflect knowledgeably on UN intelligence without first consulting all that he has written and asking about all that he plans to write.

Lieutenant Commander Andrew Chester, then serving in the Intelligence Division of Supreme Allied Commander, Atlantic, personally led my efforts in drafting the *NATO Open Source Intelligence Handbook*, and collating materials from my long-running conference for the *NATO Open Source Intelligence Reader*, and himself drafted a quite extraordinary and still enormously popular *NATO Guide to Intelligence Exploitation of the Internet*.

Afterword: Some Reflections

Since the Second World War, the international community, mainly under the auspices of the United Nations, has dispatched over a million military personnel to conflict areas around the world to serve as peacekeepers. While the "boots on the ground" and the conflicts they seek to resolve have been well publicized and studied, the air component of peacekeeping has received much less attention.

This volume has helped remedy this deficit by examining air operations in detail and seeking to answer important questions. For example, what role has air power played? What quantity and quality of aircraft were used? How effective was air power and what problems were encountered? We learned that aircraft provided strategic and tactical airlift, enforced no-fly zones, performed surveillance missions and even conducted kinetic (combat) operations to protect UN forces and civilians. Notwithstanding the harshness of the environment and the mixed results of peacekeeping and humanitarian missions, the aircraft and their aircrews conducted themselves with professionalism and often performed wonders of improvisation.

The purpose of this book, however, was not just to chronicle these contributions and commemorate the deeds of air and ground crews, but to provide lessons for practitioners. Air power will become even more critical in the future, as UN operations by the world organization and its enforcers, especially the North Atlantic Treaty Organization (NATO), will rely not only on manned aircraft but also on new technological assets such as unmanned vehicles. As has been shown in this volume, there is much to learn from in the past. Today, the United Nations has Mi-35 attack helicopters in the Democratic Republic of the Congo (DRC). Much earlier, in the 1960s, the UN's "Air Force" in the Congo included more: Swedish Saab J-29B fighter–bombers; Ethiopian F-86 Sabre fighters; and Indian Canberra bombers. The future might see the return of combat aircraft, particularly if the more technologically advanced nations re-engage in peacekeeping. In the near future, unmanned aerial vehicles, aerostats (tethered balloons) and parafoils (kites) guided by global positioning systems might also carry out the kinds of transport and surveillance missions that were once the sole preserve of crewed aircraft. The aerospace assets of the NATO peace operation in the former Yugoslavia after 1995 serve as an example to the United Nations of what can and should be brought to the field. Compared to NATO or advanced nations, UN air power is rudimentary – with a few exceptions, as shown in this volume.

Many changes in the United Nations can be made to achieve progress. As suggested, the United Nations should reform its bidding processes to allow the contracting of more capable aircraft (for example, the S-92 helicopter rather than the Mi-8MTV). Western countries can help fill the air power void by offering more capable military aircraft, including advanced UAVs and fighter jets. Robust aircraft can give peace operations a much-needed peace enforcement capability. Since many Western nations remain reluctant to commit ground forces, these countries could make a meaningful, specialized contributions by providing modern transport, surveillance, and combat aircraft to UN missions.

This volume has shown that UN aircraft are deficient not only in *quality* but also in *quantity*. Today in the DRC, a country the size of Western Europe, the United Nations has only a handful of attack helicopters – too few to do the job properly. It is recognized, however, that robust air power is not a panacea. There remains the need for peacekeepers on the ground and political processes with the major actors, though these persons can be transported and better informed through air power.

Other important lessons include the need to limit the damage of air strikes in order to maintain a mission's legitimacy. It will remain a reality that these combat operations generally will be organized on short notice and requiring agile planning. Also intelligence-gathering by air or ground necessitates proper information analysis and distribution, an area ripe for major improvements in both technology and processes. Furthermore, there is much that can be done to ensure that different missions in the same region co-operate with each other, including by sharing their air capabilities.

Finally, there is a continuing need to study this issue and to make sure that the proper lessons are learned. For example, UN officials quickly forgot that the organization had conducted kinetic air operations in the Congo in the 1960s. The loss of institutional memory meant that the United Nations was slow to implement the use of armed helicopters in the 2000s. Therefore, this study is only the first step in a broader effort to better understand the role that air power has played in peacekeeping and humanitarian operations. In the words of Senator Dallaire in the Foreword to this volume, such studies will "play an important role in illuminating the past to brighten the future".

Index

Note: Figures indexed in **bold** and tables indexed in *italics*.

CPSIA information can be obtained
at www.ICGtesting.com
Printed in the USA
BVHW081959100619
550635BV00010B/149/P